AF412713

The IMA Volumes
in Mathematics
and its Applications

Volume 130

Series Editors
Douglas N. Arnold   Fadil Santosa

Springer

New York
Berlin
Heidelberg
Barcelona
Hong Kong
London
Milan
Paris
Singapore
Tokyo

Institute for Mathematics and its Applications (IMA)

The **Institute for Mathematics and its Applications** was established by a grant from the National Science Foundation to the University of Minnesota in 1982. The primary mission of the IMA is to foster research of a truly interdisciplinary nature, establishing links between mathematics of the highest caliber and important scientific and technological problems from other disciplines and industry. To this end, the IMA organizes a wide variety of programs, ranging from short intense workshops in areas of exceptional interest and opportunity to extensive thematic programs lasting a year. IMA Volumes are used to communicate results of these programs that we believe are of particular value to the broader scientific community.

Douglas N. Arnold, Director of the IMA

* * * * * * * * * *

IMA ANNUAL PROGRAMS

1982–1983	Statistical and Continuum Approaches to Phase Transition
1983–1984	Mathematical Models for the Economics of Decentralized Resource Allocation
1984–1985	Continuum Physics and Partial Differential Equations
1985–1986	Stochastic Differential Equations and Their Applications
1986–1987	Scientific Computation
1987–1988	Applied Combinatorics
1988–1989	Nonlinear Waves
1989–1990	Dynamical Systems and Their Applications
1990–1991	Phase Transitions and Free Boundaries
1991–1992	Applied Linear Algebra
1992–1993	Control Theory and its Applications
1993–1994	Emerging Applications of Probability
1994–1995	Waves and Scattering
1995–1996	Mathematical Methods in Material Science
1996–1997	Mathematics of High Performance Computing
1997–1998	Emerging Applications of Dynamical Systems
1998–1999	Mathematics in Biology
1999–2000	Reactive Flows and Transport Phenomena
2000–2001	Mathematics in Multimedia
2001–2002	Mathematics in the Geosciences
2002–2003	Optimization
2003–2004	Probability and Statistics in Complex Systems: Genomics, Networks, and Financial Engineering
2004–2005	Mathematics of Materials and Macromolecules: Multiple Scales, Disorder, and Singularities

Continued at the back

David P. Chock   Gregory R. Carmichael
Editors

Atmospheric Modeling

With 111 Figures

Springer

David P. Chock
Chemistry and Environmental Science
  Department
Ford Research Laboratory
P.O. Box 2053, MD-3083
Dearborn, MI 48121-2053
USA
dchock@ford.com

Gregory R. Carmichael
Department of Chemical and Biochemical
  Engineering
Center for Global and Regional
  Environmental Research
204 IATL
University of Iowa
Iowa City, IA 55240
USA
gcarmich@engineering.uiowa.edu

Series Editors:
Douglas N. Arnold
Fadil Santosa
Institute for Mathematics and its
  Applications
University of Minnesota
Minneapolis, MN 55455
USA
http://www.ima.umn.edu

Mathematics Subject Classification (2000): 58J90, 76V05, 80A32, 80A50, 86A10

Library of Congress Cataloging-in-Publication Data
Atmospheric modeling / David P. Chock, Gregory R. Carmichael.
       p. cm. — (The IMA volumes in mathematics and its applications ; 130)
    Includes bibliographical references and index.
    ISBN 0-387-95497-X (alk. paper)
       1. Atmospheric models—Congresses.   I. Chock, David P., 1943–   II. Carmichael,
Gregory R.   III. IMA volumes in mathematics and its applications ; v. 130.
    QC861.3 .A86   2002
    551.51′01′1—dc21   2002070452

ISBN 0-387-95497-X   Printed on acid-free paper.

www.springer-ny.com

Springer-Verlag   New York Berlin Heidelberg
A member of BertelsmannSpringer Science+Business Media GmbH

FOREWORD

This IMA Volume in Mathematics and its Applications

ATMOSPHERIC MODELING

is a result of a very successful one-week workshop on the same title. The workshop was an integral part of the IMA annual program on Mathematics in Reactive Flow and Transport Phenomena, 1999–2000. We would like to thank David P. Chock (Ford Research Laboratory) and Gregory R. Carmichael (University of Iowa) for their excellent work as organizers of the meeting and for editing the proceedings.

We take this opportunity to thank the National Science Foundation for their support of the IMA.

Series Editors

Douglas N. Arnold, Director of the IMA

Fadil Santosa, Deputy Director of the IMA

PREFACE

During March 15–19, 2000, a workshop on Atmospheric Modeling was held at the Institute for Mathematics and Its Applications (IMA). This workshop was part of the Reactive Flow Modeling Program undertaken by IMA in the academic year 1999–2000. More than sixty scientists and mathematicians in atmospheric environmental modeling from ten countries participated in an intensive discussion of the mathematics relevant to the transport and transformation of atmospheric constituents. In atmospheric processes, a very wide range of temporal and spatial scales needs to be considered. There are a large number of chemical constituents; they can appear in gaseous, liquid and solid phases. The rates of their reactions vary considerably. In the transport processes, daytime convective mixing is distinctly different from the nighttime limited mixing and predominance of vertically stratified shear flows, and the spatial scales can range from micro-environmental to urban, regional and global. Our understanding of both the chemical and physical processes remains patchy and incomplete. Even if we can describe the atmospheric environmental system intelligibly, drastic approximations in models are needed due to incomplete knowledge and constraints in computational resources. Many interesting mathematical problems arise in the hope of reducing uncertainties in model predictions given our current state of understanding of the chemistry, the meteorology, the model input information, and the available computational tools. The present volume contains a collection of refereed papers that were either presented during the workshop or submitted separately to address many of these issues.

The range of topics covered by the papers is wide. We arranged them according to the topics, from general to specific. Operator splitting is commonly used in air pollution models. They can lead to significant errors for variables with short time scales. P.J.F. Berkvens, M.A. Botchev, M.C. Krol, W. Peters, and J.G. Verwer compare the accuracy of two splitting methods with an un-split Rosenbrock method for a 1D chemistry-transport model. A key issue in the description of atmospheric chemistry is the presence of a wide range of timescales, which results in a set of stiff ordinary differential equations. A. Sandu describes a time integration method that preserves the positivity of the species concentrations, and B. Sportisse and F. Djouad consider the mathematical aspects of multiple timescales in the numerical modeling of atmospheric chemistry. The need to describe a wide range of species spatial resolution and the need for computational efficiency lead to development of adaptive-grid methods among others. One such method is described in the paper by S. Ghorai, A. S. Tomlin, and M. Berzins. D.P. Chock, S.L. Winkler and P. Sun describe the effect of grid resolution and subgrid assumption on the model predictions of a system containing non-homogeneous chemistry. A major problem in atmospheric modeling that relies on input from meteorological models is the potential of violating mass conservation. The sources of this problem are many. Circumventing it may disturb the flow field or introduce artificial density sources or sinks. D.W. Byun and S.M. Lee compare different mass adjustment schemes for their ability to preserve the mixing ratio fields, to conserve the mass, etc. Four-dimensional variational data assimilation has also become increasingly popular to improve the forecasting ability of an air quality model. However, high memory storage and large computation time have been a disadvantage. D. Daescu and G.R. Carmichael show that use of a non-operator-splitting method eliminates the stiff transients problems and allows for large time steps and reduces storage requirements. H. Elbern and H. Schmidt use data assimilation to optimize initial values and emission rates for model performance improvement in the simulation of an ozone episode in Europe. Air pollution models require substantial computational resources. Converting a sequential code to run optimally on a parallel machine is not trivial. W. Owczarz and Z. Zlatev present an example of testing a code with two levels of parallelism on an IBM SMP parallel computer.

There are many models describing the global transport and chemistry of atmospheric constituents, but they are generally tied to specific global climate models. Here, E.P. Olaguer presents a self-contained, efficient 3D model that includes the global circulation dynamics as well as the transport and chemistry of various atmospheric species. In a similar spirit, M.Z. Jacobson developed a comprehensive coupled weather forecast and air pollution model that contains multiply nested grids to describe gaseous and aerosol processes from an urban scale to the global scale. The task of describing the fate of atmospheric aerosols or particulate matter has received considerable attention presently. In addition to the paper by Jacobson above, F.S. Binkowski, S.J. Roselle, M.R. Mebust and B.K. Eder describe a modal approach to model the fate of atmospheric particulate matter. The knowledge of secondary organic aerosol formation remains rather incomplete. R.J. Barthelmie and S.C. Pryor model the formation of secondary organic aerosols originated from the emission and oxidation of monoterpenes, which generally come from biogenic sources.

We would like to thank Patricia V. Brick of the IMA for her infinite patience in overseeing the completion of this volume. We would also like to thank the reviewers for their effort in providing helpful comments to the authors. We also want to thank Fred Dulles for planning the logistics of the workshop. We are especially grateful to Willard Miller, the then Director of the IMA, for inviting the attendees to the workshop and help make the workshop a great success.

David P. Chock
Chemistry and Environmental Science Department
Ford Research Laboratory

Gregory R. Carmichael
Department of Chemical & Biochemical Engineering
Center for Global and Regional Environmental Research
The University of Iowa

CONTENTS

SOLVING VERTICAL TRANSPORT AND CHEMISTRY IN AIR POLLUTION MODELS

P.J.F. BERKVENS[*], M.A. BOTCHEV[†], M.C. KROL[‡],

W. PETERS[‡§], AND J.G. VERWER[*¶]

Abstract. For the time integration of stiff transport-chemistry problems from air pollution modelling, standard ODE solvers are not feasible due to the large number of species and the 3D nature. The popular alternative, standard operator splitting, introduces artificial transients for short-lived species. This complicates the chemistry solution, easily causing large errors for such species. In the framework of an operational global air pollution model, we focus on the problem formed by chemistry and vertical transport, which is based on diffusion, cloud-related vertical winds, and wet deposition. Its specific nature leads to full Jacobian matrices, ruling out standard implicit integration.

We compare Strang operator splitting with two alternatives: source splitting and an (unsplit) Rosenbrock method with approximate matrix factorization, all having equal computational cost. The comparison is performed with real data. All methods are applied with half-hour time steps, and give good accuracies. Rosenbrock is the most accurate, and source splitting is more accurate than Strang splitting. Splitting errors concentrate in short-lived species sensitive to solar radiation and species with strong emissions and depositions.

Key words. Numerical time integration, stiff ODEs, splitting techniques, approximate matrix factorization, Rosenbrock methods, air pollution problems.

1991 Mathematics Subject Classification. 65M06, 65M20. Secondary: 65Y20.

1. Introduction. In global air pollution modeling the time dependent chemical composition of the atmosphere is described by a three-dimensional (3D) system of partial differential equations (PDEs),

$$(1) \qquad \frac{\partial c_i}{\partial t} = \mathcal{T}_i(c_i) + f_i(c_1, \dots, c_{m_s}), \quad i = 1, \dots, m_s.$$

The variables c_i represent the concentrations of m_s atmospheric trace gases, the linear term $\mathcal{T}_i(c_i)$ represents the transport by various atmospheric transport processes. The non-linear term f_i describes the chemistry that occurs between the trace gases; it may also contain emissions and depositions. In general, the transport term contains horizontal and vertical advection and diffusion. In global applications it also contains parameterizations of rapid vertical motions which occur on subgrid scales both in space and time. The chemical interaction term f_i is generally stiff and thus requires an implicit

---

[*]CWI, P.O. Box 94079, 1090 GB Amsterdam, The Netherlands; `patrick.berkvens@cwi.nl`.

[†]Mathematical Sciences Faculty, University of Twente, P.O. Box 217, 7500 AE Enschede, The Netherlands; `m.a.botchev@math.utwente.nl`. This work was done while this author was with the CWI, Amsterdam.

[‡]IMAU, P.O. Box 80005, 3508 TA Utrecht, The Netherlands; `m.krol@phys.uu.nl`.

[§]`w.peters@phys.uu.nl`.

[¶]`jan.verwer@cwi.nl`.

numerical solution method. The system (1) is augmented with boundary conditions, which may contain emissions and depositions. Observe that without the chemical interactions f_i, the system (1) represents m_s decoupled transport equations. In practical applications m_s varies from 20 to 100 and the global atmosphere is resolved in $\mathcal{O}(10^4)$ to $\mathcal{O}(10^6)$ grid points.

The vertical spatial discretization of the transport term on a grid of m_h grid cells yields a system of ordinary differential equations (ODEs) which fits the form

$$(2) \qquad \dot{w} = F(w) \equiv Tw + f(w), \quad w \in R^N, \quad N = m_h m_s.$$

The linear term Tw is the semi-discrete vertical transport term and the nonlinear term $f(w)$ contains the chemistry. As explained above, convection is a vertical transport process that occurs on the sub-grid scale and couples the concentrations of each species in all the layers in which the convection occurs. This gives rise to a block-diagonal matrix T composed of m_s completely full blocks of dimension m_h. The immediate consequence is that the $N \times N$ Jacobian matrix $F'(w)$ is nearly full. The Jacobian is not completely full because the Jacobian of the chemical sub-system is sparse. This sparsity, however, cannot be efficiently exploited in a direct solution of linear systems. The reason is that the sparse chemistry Jacobians of different grid cells are coupled in $F'(w)$ by the transport matrix T in such a way that the sparsity will be lost in a direct solution. Thus the straightforward use of implicit solvers is ruled out and tailored numerical integration methods are required.

An efficient and widely used method to solve such a huge system is operator splitting in which the chemistry and transport processes are advanced in time sequentially. However, splitting errors may dominate the total error for typical time step sizes used in long-term simulations. Splitting errors are caused by splitting the time step advance for a number of processes that act simultaneously, into steps for separate processes which are then carried out consecutively. For example, when the processes of vertical mixing, emissions, depositions, and chemical reactions — which act simultaneously in reality — are time-stepped one after another, this leads to splitting errors. There is one exception: if the mathematical operators linked to the processes *commute*, then no splitting error arises [18]. In practice, though, this never occurs.

In this paper several methods will be studied that aim at the reduction of splitting errors. Our starting point is a Strang-type operator splitting method that is widely used in air pollution modeling [3, 8, 18, 21, 24, 26, 30]. Two alternative methods will be compared to Strang splitting and their effect on the numerical solution will be investigated. The two methods are the so-called source-splitting method and the Rosenbrock method with approximate matrix factorization. The methods will be applied to a vertical column of a realistic atmospheric chemistry model [22]. Advection is left out of system (1), but the vertical mixing processes, emission and deposition processes, and chemistry are included.

All three methods that are studied here have in common that they avoid the use of the huge and nearly full Jacobian $F'(w)$ in the linear system solutions. The methods will be outlined in more detail in Section 3.

In a previous study [3] it was found that splitting errors are of minor importance for longer lived species like ozone (O_3). However, for rapidly varying short-lived species (like the N_2O_5) standard operator splitting completely fails occasionally. These large deviations from the reference solution occurred mainly near strong emission peaks and at sunrise and sunset. The Rosenbrock method was found to provide the best solution. The model used in [3] contained only vertical diffusion as a transport mechanism. In the current study a different chemical mechanism is used and also convection type transport is included, but, different from [3], advection is left out. One aim of this paper is to reexamine the splitting errors in this different model set-up.

We address the following questions:

- How accurate are the three methods Strang splitting, source splitting, and Rosenbrock with approximate matrix factorization?
- Are these accuracies acceptable for the application field?
- How do the three methods compare among themselves?
- What are the effects of operator splitting? Particularly interesting is the question, whether for the current model, based on a vertical transport mechanism and set of chemical reactions different from [3], unacceptably large splitting errors again remain restricted to the short-lived species.
- What are the computational costs?

The contents are as follows. In Section 2 we describe the systems (1) and (2) in greater detail. Section 3 is devoted to the integration algorithms for the semi-discrete system (2). The methods are compared in a numerical experiment in Section 4 and the main conclusions are summarized in Section 5.

2. Model definition.

2.1. Meteorological model. In this study we use a one-column version of the Chemistry Transport Model version 3 (TM3). This model was used for studies of the photochemistry of the atmosphere on numerous occasions, e.g. [19, 7]. The same processes as included in the 3D global CTM are simulated, with the exception of advection. Processes included in the one-column model are summarized in Table 1, together with the operator notation adopted in this work.

Vertical diffusion and convection together constitute the vertical transport in the model. Diffusion [20] represents transport of trace gases by large turbulent eddies in the atmosphere due to differential heating of the surface and wind shear. Since these eddies are typically smaller than 1 km, dif-

TABLE 1
Operator notation.

Process	Operator	
Vertical diffusion	$\mathcal{U}$	$\mathcal{T}$
Convection	$\mathcal{V}$	
Wet deposition	$\mathcal{W}$	
Deposition	$\mathcal{D}$	f
Gas phase chemistry	r	
Emission sources	ς	

fusion is a local process redistributing the air over several adjacent layers. Diffusion is strongest near the surface, and near the wind maxima in the upper troposphere (jet streams). Convection [25] represents vertical transport resulting from large-scale instability in the atmosphere. Due to condensation of water vapour, air parcels become buoyant and can quickly loft several kilometers, forming cumulus clouds. During this lofting, exchange with the air surrounding the cloud occurs. Convection can redistribute trace gases over large vertical distances. Vertical diffusion and convection are non-divergent and mass conservative, and only cause spatial coupling in the vertical direction. In the cumulus clouds, soluble trace gases can deposit on cloud droplets. When these cloud droplets rain out of the cloud, trace gases are removed from the atmosphere. This process is called wet deposition [10] and is proportional to the solubility, the scavenging efficiency of the cloud, and the trace gas concentration. Wet deposition thus removes tracer mass from the atmosphere. Likewise, dry deposition [29] removes trace gases at the surface. Soils and vegetation have the ability to absorb some trace gases. The efficiency depends on many factors such as sunlight, soil type, solubility. The gas phase photo-chemistry [13] consists of 29 trace gases, that react under the influence of temperature and sunlight. The strong dependence of some reactions on the intensity of sunlight makes this system very stiff. The tendency of photochemistry to converge to an equilibrium state is constantly perturbed by the changing photolysis rates, but also by transport, emission, and removal processes. Finally, emissions in the model are present mostly in the lowermost layer, where anthropogenic activities and natural processes occur. The only exceptions are the production of NO by lightning and airplane emissions, which occurs at higher model levels.

In the rest of the paper, operator $\mathcal{T}_i$ represents the first three processes from Table 1:

$$\mathcal{T}_i = \mathcal{U} + \mathcal{V} + \mathcal{W}_i \quad . \tag{3}$$

The diffusion $\mathcal{U}_i$ is defined as:

$$(4) \qquad \mathcal{U}(z, c_i(z)) \equiv \frac{\partial}{\partial z}\left(K(z)\frac{\partial c_i(z)}{\partial z}\right) ,$$

where $z \in [0, H]$ is the vertical coordinate and K is the diffusivity.
The convection operator is defined as:

$$(5) \qquad \mathcal{V}(z, c_i(z)) \equiv \int_0^H [M(z, \zeta)c_i(\zeta) - M(\zeta, z)c_i(z)]d\zeta ,$$

where $M(z, \zeta)$ gives the rate at which tracer mass is transported from height ζ to height z, and H is the column height. The first term under the integral gives the gain (always ≥ 0) at height z by transport from other heights, the second term gives the loss (always ≤ 0) at height z to other heights. The operator conserves tracer mass within the column. More specifically, M consists of an updraft, a subsidence, and a downdraft contribution, thus $M = M_u + M_s + M_d$. For $\zeta \leq z$ we have $M_u \geq 0$, $M_s = 0 = M_d$, whereas for $\zeta > z$ we have $M_u = 0$, $M_s \leq 0$, and $M_d \leq 0$. Note that the convection operator directly couples any height z to all other heights ζ. This was done in the TM3 model because convection has very short time-scales for the coupling along the vertical direction compared with the timesteps used.

Finally, the wet-deposition operator $\mathcal{W}_i$ is defined as:

$$(6) \qquad \mathcal{W}_i(z, c_i(z)) \equiv \beta_i(z)\varepsilon(z)\int_0^z M_u(z, \zeta)c_i(\zeta)d\zeta ,$$

where β_i denotes the solubility of tracer i and ε is the scavenging efficiency of the convective precipitation. So, wet deposition is proportional to the updraft flux.

The remaining processes are described by operator f:

$$(7) \qquad f_i(c_1, \dots, c_{m_s}) = r_i(c_1, \dots, c_{m_s}) + \mathcal{D}_i(c_i) + \varsigma_i .$$

The upper boundary conditions of the system are that no fluxes are allowed. At the lower boundary the fluxes are determined by the emission and deposition processes. At the surface and in the top model layers, the concentrations of some longer lived species is constrained by relaxation of the concentration to a prescribed value that represents a climatology. This is done by modifying the terms $\mathcal{D}_i(c_i) + \varsigma_i$ in (7).

In this study a column of air over the Amazonian rainforest ($60°W$, $0°N$) is selected. This location is of particular interest because of the high solar radiation, the strong vertical diffusion and frequent occurrence of convection. Furthermore, dry deposition and natural emissions from the forest are abundant in this column, and the photochemistry is very stiff. Under these conditions splitting errors are expected to maximize. Transport is based on 6-hourly meteorological data from the European Center for Medium Range Weather Forecasting (ECMWF) model for March 1997, while initial tracer concentrations and surface emissions are taken from a full 3D TM3 simulation. Since we do not include advection in the simulations, concentrations of species with strong sources can become unrealistically high. In order to prevent this, and to retain as much resemblance to the observed atmospheric situation as possible, a relaxation (or nudging) of the trace gas concentrations is applied. The concentrations are relaxed to the clean background concentrations c_i^{ocean} observed over the oceans surrounding the continent by modifying $\mathcal{D}_i$ and s_i according to:

$$\begin{aligned} \mathcal{D}_i &:= \mathcal{D}_i + \tilde{\mathcal{D}}, \\ s_i &:= s_i + \tilde{s}_i, \end{aligned} \tag{8}$$

where $\tilde{\mathcal{D}}$ represents the reciprocal of the turnover time of a grid box with respect to advection (taken to be 10 days), and $\tilde{s}_i = -\tilde{\mathcal{D}} c_i^{\text{ocean}}$.

2.2. Semi-discretization. The vertical column is divided into m_h cells, indexed with k ($k = 1$ at the surface and $k = m_h$ at the top), with volumes v_k, assembled in the vector v. In the present test, v does not change with time, since the advection and the related pressure changes are zero. The approximations of the masses of tracer i in the cells, which are the cell-integrals of the concentrations, are assembled in μ_i with elements $\mu_{i,k}$, where i indicates the species.

The semi-discretization of the vertical-transport operator $\mathcal{T}_i$ yields a matrix T_i operating on column vectors of tracer masses, thus $T_i\mu_i$. The vertical-transport matrix can be split into a diffusion matrix U (corresponding to $\mathcal{U}$), a convection matrix V (corresponding to $\mathcal{V}$) and a convective wet deposition matrix W_i (corresponding to $\mathcal{W}_i$), thus $T_i = U + V + W_i$. For the diffusion standard central differences on a three-point stencil are used. The semi-discretization of the convection [12] results in a full matrix. This reflects the non-local nature of the vertical coupling in the parameterization of the convection [25]. The semi-discretized submodel for vertical mixing reads:

$$\dot{\mu}_i = (U + V)\mu_i \ . \tag{9}$$

Element $(u + v)_{k\ell}$ of V, $\ell \neq k$, indicates the net fraction of tracer mass $\mu_{i,\ell}$ in cell ℓ which moves from cell ℓ to cell k per time unit. $U + V$ conserves the tracer mass in a column.

The vertical-mixing matrix V can be split into an updraft part V_u, a subsidence part V_s and a downdraft part V_d, thus $V = V_u + V_s + V_d$, and the mass-conservation property holds for both of these matrices separately. V_u is a lower-triangular matrix, which expresses the fact that the updraft only transports tracer upward. (Similarly, V_s and V_d are upper triangular.) The semi-discretization of the convective wet deposition yields, for each species, matrices W_i which are closely related to V_u. They read:

$$(10) \qquad W_i = B_i E(V_u - \text{diag}(V_u)) \ ,$$

where $B_i = \text{diag}_{k=1}^{m_h}\{\beta_{i,k}\}$ and $\beta_{i,k} \in [0,1]$ gives the solubility of tracer i, $E = \text{diag}_{k=1}^{m_h}\{\varepsilon_k\}$, with $\varepsilon_k \in [0,1]$, is a diagonal matrix expressing the scavenging efficiency, and $\text{diag}(V_u)$ is the diagonal matrix containing the main diagonal of V_u.

The deposition operators $\mathcal{D}_i$ (containing both non-convective wet deposition and dry deposition) are discretized into diagonal matrices D_i with $d_{i,kk} \leq 0$. The diagonal elements give the cell-averaged drain rates. The volume sources ς_i are discretized into vectors s_i whose elements contain the cell-integrated values of the sources.

With regard to the boundary conditions at the surface, after discretization the dry-depositions Δ_i are incorporated into the elements $d_{i,11}$ of D_i, and the surface emissions σ_i are incorporated into the element $s_{i,1}$ of s_i, so they are attributed to the lowest cell.

Finally, the chemical reaction operator $r_i(c_1,\dots,c_{m_s})$ is semi-discretized into $r_i(c_1,\dots,c_{m_s})$ by taking cell-averaged photochemical and reaction rates operating on cell-averaged concentrations.

The above semi-discretization yields the following system of ODE's for the concentrations:

$$(11) \qquad \frac{\partial c_i}{\partial t} = (T(c_i \odot v)) \oslash v + D_i c_i + s_i \oslash v + r_i(\vec{c}) \ , \quad i = 1,\dots,m_s \ ,$$

where $\odot$ means elementwise multiplication. We have to transform back and forth between concentrations c_i and tracer masses $\mu_i = c_i \odot v$, since T operates on tracer masses rather than on concentrations. The average concentrations for all species and cells are assembled in $\vec{c}$, where $\oslash$ means entrywise division. $\vec{c}$ has elements $c_{i,k}$ for species numbered by i and cells numbered by k. (We use boldface notation for 'column' vectors of quantities varying with the computational cells, arrow notation for 'species' vectors of quantities varying with the chemical species, and a combination of boldface and arrow notation for vectors of quantities varying with *both* the cells *and* the species.)

In Section 3 we describe three numerical time-integration methods for solving this system of ODE's. Two of the methods use operator splitting, where for efficiency reasons the two operators T and r_i are applied separately. This leaves us with the following choice: with which of these two operators will we combine D and s? For accuracy reasons, one would choose to combine D and s with the operator which causes the largest splitting error if it is separated from D and s. Because the (photo)chemical reactions r_i are usually more stiff than the vertical mixing T for most of the species, we choose to combine both D_i and s_i with r_i, and define f_i as follows:

$$(12) \qquad f_i(\vec{c}) = D_i c_i + s_i \oslash v + r_i(\vec{c}) \ , \quad i = 1,\dots,m_s \ .$$

Equation (11) is now written as (cf. (2)):

$$(13) \qquad \frac{\partial c_i}{\partial t} = F_i(\vec{c}) \equiv (T(c_i \odot v)) \oslash v + f_i(\vec{c}) \ , \quad i = 1, \ldots, m_s \ .$$

3. Integration methods. The chemistry processes are continuously driven by rapidly changing photolysis rates. As a result, chemistry is very stiff and implicit integration is indispensible. A typical time step that is desirable for the other processes is in the order of 30 minutes. This time step enables the simulation of long term changes in the atmospheric composition which is one of the main application areas of the TM3 model. For the time step $\tau = 30$ minutes one normally has

$$(14) \qquad \tau \|f'\|_2 \sim \mathcal{O}(10^6), \qquad \tau \|T\|_2 \sim \mathcal{O}(10),$$

whereas the smallest in modulus eigenvalues are of order $\mathcal{O}(10^{-5})$.

All three methods for solving (2) are derived from a particular Rosenbrock method (which we call ROS2) from the stiff ODE field [6, 11]. Let $\gamma = 1 + \frac{1}{2}\sqrt{2}$ and denote $A = F'(w_n)$. The Rosenbrock method reads

$$
\begin{aligned}
w_{n+1} &= w_n + \frac{1}{2}k_1 + \frac{1}{2}k_2, \\
(I - \gamma\tau A)\,k_1 &= \tau F(w_n), \\
(I - \gamma\tau A)\,k_2 &= \tau F(w_n + k_1) - 2\gamma\tau A k_1,
\end{aligned}
\qquad (15)
$$

where $w_n \approx w(t_n)$ and $\tau = t_{n+1} - t_n$ denotes the step size. An equivalent formulation, more attractive for implementation, is

$$
\begin{aligned}
w_{n+1} &= w_n + \frac{3}{2}k_1 + \frac{1}{2}k_2, \\
(I - \gamma\tau A)\,k_1 &= \tau F(w_n), \\
(I - \gamma\tau A)\,k_2 &= \tau F(w_n + k_1) - 2k_1.
\end{aligned}
\qquad (16)
$$

With this formulation we economize on a matrix-vector multiplication. This two-stage method is 2nd-order consistent and it has the stability function $R(z) = (1 + (1 - 2\gamma)z)/(1 - \gamma z)^2$ which is L-stable and positive for all $z \leq 0$. In [3, 27] it is argued that this low order, linearly implicit method is suitable for solving stiff atmospheric chemical kinetics problems (box models), provided the costs for the linear system solutions and the Jacobian updates are minimized by using optimal sparsity routines. Such routines are provided by the chemical kinetics preprocessor KPP [4, 5, 23]. ROS2 remains order two consistent with an arbitrary matrix A, $A \neq F'(w_n)$. However, stability of the scheme is corrupted when A is not the exact Jacobian.

As is explained in Section 1, straightforward application of method (16) to the coupled vertical-transport-chemistry problem is expensive. In full three-dimensional computations the costs for the linear systems solutions with the nearly full and large matrices $F'(w_n) = I - \gamma\tau(T + f'(w_n))$ would be unacceptably large. But solving linear systems with the submatrices $I - \gamma\tau T$ and $I - \gamma f'(w_n)$ is economical, as both are block-diagonal (upon reordering, cf. Section 2). This property, with a few others, largely determines the choice for the three integration methods discussed next.

3.1. Strang splitting. Let $\Phi_T(t;\tau)$ and $\Phi_f(t;\tau)$ denote the ROS2 integrator (16) applied to $\dot{w} = Tw$ and $\dot{w} = f(w)$, respectively. The Strang-type operator splitting method for system (2) using a stepsize τ and ending with a chemistry step [24], can then be formulated as

$$(17) \quad w_{n+1} = \Phi_f(t_{n+1/2}; \tfrac{1}{2}\tau)\, \Phi_T(t_{n+1/2}; \tfrac{1}{2}\tau)\, \Phi_T(t_n; \tfrac{1}{2}\tau)\, \Phi_f(t_n; \tfrac{1}{2}\tau)w_n.$$

This method is 2nd-order consistent. The leading local error term is the sum of the leading local errors made by the Rosenbrock method and the leading local splitting error. The latter is zero in case Tw and $f(w)$ commute [8, 18]. This, however, is in general not true. In what follows, we refer to method (17) as STRANG.

The STRANG method is an improved version of a splitting method used in the TM3 model [22]. The essential difference from (17) is that the vertical transport is integrated with larger step sizes of order several hours alternatively with series of still small chemistry substeps. Also, in the splitting method used in TM3 the deposition and source operators are split from the chemistry operator. Since the vertical transport matrix changes only every 6 hours, its inverse is also recomputed only every 6 hours, the same number of times in both methods (17) and the splitting method used in TM3. Thus, with a relatively small computational overhead, (17) appears an attractive alternative to the splitting method in TM3.

The stability properties of the STRANG method are largely determined by those of each of its substeps: since at each substep the L-stable ROS2 method is applied, the STRANG method is L-stable.

The splitting in both methods has adverse effects on especially the integration of the stiff chemistry: initial values for the chemistry substeps are not always the results of the previous chemistry substeps, they are 'discontinuous' for the chemistry integration. This leads to so-called stiff transients. The methods considered in the next two sections are aimed to avoid this artifact.

3.2. S-SPLIT: source splitting method. In the source splitting method [16, 15], one of the split processes is approximated at each time step as a constant source to be applied with the other split process. In fact, we replace (2) with

$$(18) \quad \dot{w} = \hat{F}(w) = f(w) + S, \qquad t \in [t_n, t_n + \tau],$$

where S is the source term approximating vertical transport. The scheme can be written as

$$(19) \qquad \tilde{w}_{n+1} = \Phi_T(t_n; \tau) w_n,$$

$$S = \frac{\tilde{w}_{n+1} - w_n}{\tau},$$

$$(20) \qquad w_{n+1} = \Phi_{\hat{F}}(t_n, \tau) w_n.$$

Formally, splitting has been removed because the step $\Phi_{\hat{F}}$ is in fact done for the unsplit system (18) that approximates the full system (2). Independently of the order of the methods used per substep in (19) and (20), this S-SPLIT method is only first order consistent. Observe that $S = Tw + \mathcal{O}(\tau)$. Just as in standard splitting, in S-SPLIT there is essentially the same freedom left to use one's favorite combination of algorithms. Higher order generalizations of the source splitting are considered in [17].

We note in passing that it is profitable to have the substep corresponding to the stiffest process, chemistry, at the end of the splitting sequence [24]. This is the case for methods STRANG and S-SPLIT.

To get an impression of the stability of the S-SPLIT method, we consider the standard scalar test equation

$$\dot{w} = \lambda_f w + \lambda_T w.$$

Denoting $z_f = \tau \lambda_f$ and $z_T = \tau \lambda_T$, we can write the stability function of the S-SPLIT method as

$$R(z_f, z_T) = R_0(z_f) + R_1(z_f) \left[R_0(z_T) - 1 \right],$$

where

$$R_0(z) = \frac{1 + (1 - 2\gamma)z}{(1 - \gamma z)^2}, \quad R_1(z) = \frac{1 + (\frac{1}{2} - 2\gamma)z}{(1 - \gamma z)^2}.$$

Here $R_0(z)$ is the stability function of the ROS2 method and function $R_1(z)$ is such that a ROS2 step for the equation $\dot{w} = \lambda w + \sigma$ can be written as

$$w_{n+1} = R_0(z) w_n + R_1(z) \tau \sigma.$$

Simple analysis leads us to the following observations:
(a) Fix $z_T = z_T^* \leq 0$. Then function $R_f(z_f) = R(z_f, z_T^*)$ is L-stable, but not positive for $z_f \leq 0$.
(b) Fix $z_f = z_f^* \leq 0$. Then function $R_T(z_T) = R(z_f^*, z_T)$ is A-stable, but not positive for $z_T \leq 0$.

Property (a) quarantees good damping for large chemistry eigenvalues λ_f, whereas the lack of L-stability in (b) is hardly harmful taking into account that vertical transport eigenvalues are normally several orders less than those of chemistry. Positivity of R is argued to be important for chemistry kinetic problems in [27], so, for S-SPLIT lacking this property, we can expect positivity problems. Indeed, in our numerical experiments we have observed that negative values in chemistry computations were encountered in the S-SPLIT scheme sligthly more often than in STRANG (unphysical negative concentrations occuring occasionally in computations were set to zero).

3.3. ROS2-AMF: approximate factorization. The ROS2-AMF method is the ROS2 scheme (16) applied to the whole system (2) with an inexact matrix $I - \gamma\tau A$. For $I - \gamma\tau A$ we take Approximate Matrix Factorization (AMF) in the form:

$$(21) \qquad I - \gamma\tau A = (I - \gamma\tau T)(I - \gamma\tau f').$$

This ROS2-AMF method was introduced in [27] and recently tested in [3]. The idea of approximate matrix factorization is known at least since papers [2, 9, 28] and still widely used (see e.g. [1, 14]).

The ROS2-AMF method is split-free because the 'splitting' appears only at the level of the linear solves with matrix $I - \gamma\tau A$. However, the method has retained the ease of implementation of the standard splitting: the algorithmic ingredients (among which the linear solves with matrices $I - \gamma\tau T$ and $I - \gamma\tau f'$ are the most essential) remain basically the same as in conventional splitting. Taking an inexact Jacobian in (21) leads to only very moderate loss of stability: in fact, the scheme maintains its A-stability, only the L-stability is lost [27].

3.4. Computational costs. Methods S-SPLIT and ROS2-AMF require the same work per time step and, since the whole step of STRANG consists of two substeps for each process (cf. (17)), two steps of either of the methods are equal in cost to one step of STRANG. In our experiments we use the same constant step size τ for S-SPLIT and ROS2-AMF and twice as large a step size for STRANG. This makes these three methods equal in costs.

The work per step of S-SPLIT or ROS2-AMF consists essentially of one evaluation of the chemistry Jacobian f', one sparse LU factorization of $I - \gamma\tau f'$, two linear system solves with each of the matrices $I - \gamma\tau f'$ and $I - \gamma\tau T$, two chemistry function evaluations and two matrix-vector multiplication with the vertical transport matrix T. Matrices T and $(I - \gamma\tau T)^{-1}$ are updated every 6 hours.

4. Numerical comparisons and discussion. We have run the S-SPLIT and ROS2-AMF methods with a step size $\tau = 30$ min and the STRANG method with a step size $\tau = 60$ min. With these step sizes all three methods are equal in computational costs (see Section 3.4). These step sizes are quite large for the chemistry reaction set used (cf. (14)).

We compare the accuracy of the methods against a reference solution. The reference solution was obtained by the STRANG method run with a

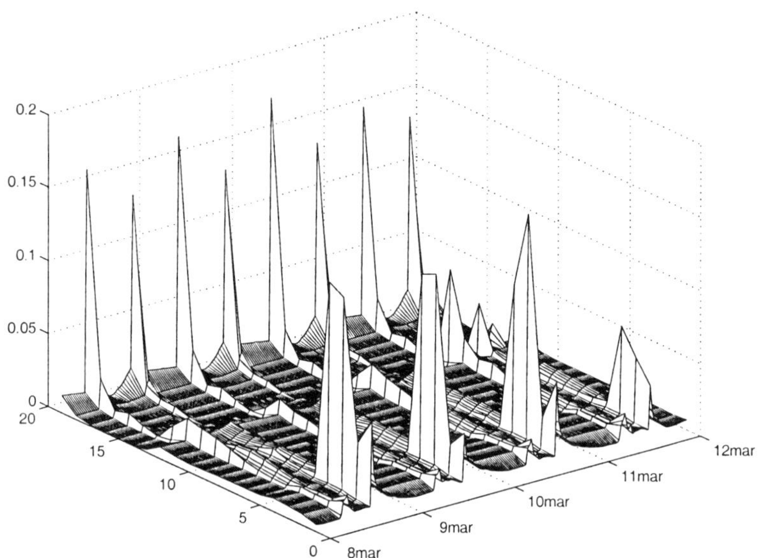

FIG. 1. *The error $\epsilon_{k;n}$ for the method* ROS2-AMF *as a function of time and model level.*

tiny step size $\tau = 3$ sec. With such a step size, all three methods converged to the same solution.

Despite the large time steps chosen, all three methods perform well delivering notably small errors (mostly less than 10% of the daily variations). Since all the methods are based on the ROS2 time stepping scheme, one may conclude that this integrator is stable and accurate.

In Figure 1 we plot the error $\epsilon_{k;n}$ as a function of time level n and of height k for the method ROS2-AMF. In Figure 2, we plot the height-average of the error $\epsilon_{k;n}$ of the three methods as functions of local time. In these figures the local-time interval ranges from 9 to 12 March. In Figure 3, we plot the time-average of the error $\epsilon_{k;n}$ (for 6 to 15 March) of the three methods as functions of the vertical level. The error $\epsilon_{k;n}$ in cell k and at time n is defined as the scaled distance in the m_s-dimensional space of the concentration vectors $\vec{c}_{k;n}$ of a particular method and $\vec{c}_{k;n}^{\,\mathrm{ref}}$ of the reference solution. The formal definition is:

$$(22) \qquad \epsilon_{k;n} \equiv \|(\vec{c}_{k;n} - \vec{c}_{k;n}^{\,\mathrm{ref}}) \oslash (\max_{k;n} \vec{c}_{k;n}^{\,\mathrm{ref}} - \min_{k;n} \vec{c}_{k;n}^{\,\mathrm{ref}})\|_2 / \sqrt{m_s} \ ,$$

where the max and min operators work elementwise, i.e. per species. The distances are scaled elementwise with the total range of the concentration in the reference solution. They are also scaled with $\sqrt{m_s}$. This guarantees that $\epsilon_{k;n}$ lies between the smallest and the largest relative error per species.

As we see in Figure 1, ROS2-AMF produces larger errors in the morning and in the evening. Since the chemistry is driven by photolysis rates that

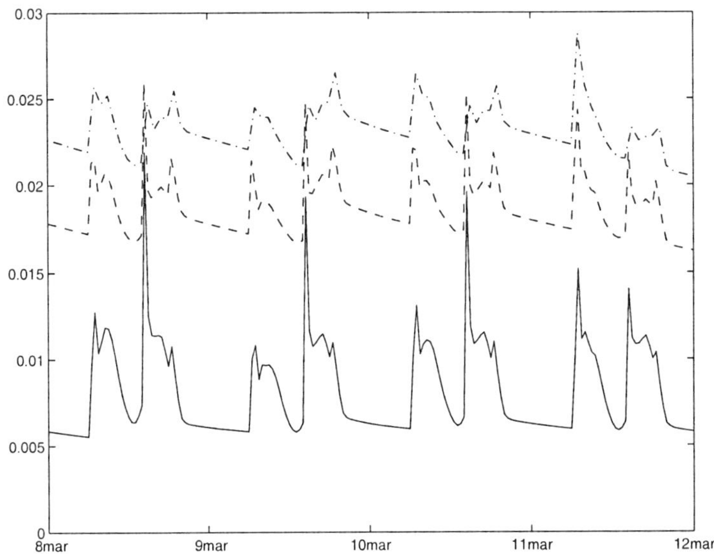

FIG. 2. *The height-averaged error as a function of time (solid line for* ROS2-AMF, *dashed line for* S-SPLIT, *and dash-dotted line for* STRANG*).*

change rapidly with the solar zenith angle, the reference solution (time step of 3 seconds) is able to capture this rapid change better than the other solutions, which use much larger time steps. The errors also become larger at the times when the vertical transport update occurs during daytime, notably at 8h and 14h local time. However, the errors of all the methods remain small and do not grow in time. Quite similar behaviour also occurs for S-SPLIT and STRANG (not shown). In Figures 6 and 7 of [3], it is also seen that large errors in the solution for the short-lived species N_2O_5 mainly occur — or start and end — at sunrise and sunset.

In general ROS2-AMF appears more accurate than both STRANG and S-SPLIT, and S-SPLIT is slightly more accurate than STRANG. However, this is not generally true for every tracer in each layer. However, it can be observed in Figure 3 that ROS2-AMF performs better than both STRANG and S-SPLIT in most of the layers. Since ROS2-AMF does not split the emission and deposition processes to the vertical transport, these differences are due to split errors that occur in S-SPLIT and STRANG. This observation is emphasized in Figures 4 to 8, which show the concentration as a function of time for a few selected tracers in the lowest model layer. Their lifetimes range from seconds (OH) to several hours (O_3). Due to reduced splitting errors, ROS2-AMF performs best compared to the reference solution.

For each method we have also calculated the time and layer averaged relative contributions per species to the total error defined in (22). The results for STRANG and S-SPLIT are almost identical, with very large contri-

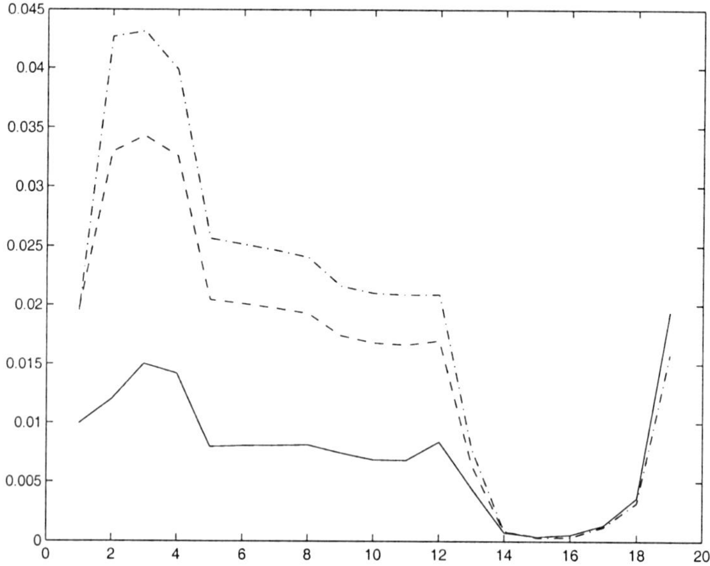

FIG. 3. *The time-averaged error as a function of model level (solid line for* ROS2-AMF, *dashed line for* S-SPLIT, *and dash-dotted line for* STRANG*).*

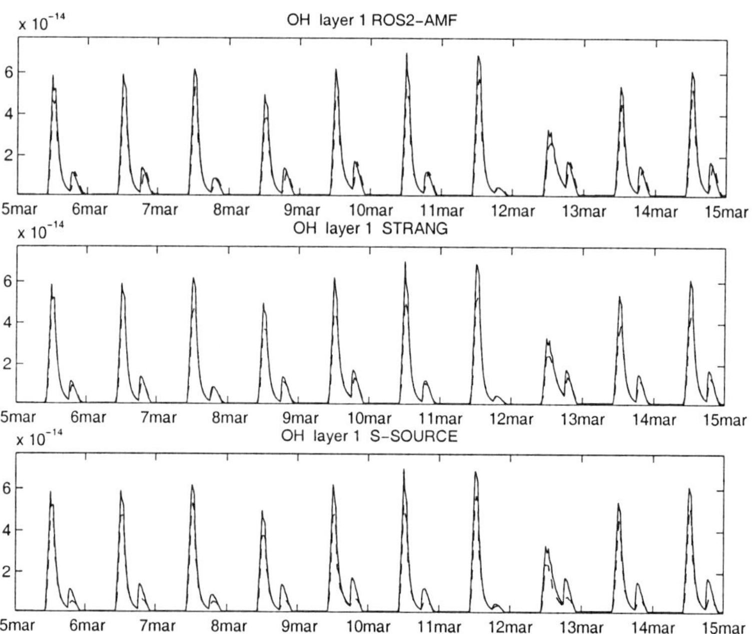

FIG. 4. *The OH solution in the surface cell as a function of time in molecules per air molecules (volume mixing ratio).*

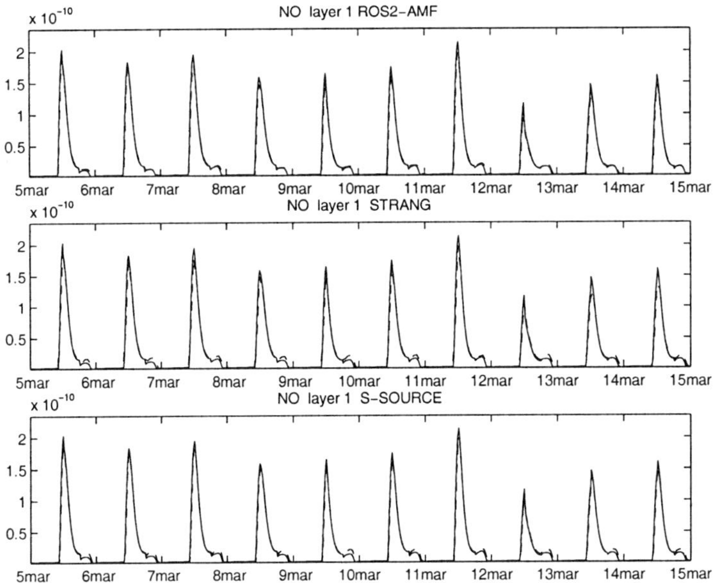

FIG. 5. *The NO solution in the surface cell as a function of time in molecules per air molecules (volume mixing ratio).*

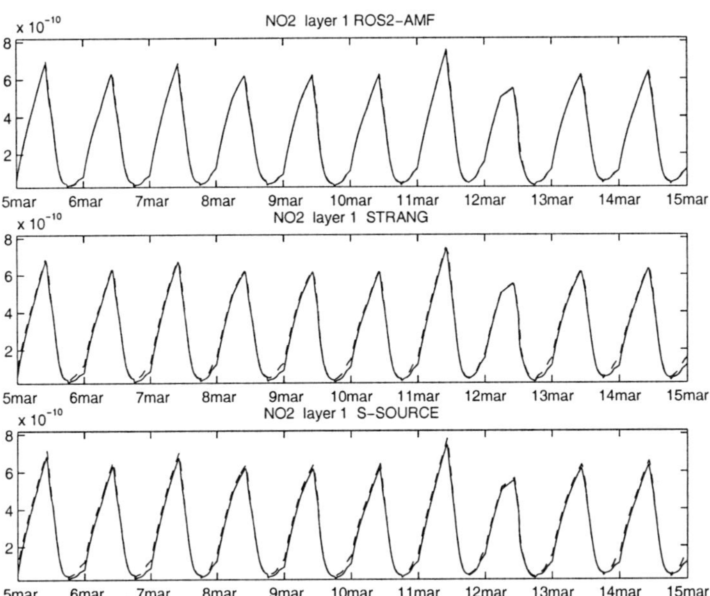

FIG. 6. *The NO_2 solution in the surface cell as a function of time in molecules per air molecules (volume mixing ratio).*

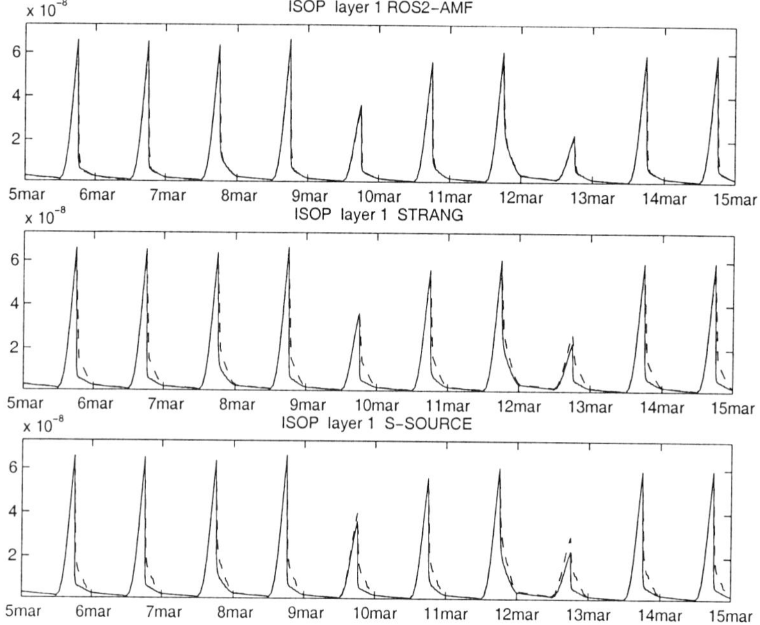

FIG. 7. *The ISOP solution in the surface cell as a function of time in molecules per air molecules (volume mixing ratio).*

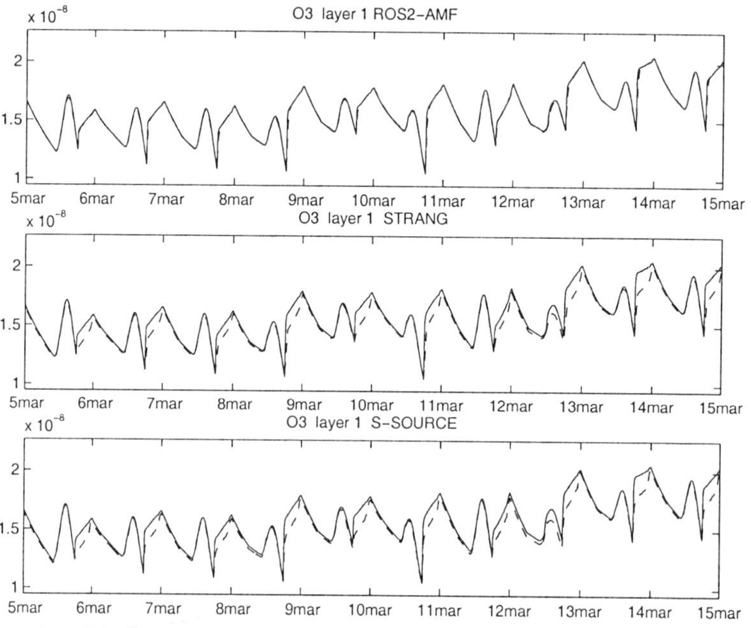

FIG. 8. *The O_3 solution in the surface cell as a function of time in molecules per air molecules (volume mixing ratio).*

butions for NH_3 and HNO_4. For ROS2-AMF they differ considerably in that the errors are more evenly distributed over the species. For all methods, the species that contribute most to the total error are those species which react strongly to changes in solar radiation (the only exception being NH_3). This happens at sunrise and sunset, as well as at the times of the vertical transport updates, when the cloudiness changes also. Occasionally these errors are larger than 10% of the daily variations. This confirms the finding in [3] that N_2O_5 has large errors, which are related to sunrise and sunset. As mentioned, the only exception is NH_3, which is not (photo)chemically active in this model. However, it has strong emissions and (solar radiation dependent) depositions, which are split from the vertical transport in the numerical methods, see the argumentation and equation (12) in Section 2.2.

In general, both S-SPLIT and STRANG, based on the standard operator splitting, perform well. However, it seems to be crucial for the performance of STRANG that the step size for the vertical substep is not taken larger than for the chemistry substep. Our experiments showed that alternating two 3-hour vertical-transport substeps $(T^{(3)})$ with three 2-hour groups of half an hour chemistry substeps $(R^{(2)} = \tilde{R}^{(2)} \circ \tilde{R}^{(2)} \circ \tilde{R}^{(2)} \circ \tilde{R}^{(2)})$ within a 6-hour timestep (i.e. the splitting method used in TM3 with a 6-hour time step with operator sequence $R^{(2)} \circ T^{(3)} \circ R^{(2)} \circ T^{(3)} \circ R^{(2)}$) leads to dramatical loss of accuracy. Errors may reach the order of magnitude of the daily concentration variations. On the other hand, STRANG seems to be not very dependent on whether emissions and depositions are included in the chemistry substep or handled separately as an extra operator in the splitting. We note that S-SPLIT, though being a first order method, compares well with the second order STRANG and ROS2-AMF. Again, it is likely that this is caused by reduced splitting in the S-SPLIT method. The splitting is reduced because the effect of vertical transport has direct influence on the chemistry (but not the other way around) through the term S in equation (18).

As noted earlier, the time-accurate reference solution accurately follows the rapid transitions in the morning and evening period, as at other points in time where the ODEs are strongly non-autonomous due to changes in model parameters (cloudiness, turbulence, etc.). ROS2-AMF is less accurate at those points. So it might be the case that the AMF in ROS2-AMF method has some difficulties with the non-autonomous nature of the problem. In order to investigate this, we compared the ROS2-AMF scheme with ROS2 applied to the whole unsplit problem with a time step of 30 minutes. We already noted that this scheme is by far more expensive than the other schemes. The solution produced by this fully implicit ROS2 scheme was hardly distinguishable from the cheap ROS2-AMF solution. This indicates that, at least in this particular case, the AMF provides a good approximation to the whole Jacobian linear system (cf. equation (21)). A more expensive, non-autonomous version of ROS2 might give somewhat better results. This was not investigated.

5. Conclusions. In a column submodel of the global dispersion model TM3, with strong mixing and stiff chemistry, we have compared the standard operator splitting method STRANG versus two other methods S-SPLIT and ROS2-AMF with (partially) eliminated splitting.

Our conclusions are:

- With relatively large time steps (of half an hour), all three methods perform well. Generally the errors in the concentrations are only a few percent of the total daily variations.
- Such accuracy is by far sufficient for the application area. An error limit of 10% is acceptable in atmospheric chemistry modelling, since other error sources, e.g. emissions, cause errors of similar or larger magnitude.
- In general the ROS2-AMF scheme appears to be the most accurate scheme, giving the smallest errors for most tracers and layers. In particular this is the case for the important surface layer where ROS2-AMF clearly gives the best solution.
- Since the only differences between the three methods are in the degree and the way of splitting, the differences in the errors must be caused by differences in operator splitting. In the present experiment this implies that splitting errors result from splitting chemistry (including emission and deposition) and transport. The higher accuracy of ROS2-AMF as compared to STRANG and S-SPLIT is therefore attributed to the absence of operator splitting in the former method. We note that the first-order S-SPLIT compares favorably with the second-order STRANG. This is attributed to the fact that splitting has been formally removed in S-SPLIT. Unacceptably large splitting errors remain restricted to short-lived species sensitive to solar radiation and those with strong emissions and depositions. For the accuracy of a standard operator splitting method (notably the splitting method used in TM3) we found that it is crucial not to interchange the relatively small substeps of the stiff process (chemistry) with much larger timesteps for the not-so-stiff process (transport), even if stability allows this.
- The computational costs of all three methods are essentially the same. Approximate Matrix Factorization (AMF) allows to apply the ROS2 scheme to the whole unsplit problem in an efficient way, thus avoiding splitting *at no extra cost.*
- Since all the schemes are based on the ROS2 integrator, we conclude that ROS2 proves successful for this difficult problem, and, based also on conclusions in [3], we recommend ROS2 as a viable method for use in the field of air pollution modelling.

Acknowledgements. This work has been sponsored by the Netherlands Organization for Scientific Research (NWO) under grants no. 613-

302-040, 750-298-02, by the Space Research Organization Netherlands (SRON), and by the Netherlands Institute for Computer Facilities (NCF).

REFERENCES

[1] I. Ahmad and M. Berzins. An algorithm for ODEs from atmospheric dispersion problems. *Appl. Num. Math.*, **25**: 137–150, 1997.

[2] R.M. Beam and R.F. Warming. An implicit finite-difference algorithm for hyperbolic systems in conservation-law form. *J. Comp. Phys.*, **22**: 87–110, 1976.

[3] J.G. Blom and J.G. Verwer. A comparison of integration methods for atmospheric transport-chemistry methods. *J. Comp. Appl. Math.*, to appear in 2001.

[4] V. Damian-Iordache. KPP – chemistry simulation development environment. Master's thesis, University of Iowa, 1996.

[5] http://www.cgrer.uiowa.edu/people/vdamian/kpp/.

[6] K. Dekker and J.G. Verwer. *Stability of Runge-Kutta methods for stiff non-linear differential equations.* North-Holland Elsevier Science Publishers, 1984.

[7] F.J. Dentener, J. Feichter, and A. Jeuken. Simulation of transport of ^{222}Rn using on-line and off-line global models at different horizontal resolutions: A detailed comparison with measurements. *Tel. Ser. B*, **51**: 573–602, 1999.

[8] I. Dimov, I. Faragó, and Z. Zlatev. Commutativity of the operators in splitting methods for air pollution models. Technical Report 04/99, Central Laboratory for Parallel Processing, Bulgarian Academy of Sciences, Sofia, 1999.

[9] E.G. D'yakonov. Difference systems of second order accuracy with a divided operator for parabolic equations without mixed derivatives. *USSR Comput. Math. Math. Phys.*, **4**(5): 206–216, 1964.

[10] W. Guelle, Y.J. Balkanski, M. Schulz, F. Dulac, and P. Monfray. Wet deposition in a global size-dependent aerosol transport model, 1, comparison of a 1 year ^{210}Pb simulation with ground measurements. *J. Geophys. Res.*, **103**: 11429–11445, 1998.

[11] E. Hairer and G. Wanner. *Solving Ordinary Differential Equations II. Stiff and Differential-Algebraic Problems.* Springer-Verlag, Berlin, 2nd edition, 1996.

[12] M. Heimann. The global atmospheric tracer model TM2. Technical report, DKRZ (Deutsches Klimarechenzentrum), Hamburg, Germany, 1995.

[13] S. Houweling, F.J. Dentener, and J. Lelieveld. The impact of non-methane hydrocarbon compounds on tropospheric photochemistry. *J. Geophys. Res.*, **103**: 10673–10696, 1998.

[14] P.J. v.d. Houwen and B.P. Sommeijer. Approximate factorization for time-dependent partial differential equations. *J. Comp. Appl. Math.*, 2000. Report MAS-R9915, CWI (Centre for Mathematics and Computer Science), Amsterdam, The Netherlands.

[15] C. Kessler. *Entwicklung eines effizienten Lösungsverfahrens zur modellmäßigen Beschreibung der Ausbreitung und chemischen Umwandlung reaktiver Luftschadstoffe.* PhD thesis, Fakultät für Maschinenbau der Universtät Karlsruhe, 1995.

[16] O. Knoth and R. Wolke. A comparison of fast chemical kinetic solvers in a simple vertical diffusion model. In S.-V. Gryning and M.M. Millán, editors, *Air Pollution Modelling and Its Application*, volume X, pp. 287–294. Plenum Press, 1994.

[17] O. Knoth and R. Wolke. Implicit-explicit Runge-Kutta methods for computing atmosperic reactive flows. *Appl. Num. Math.*, **28**: 327–341, 1998.

[18] D. Lanser and J.G. Verwer. Analysis of operator splitting for advection-diffusion-reaction problems from air pollution modelling. *J. Comp. Appl. Math.*, **111**: 201–206, 1999.

[19] J. Lelieveld and F.J. Dentener. What's controlling tropospheric ozone. *J. Geophys. Res.*, **105**: 3531, 2000.

[20] J.F. Louis. A parametric model of vertical eddy fluxes in the atmosphere. *Bound. Layer Met.*, **17**: 178–202, 1979.

[21] G.J. McRae, W.R. Goodin, and J.H. Seinfeld. Numerical solution of the atmospheric diffusion equation for chemically reacting flows. *J. Comp. Phys.*, **45**: 1–42, 1982.

[22] Url http://www.phys.uu.nl/~peters/TM3/TM3S.html.

[23] A. Sandu, F.A. Potra, G.R. Carmichael, and V. Damian. Efficient implementation of fully implicit methods for atmospheric chemical kinetics. *J. Comp. Phys.*, **129**: 101–110, 1996.

[24] B. Sportisse. An analysis of operator splitting techniques in the stiff case. *J. Comp. Phys.*, **161**: 140–168, 2000.

[25] M. Tiedtke. A comprehensive mass-flux scheme for cumulus parameterization in large-scale models. *Mon. Weather Rev.*, **117**: 1779–1800, 1989.

[26] J.G. Verwer, W. Hundsdorfer, and J.G. Blom. Numerical time integration for air pollution models. *Surv. Math. Ind.*, to appear in 2001. Report MAS-R9825, CWI (Centre for Mathematics and Computer Science), Amsterdam, The Netherlands.

[27] J.G. Verwer, E.J. Spee, J.G. Blom, and W. Hundsdorfer. A second order Rosenbrock method applied to photochemical dispersion problems. *SIAM J. Sci. Comp.*, **20**: 1456–1480, 1999.

[28] R.F. Warming and R.M. Beam. An extension of A-stability to alternating direction methods. *BIT*, **19**: 395–417, 1979.

[29] M.L. Wesely. Parameterization of surface resistance to gaseous dry deposition in regional-scale numerical models. *Atm. Env.*, **23**: 1293–1304, 1989.

[30] Z. Zlatev. *Computer Treatment of Large Air Pollution Models*. Kluwer Academic Publishers, 1995.

TIME-STEPPING METHODS THAT FAVOR POSITIVITY FOR ATMOSPHERIC CHEMISTRY MODELING

ADRIAN SANDU*

Abstract. Chemical kinetics conserves mass and renders non-negative solutions; a good numerical simulation would ideally produce mass balanced, positive concentration vectors. Many time stepping methods are mass conservative; however, unconditional positivity restricts the order of a traditional method to one. The projection method presented in [13] post-processes the solution of a one-step integration method to ensure mass conservation and positivity. It works best when the underlying time-stepping scheme favors positivity. In this paper several Rosenbrock-type methods that favor positivity are presented; they are designed such that their transfer functions together with several derivatives are nonnegative for all real, negative arguments.

Key words. Chemical kinetics, linear invariants, positivity, numerical time integration.

1. Introduction. Air quality models [3, 11] solve the convection-diffusion-reaction set of partial differential equations which describe the atmospheric physical and chemical processes. Usually an operator-split approach is taken: chemical equations and convection-diffusion equations are solved in alternative steps. In this setting the integration of chemical kinetic equations is a demanding computational task. The chemical integration algorithm should be stable in the presence of stiffness; ensure a modest level of accuracy, typically 1%; preserve mass; and keep the concentrations positive.

Most popular ODE integrators (multistep, Runge-Kutta, Rosenbrock) preserve mass, but positivity is more difficult to achieve. Unconditional positivity restricts the order of a numerical method to one [2]. Clipping (setting the negative concentrations to zero) enhances stability but artificially adds mass to the system. Solution projection and stabilization [13] are postprocessing techniques which allow a positive, mass balanced solution without restrictions on the order or time step of the integration method.

In order to be effective, projection and stabilization must be paired with time stepping methods that favor positivity. Although positivity is not guaranteed, these methods tend to keep nonnegative concentrations and reduce the postprocessing overhead. One example of positivity favorable method is Ros2 (advocated by Verwer et al. [15]).

In this paper we explore several methods that favor positivity. These methods are developed based on the conjecture that good positivity behavior is related to the positivity of the transfer function and its first several

---

*Department of Computer Science, Michigan Technological University, 205 Fisher Hall, 1400 Townsend Drive, Houghton, MI 49931 (`asandu@mtu.edu`).

derivatives for negative real arguments. All methods are of Rosenbrock type. Numerical experiments show that the new methods paired with solution projection produce smaller overheads.

The paper is organized as follows. Section 2 reviews the properties of chemical kinetic systems. Positive time stepping methods are discussed in Section 3 (following Bolley and Crouzeix [2]), and in Section 4 (solution postprocessing techniques). Section 5 presents new time stepping methods that favor positivity. A test case from stratospheric chemistry is considered in Section 6, where different numerical results are presented. Finally, the findings and conclusions of the paper are summarized in Section 7. Appendix A defines the numerical integration methods and Appendix B the test chemical mechanism.

2. Mass action kinetics, linear invariants and positivity. Consider a chemical kinetic system with N species $y_1, \cdots, y_N$ interacting in r chemical reactions. The time evolution of the chemical concentrations is governed by the "mass action kinetics" differential law[1]

$$(2.1) \qquad\qquad y' = S \cdot \omega(y) , \quad y(t_0) = y^0 ,$$

where $S \in \Re^{N \times r}$ is the matrix of stoichiometric coefficients and $\omega \in \Re^r$ is the vector of reaction speeds. Any solution of (2.1) satisfies

$$(2.2) \qquad A^T y(t) = A^T y^0 = b = \text{const.} \qquad \text{for all } t \geq t_0 ,$$

where $A \in \Re^{N \times m}$ is a matrix whose columns span $\ker\left(S^T\right)$, i.e. $A^T S = 0$. This implies $A^T y'(t) = 0$ and $A^T y(t)$ is invariant in time. The vector $b \in \Re^m$ contains the m invariant values. Simply stated, the existence of linear invariants ensures that mass is conserved during chemical reactions.

Let us now separate the production terms $P(y)$ from the destruction terms $D(y)$ in (2.1)

$$y_i' = P_i(y) - D_i(y)\, y_i , \qquad 1 \leq i \leq N .$$

The special form of the reaction speeds ensures that $P_i(y)$ and $D_i(y)$ are polynomials in y with positive coefficients. If at time moment τ all concentrations are nonnegative, $y(\tau) \geq 0$, and the concentration of species i is zero, $y_i(\tau) = 0$, then the corresponding derivative is nonnegative, $y_i'(\tau) = P_i(\tau) \geq 0$, which implies that

$$(2.3) \qquad\qquad y(t_0) \geq 0 \quad \Longrightarrow \quad y(t) \geq 0 \quad \text{for all } t \geq t_0 .$$

In short, the concentrations cannot become negative during chemical reactions.

---

[1] We denote by y_i both the chemical species i and its mass concentration.

Linear invariants (2.2) and positivity (2.3) imply that the solution of (2.1) remains all the time within the reaction simplex

$$(2.4) \quad y(t) \in \mathcal{S} \text{ for all } t \geq t_0 , \quad \mathcal{S} = \left\{ y \in \Re^s \text{ s.t. } A^T y = b \text{ and } y \geq 0 \right\} .$$

A general principle in scientific computing says that the numerical solution must capture (as much as possible) the qualitative behavior of the true solution. Good numerical methods for integrating chemical reaction models (2.1) should therefore *be unconditionally stable*, as the system is usually stiff (this requires implicit integration formulas); should *preserve the linear invariants*, otherwise artificial mass sources (or sinks) are introduced; and should *preserve solution positivity*.

Negative concentrations are non-physical. In addition, the kinetic system may become unstable for negative concentrations, as shown by Verwer et al. [15]. An operator-split solution of convection-diffusion-reaction atmospheric equations alternates chemical integration steps with advection steps; negative concentrations from chemical integration can hurt the positivity of the following advection step, which will perturb the next chemical step etc. leading to poor quality results.

It is well known that the most popular integration methods (Runge-Kutta, Rosenbrock and Linear Multistep) preserve exactly[2] all the linear invariants of the system [15]. Moreover, the (modified) Newton iterations used to solve for implicit solutions also preserve the linear invariants at each iteration. With linear-preserving integration methods the accuracy of the individual components is given by the truncation errors (e.g. having a magnitude 10^{-4}), while the accuracy of the linear invariants is only affected by the roundoff errors (having a much smaller magnitude, e.g. 10^{-14}).

3. Positive time-stepping methods. Positivity of the numerical solution is more difficult to achieve. We would like to obtain higher order methods that unconditionally preserve positivity. The results of Bolley and Crouzeix [2] show that unconditional positivity limits the order of the numerical method to one. For convenience we briefly review their analysis. Bolley and Crouzeix [2] focus on linear systems

$$(3.1) \qquad y' = A y \quad \text{with} \quad A \in \mathcal{M}_\alpha , \quad y(t_0) = v \geq 0 ,$$

where

$$\mathcal{M}_\alpha = \{ A \in \Re^{n \times n} \; : \; A_{ij} \geq 0 \text{ for } i \neq j \text{ and } A_{ii} \geq -\alpha , \forall i \} .$$

A one step method of order p, applied to (3.1) gives

$$y^{n+1} = R(hA) y^n ,$$

---

[2]If the computations are performed in infinite arithmetic precision.

where $R(z)$ is a rational function whose Taylor expansion around zero matches the expansion of e^z up to z^p. A necessary and sufficient condition for the scheme to be positive is that $R(hA) \geq 0$.

Let $M = -h\alpha I$ and $N = h(\alpha I + A)$, such that $M + N = hA$ and $N \geq 0$. Bolley and Crouzeix established that the following Taylor type expansion holds

$$(3.2) \quad R(M + N) = \sum_{j \geq 0} \frac{1}{j!} R^{(j)}(M) \cdot N^j = \sum_{j \geq 0} \frac{h^j}{j!} R^{(j)}(-h\alpha) \cdot (\alpha I + A)^j \ .$$

Let γ_R be the largest positive number such that $R(z)$ and all its derivatives $R'(z)$, $R''(z)$, $\cdots$ are nonnegative for $z \in [-\gamma_R, 0]$. We say that R is *absolutely monotonic* on $[-\gamma_R, 0]$. Since $\alpha I + A \geq 0$, it is clear that

$$(3.3) \qquad R(hA) \geq 0 \ \text{ for all } \ A \in \mathcal{M}_\alpha \quad \Leftrightarrow \quad h \leq \frac{\gamma_R}{\alpha} \ .$$

For (linearized) chemical kinetics $1/\alpha$ is roughly the lifetime of the fastest species in the system. Consequently, the positivity step size restriction (3.3) is similar to the upper bound required for the stability of an explicit integration scheme. This leads to impractically small steps for stiff chemical systems.

In practice we need a numerical method which unconditionally preserves positivity: $A \in \mathcal{M}_\alpha \Rightarrow R(hA) \geq 0$ for all h. From (3.3) this method must have a stability function with $\gamma_R = \infty$, i.e. $R(z)$ must be absolutely monotonic on $(-\infty, 0]$. But this condition prohibits R of approximating e^z to more than first order (see [2]). Therefore there is no one-step method which unconditionally conserves positivity of order greater than or equal to two.

Hundsdorfer [8] proved that the (first order) implicit Euler method is unconditionally positive. In practice, even implicit Euler may produce negative values since the iterative solution process is halted after a finite number of steps; while the exact solution is non-negative, the successive approximations computed by (modified) Newton are not.

4. Positivity by solution postprocessing. Postprocessing corrects the computed solution at each step such that negative concentrations are avoided. The most popular method is *clipping*, which simply sets the negative components to zero. Clipping destroys the preservation of linear invariants. Moreover, all clipping errors act in the same direction, namely increase mass (artificially); therefore they accumulate over time and may lead to significant global errors over longer simulation intervals.

In [13] two methods for preserving both mass balance and positivity were developed. *Solution projection* method is based on a linear invariant-preserving one-step integration method Φ (e.g. Runge-Kutta or Rosenbrock)

$$z^{n+1} = \Phi_h^f(y^n) \ .$$

Here t^n denotes the discrete time value at n^{th} step, y^n the nth step solution, $h = t^{n+1} - t^n$ the step size, and $f(t, y) = S\omega(t, y)$.

If some of the computed concentrations are negative

$$z_{i1}^{n+1} < 0 \quad \cdots \quad z_{ip}^{n+1} < 0 \, ,$$

the next step approximation y^{n+1} is the solution of the following quadratic optimization problem

$$(4.1) \quad \min \ \frac{1}{2} \left\| y^{n+1} - z^{n+1} \right\|_G^2 \quad \text{subject to} \quad A^T y^{n+1} = b \, , \ y^{n+1} \geq \epsilon \, .$$

If there are no negative components then $y^{n+1} = z^{n+1}$. The solution y^{n+1} is the projection of z^{n+1} onto the reaction simplex $\mathcal{S}$; ϵ are small numbers which ensure a positive solution in the presence of roundoff. The norm $\|w\|_G = \sqrt{w^T G w}$ is the typical error norm used by the ODE step size controller, with the positive matrix G given at each step by

$$(4.2) \qquad\qquad G = \operatorname*{diag}_{1 \leq i \leq s} \left[\frac{1}{s \left(atol + rtol \, |z_i^{n+1}| \right)^2} \right] .$$

In [13] it was shown that the projected vector is a better (G-norm) approximation to the true solution then is the computed vector,

$$\left\| y^{n+1} - y(t^{n+1}) \right\|_G \leq \left\| z^{n+1} - y(t^{n+1}) \right\|_G \, .$$

Any algorithm for quadratic programming can be employed to solve (4.1). We found the primal-dual algorithm of Goldfarb and Idnani [5] to be a suitable solution method.

Solution stabilization is a simpler alternative to projection. The next step solution is computed as

$$
\begin{aligned}
z^{n+1} &= \Phi_h^f(y^n) \quad \text{with} \quad z_{i1}^{n+1}, \cdots, z_{ip}^{n+1} < 0 \, , \\[2mm]
B &= [A \,|\, e_{i1} \,|\, \cdots \,|\, e_{ip}] \, , \\[2mm]
(4.3) \qquad y^{n+1} &= z^{n+1} - G^{-1} B \left(B^T G^{-1} B \right)^{-1}
\begin{bmatrix}
A^T z^{n+1} - b \\
z_{i1}^{n+1} - \epsilon_{i1} \\
\vdots \\
z_{ip}^{n+1} - \epsilon_{ip}
\end{bmatrix} .
\end{aligned}
$$

It can be directly verified that the solution satisfies $A^T y^{n+1} = b$, $y_{i1}^{n+1} = \epsilon_{i1}, \cdots, y_{ip}^{n+1} = \epsilon_{ip}$. The method does not guarantee positivity, since the projection step may render other components negative (i.e. $y_j^{n+1} < 0$ for $j \neq i1...ip$), but can have a beneficial effect on maintaining positivity.

In [13] it was shown that projection and stabilization work best when the underlying numerical method favors positivity. Such methods are discussed next.

5. Methods that favor positivity. In the view of the theory developed in [2] we do not require the methods to be unconditionally positive, but we relax the requirements to methods which favor positivity. Although positivity is not guaranteed, these methods tend to keep nonnegative concentrations and reduce the postprocessing overhead.

In [15, 16, 17] it was noted that the second order Rosenbrock method Ros2 (A.5) has favorable positivity properties, and the method is stable for nonlinear problems even with large fixed step sizes. It was also noted that Ros2 provides positive solutions for the scalar problems $C' = -kC$ and $C' = -k\,C^2$, $C(t_0) \geq 0$.

In [13] we conjectured that a possible explanation for the good observed behavior is that not only the transfer function of this method, but also its first two derivatives are nonnegative for real, negative arguments (i.e. $R(z), R'(z), R''(z) \geq 0$ for any $z \leq 0$). Therefore the first (most significant) terms in the Taylor series (3.2) are nonnegative. Higher order terms may be negative, but they are weighted by higher powers of the step size h, therefore the negative part will hopefully be small.

Based on this conjecture we develop several methods whose transfer functions and their first several derivatives are nonnegative for any real, negative argument. Specifically, we consider one-step, second order, s-stage methods ($s \geq 2$) with a stability function of the form

$$R(z) = \frac{1 - az}{(1 - \gamma z)^s} \quad \text{with} \quad a = \frac{1 + \sqrt{s}}{s - 1}, \quad \gamma = \frac{s + \sqrt{s}}{s(s - 1)}.$$

The relations for a and γ follow from imposing R to approximate the exponential to second order. We have

$$R'(z) = \frac{1 - \gamma\left(\sqrt{s} + 1\right)z}{(1 - \gamma z)^{s+1}},$$

$$R''(z) = \frac{\gamma s\left(2 + (s - 1)\gamma\right) - \gamma^2 s\left(\sqrt{s} + 1\right)z}{(1 - \gamma z)^{s+2}},$$

and we see that $R(z), R'(z), R''(z) \geq 0$ for any $z \leq 0$. Ros2 falls into this category of methods for $s = 2$. For more stages these methods have the added benefit of higher damping of transients, and faster steady-state introduced for short-lived species.

We developed four such methods which, to our knowledge, have not been proposed elsewhere. The methods are second order accurate and require only two function evaluations:

- Method A (A.6) is three-stage, second-order, stiffly accurate. It preserves its order for inexact Jacobians;
- Method B (A.7) is three-stage, second-order, stiffly accurate. One of the order 3 conditions is also satisfied;
- Method C (A.8) is three-stage, second-order (but not stiffly accurate). It works with inexact Jacobians; One of the order 3 conditions is also satisfied.

- Method D (A.9) is four-stage, second-order, stiffly accurate. It allows inexact Jacobians and has an embedded method of order 3.

The coefficients for all methods are given in Appendix A.

6. Numerical Results. We consider the basic stratospheric reaction mechanism presented in Appendix B (and adapted from NASA HSRP/ AESA [9]). The numerical examples are implemented in MATLAB. The simulation starts at noon with the initial concentrations of Table 1 and continues for 72 hours. The computation of G-norms was done with $rtol = 10^{-5}$ and $atol = 10^{-3}$. Throughout the tests the minimal values were set to $\epsilon_i = 1$ molec/cm^3. Reference solutions were obtained with the MATLAB integration routine ODE15S (variable order numerical differentiation formula); the control parameters were $RelTol = 10^{-8}$, $AbsTol = 10^{-8}$, with analytic Jacobian.

TABLE 1

Initial concentrations for the simulation (molec/cm^3).

O^{1D}	O	O_3	O_2	NO	NO_2
9.906E01	6.624E08	5.326E11	1.697E16	8.725E08	2.240E08

The integration algorithms used are BDF2 (A.1), Ros2 (A.5), Rodas3 (A.4), RK2 and RK2+ (A.2), as well as the new methods A (A.6), B (A.7), C (A.8), and D (A.9). For the Rosenbrock methods we use the non autonomous forms. RK2 and RK2+ differ by the value of the coefficient γ; the classical formula RK2 is more accurate, while RK2+ has the same transfer function as Ros2 (hence falls in the previously discussed category of positive favorable methods).

To compare the performance of different methods we measured the solution accuracy at the end of the integration interval ($t = T_F$). With y^R the reference solution and y the computed solution, the error measure reads

$$(6.1) \qquad E = \sqrt{\frac{1}{s} \sum_{i=1}^{s} \left(\frac{y_i\,(T_F) - y_i^R\,(T_F)}{y_i^R\,(T_F)} \right)^2 }.$$

Figure 1 shows the solution accuracies (6.1) versus computational work (Kflops) for different methods (plain versions). All methods perform similarly, with methods A and B slightly better at low accuracies and C more performant at high accuracies. RK2+ and BDF2 are not competitive at low accuracies, presumably due to convergence problems at large step sizes. The conclusions of this diagram are limited, however, since the system is small and sparsity is not accounted for.

Figures 2, 3, and 4 show the average negative values and the percent of time negative values were obtained for O^{1D}, O and NO respectively.

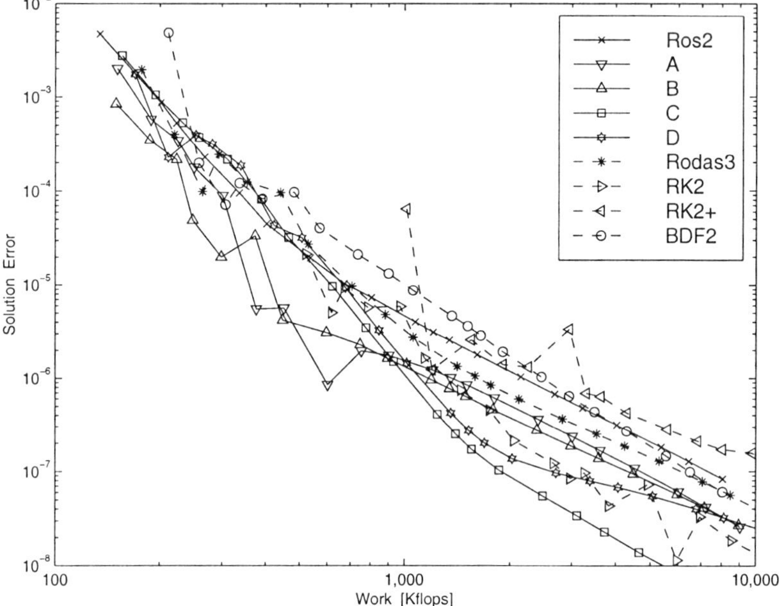

FIG. 1. *Work-precision diagram shows the solution accuracy versus the computational work. All methods perform similarly.*

For O^{1D} (Figure 2) methods A, B, D give reasonably small negative concentrations. Even smaller concentrations (in absolute value) are produced by BDF2 and Rodas3. The percent of steps which produced negative O^{1D} concentrations changes for different step sizes for each of the methods, so there is no "clear winner" here. For O (Figure 3) method D shows negligible percent of negative values for steps up to 30 min. Methods A and B show small percent for small and large steps, while C and Ros2 show large percents of negative values. The mean negative values are smallest for methods A, B and D. Also RK2+ shows better results than RK2 for small step sizes. For NO (Figure 4) the methods A, B (up to 15 min), C (up to 10 min), and D (up to 30 min) show a negligible percent of negative steps. The mean negative concentrations are smallest for A, B and D. RK2+ again produces fewer times negative concentrations than Ros2, but the average negative concentrations are large.

Interestingly, the Ros2 performance is rather modest with regard to positivity. We assume that this is due to the fact that, once a concentration c^n accidentally becomes negative, the next step approximation (of the form $c^{n+1} \approx R(hJ)c^n$) will tend to remain negative precisely because of the positivity of R.

Figure 5 shows the computational overhead (percentage) of the projected versions versus the plain versions. At small time steps (large compu-

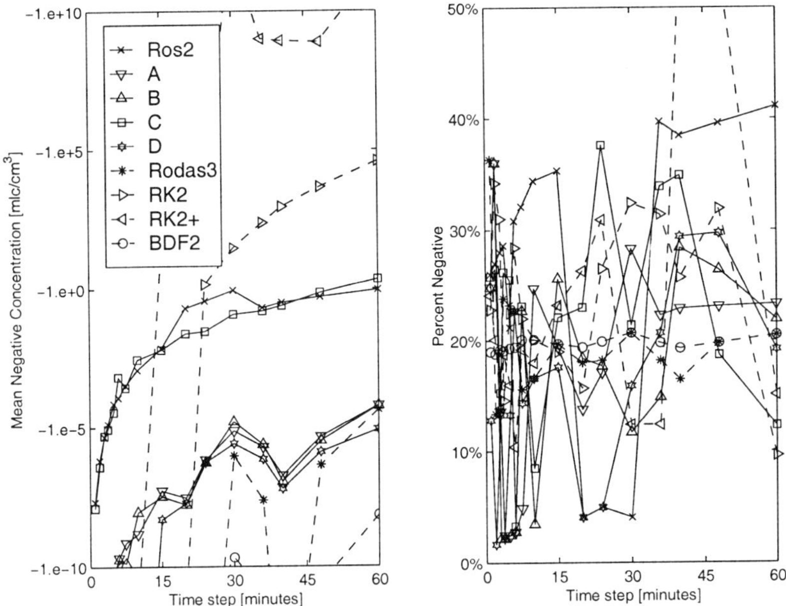

FIG. 2. *Mean negative values of O^{1D} (left) and the percent of steps which produced negative values (right). Different methods and different step sizes are considered.*

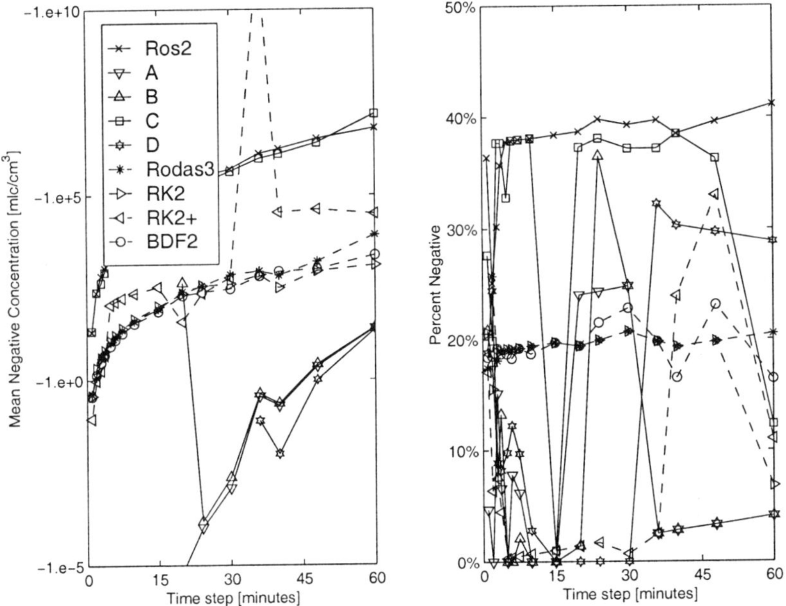

FIG. 3. *Mean negative values of O (left) and the percent of steps which produced negative values (right). Different methods and different step sizes are considered.*

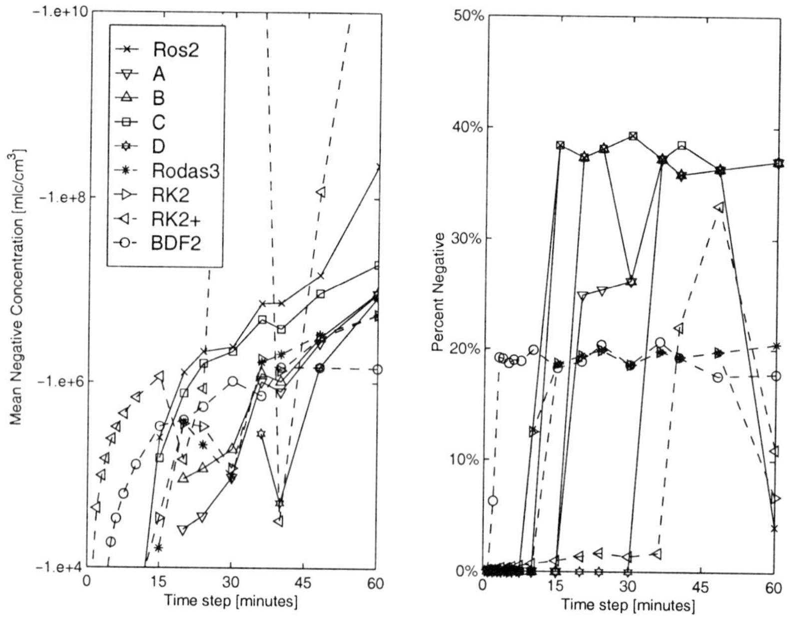

FIG. 4. *Mean negative values of NO (left) and the percent of steps which produced negative values (right). Different methods and different step sizes are considered.*

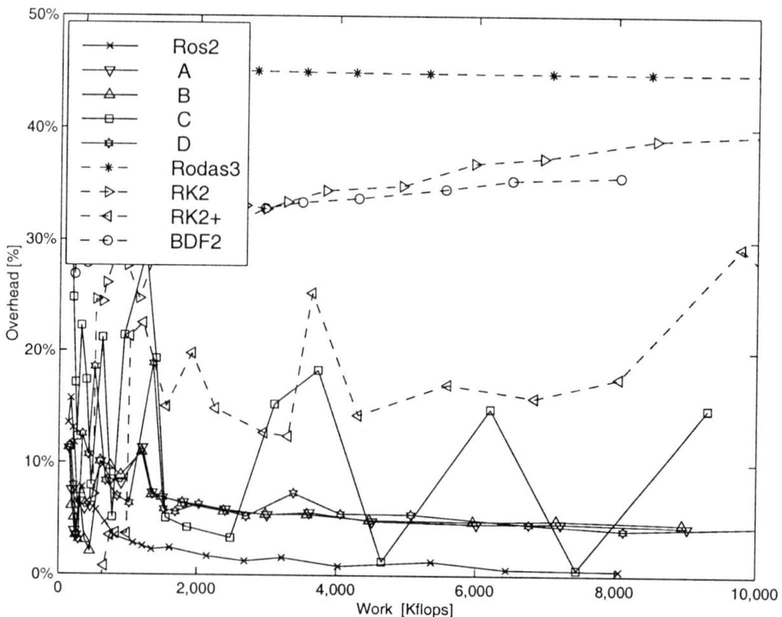

FIG. 5. *Overheads incurred by solution projection versus computational work of the plain version.*

tational work and high accuracies) Ros2 has negligible overheads; methods A, B and D have overheads around 5%, while the overheads of method C are different at different step sizes. RK2+ has half the overhead of RK2. BDF2, RK2 and Rodas3 have higher overheads. At large steps (small computational work and low accuracies) Ros2 is the clear winner, with small overheads.

These experiments show a clear difference between the plain versions and the projected versions. While the plain Ros2 performance is modest with regard to positivity, the corrected version performance is excellent. This can be explained by the fact that once the negative concentration c^n are corrected, then the next step value c^{n+1} will tend to remain positive.

7. Conclusions. Projection and stabilization ensure mass conservation and positivity for the numerical solutions of chemical kinetic problems. The techniques are based on postprocessing the next-step approximations given by linear-preserving methods, should negative concentrations develop.

Both techniques have to be paired with a positivity-favorable numerical integration methods, in order to reduce the overheads. Although such methods do not guarantee positivity, they seldom produce non-positive results, which minimizes the overheads incurred by projection or stabilization. Such a method is for example Ros2 [16, 17]; Verwer et al. explained the good properties of Ros2 by the fact that its stability function $R(z) \geq 0$ for all $z \leq 0$, and also by the fact that the result of integrating $c' = -c^2$, $c(t_0) \geq 0$ is unconditionally positive.

Based on the theory of Bolley and Crouzeix [2] we conjectured that the positivity of the first and second derivatives of the transfer function $R'(z), R''(z) \geq 0$ for all $z \leq 0$ play a role in making a method positivity favorable. Following this conjecture we derived four new Rosenbrock methods which fulfill this condition. In addition, we tested a a Runge-Kutta scheme (RK2+) whose transfer function has similar properties.

The results seem to confirm the conjecture. Solution projection gives low overheads when paired one of these positivity favorable techniques. Also, the modified RK2+ method has consistently given lower overheads than the original method RK2, even if the latter method is more accurate.

The conclusion, however, is not as clear cut as one hoped. First, method C does not behave as well as the other methods, suggesting that extra factors besides the transfer function determine positivity. Second, there is a significant difference between the behavior of the plain and the corrected methods. For example the Ros2 results are rather dissapointing (in terms of positivity) for the noncorrected version, but are excellent for the corrected version. The observed behavior is due to factors not captured by our linearized analysis. One important factor is the nonlinearity of the underlying chemical system, not considered by the transfer function analysis. Another factor is the stifness of the chemical system, which implies that terms of the form $h^k J$ are not necessarily small, therefore higher negative derivatives can influence significantly the solution and affect positivity.

APPENDIX

A. Numerical integration methods. The second order backward differentiation formula BDF2 [6, Section III.1] is

$$(\text{A.1}) \quad y^{n+1} = Y^n + \frac{2}{3} h f \left(t^{n+1}, y^{n+1} \right) , \quad Y^n = \frac{4}{3} y^n - \frac{1}{3} y^{n-1} .$$

Here $t^{n+1} = t^n + h = t^{n-1} + 2h$; for variable time steps the coefficients change. The very first step requires both y^1 and y^0; the former is given, while the latter is obtained with one backward Euler step.

The second order Runge-Kutta method RK2 [10] is

$$
\begin{aligned}
(\text{A.2}) \qquad y^{n+1} &= y^n + (1 - \gamma)\, k_1 + \gamma\, k2 \\
k_1 &= h f \left(t^n + \gamma h, y^n + \gamma k_1 \right), \\
k_2 &= h f \left(t^n + h, y^n + (1 - \gamma) k_1 + \gamma k_2 \right),
\end{aligned}
$$

with $\gamma = 1 - \sqrt{2}/2$. For RK2+ the value is $\gamma = 1 + \sqrt{2}/2$.

An s-stage Rosenbrock method reads

$$
\begin{aligned}
(\text{A.3}) \qquad y^{n+1} &= y^n + \sum_{i=1}^{s} m_i k_i , \\
\hat{y}^{n+1} &= y^n + \sum_{i=1}^{s} \hat{m}_i k_i , \\
\left(\frac{1}{\gamma h} I - J \right) k_i &= f \left(t^n + \alpha_i h, y^n + \sum_{j=1}^{i-1} a_{ij} k_j \right) \\
&\quad + \sum_{j=1}^{i-1} \left(\frac{c_{ij}}{h} \right) k_i + g_i h f_t , \quad i = 1, \cdots, s .
\end{aligned}
$$

where the Jacobian matrix $J = \partial f(t, y) / \partial y$ and the time partial derivative $f_t = \partial f(t, y) / \partial t$ are evaluated at $t = t^n$. A specific method is defined by its coefficients.

The Rodas3 method [12] is third order accurate and reads

$$\gamma = 1/2 \,, \qquad A = \begin{bmatrix} 0 & 0 & 0 & 0 \\ 0 & 0 & 0 & 0 \\ 2 & 0 & 0 & 0 \\ 2 & 0 & -1 & 0 \end{bmatrix} ,$$

$$(\text{A.4}) \quad C = \begin{bmatrix} 0 & 0 & 0 & 0 \\ 4 & 0 & 0 & 0 \\ 1 & -1 & 0 & 0 \\ 1 & -1 & -8/3 & 0 \end{bmatrix} , \qquad \alpha = \begin{bmatrix} 0 \\ 0 \\ 1 \\ 1 \end{bmatrix} ,$$

$$g = \begin{bmatrix} 1/2 \\ 3/2 \\ 0 \\ 0 \end{bmatrix} , \qquad m = \begin{bmatrix} 2 \\ 0 \\ 1 \\ 1 \end{bmatrix} , \qquad \hat{m} = \begin{bmatrix} 0 \\ 0 \\ 0 \\ 0 \end{bmatrix} .$$

The second order Rosenbrock scheme Ros2 [16, 17] is defined as

$$\gamma = 1 + \sqrt{2}/2 \,, \quad A = \begin{bmatrix} 0 & 0 \\ 2 - \sqrt{2} & 0 \end{bmatrix} ,$$

$$(\text{A.5}) \qquad C = \begin{bmatrix} 0 & 0 \\ -4 + 2\sqrt{2} & 0 \end{bmatrix} , \quad \alpha = \begin{bmatrix} 0 \\ 1 \end{bmatrix} ,$$

$$g = \begin{bmatrix} 1 + \sqrt{2}/2 \\ -1 - \sqrt{2}/2 \end{bmatrix} , \quad m = \begin{bmatrix} (6 - 3\sqrt{2})/2 \\ 1 - \sqrt{2}/2 \end{bmatrix} , \quad \hat{m} = \begin{bmatrix} 2 - \sqrt{2} \\ 0 \end{bmatrix} .$$

The vector $y^n + (1/\gamma)k_1$ is a consistent approximation at t^{n+1} and was used to implement the error estimator in the variable step formulation.

Method A is second-order, stiffly accurate, and is defined by

$$\gamma = (3 + \sqrt{3})/6 \,, \qquad A = \begin{bmatrix} 0 & 0 & 0 \\ 3 - \sqrt{3} & 0 & 0 \\ 3 - \sqrt{3} & 0 & 0 \end{bmatrix} ,$$

$$(\text{A.6}) \quad C = \begin{bmatrix} 0 & 0 & 0 \\ -6 + 4\sqrt{3} & 0 & 0 \\ 3 - 2\sqrt{3} & 3 - 2\sqrt{3} & 0 \end{bmatrix} , \qquad \alpha = \begin{bmatrix} 0 \\ 1 \\ 1 \end{bmatrix} ,$$

$$g = \begin{bmatrix} (3 + \sqrt{3})/6 \\ (1 + \sqrt{3})/2 \\ 0 \end{bmatrix} , \qquad m = \begin{bmatrix} 3 - \sqrt{3} \\ 0 \\ 1 \end{bmatrix} .$$

It requires only 2 function evaluations and preserves its order for inexact Jacobians.

Method B is second-order, stiffly accurate, and is defined by

$$\gamma = (3 + \sqrt{3})/6 \,, \qquad A = \begin{bmatrix} 0 & 0 & 0 \\ 3 - \sqrt{3} & 0 & 0 \\ 3 - \sqrt{3} & 0 & 0 \end{bmatrix} \,,$$

$$(\text{A.7}) \quad C = \begin{bmatrix} 0 & 0 & 0 \\ 0 & 0 & 0 \\ -4 + 2\sqrt{3} & 1 - \sqrt{3} & 0 \end{bmatrix} \,, \quad \alpha = \begin{bmatrix} 0 \\ 1 \\ 1 \end{bmatrix} \,,$$

$$g = \begin{bmatrix} (3 + \sqrt{3})/6 \\ (3 + \sqrt{3})/6 \\ 0 \end{bmatrix} \,, \qquad m = \begin{bmatrix} 3 - \sqrt{3} \\ 0 \\ 1 \end{bmatrix} \,.$$

One of the order 3 conditions is also satisfied.

Method C is second-order (but not stiffly accurate) and is defined by

$$A = \begin{bmatrix} 0 & 0 & 0 \\ 0 & 0 & 0 \\ (1272 - 823\sqrt{3})/354 & 3(-51 + 49\sqrt{3})/59 & 0 \end{bmatrix} \,,$$

$$C = \begin{bmatrix} 0 & 0 & 0 \\ -1 + 7\sqrt{3}/18 & 0 & 0 \\ (25 - 13\sqrt{3})/6 & 12 - 6\sqrt{3} & 0 \end{bmatrix} \,,$$

$$(\text{A.8})$$

$$\gamma = (3 + \sqrt{3})/6 \,, \quad \alpha = \begin{bmatrix} 0 \\ 0 \\ 2/3 \end{bmatrix} \,,$$

$$g = \begin{bmatrix} (3 + \sqrt{3})/6 \\ (39 + 14\sqrt{3})/108 \\ (9 + \sqrt{3})/6 \end{bmatrix} \,, \quad m = \begin{bmatrix} (-12089 + 5037\sqrt{3})/472 \\ 9(344 - 135\sqrt{3})/118 \\ 3(3 - \sqrt{3})/4 \end{bmatrix} \,.$$

It also requires only 2 function evaluations and is more accurate than method A, since one of the order 3 conditions is also satisfied. It works with inexact Jacobians.

The stability function for both methods A and B is

$$R(z) = \frac{1 - \dfrac{1 + \sqrt{3}}{2} z}{\left(1 - \dfrac{3 + \sqrt{3}}{6} z\right)^3}$$

and we have $R(z), R'(z), R''(z) \geq 0$ for $z \leq 0$.

Method D is second-order, stiffly accurate,

$$\gamma = \frac{1}{2}, \qquad A = \begin{bmatrix} 0 & 0 & 0 & 0 \\ 2 & 0 & 0 & 0 \\ 2 & 0 & 0 & 0 \\ 2 & 0 & 0 & 0 \end{bmatrix},$$

$$\text{(A.9)} \quad C = \begin{bmatrix} 0 & 0 & 0 & 0 \\ -4/3 & 0 & 0 & 0 \\ -10/3 & -2 & 0 & 0 \\ -1/2 & 0 & 3/2 & 0 \end{bmatrix}, \qquad \alpha = \begin{bmatrix} 0 \\ 1 \\ 1 \\ 1 \end{bmatrix},$$

$$g = \begin{bmatrix} 1/2 \\ 1/6 \\ -1/2 \\ 0 \end{bmatrix}, \qquad m = \begin{bmatrix} 2 \\ 0 \\ 0 \\ 1 \end{bmatrix}, \qquad \hat{m} = \begin{bmatrix} 8/3 \\ 1 \\ 1 \\ -1/3 \end{bmatrix}.$$

It requires 2 function evaluations. It allows inexact Jacobians and the embedded method is order 3. The stability function is

$$R(z) = \frac{1-z}{\left(1 - \dfrac{z}{2}\right)^4}$$

and we have $R(z), R'(z), R''(z), R'''(z) \geq 0$ for $z \leq 0$.

B. The test mechanism. Consider the basic stratospheric reaction mechanism (adapted from NASA HSRP/AESA [9])

$$\text{(B.1)} \quad \begin{aligned} r1) \quad & O_2 + h\nu \xrightarrow{k_1} 2\,O && (k_1 = 2.643 \times 10^{-10} \cdot \sigma^3) \\ r2) \quad & O + O_2 \xrightarrow{k_2} O_3 && (k_2 = 8.018 \times 10^{-17}) \\ r3) \quad & O_3 + h\nu \xrightarrow{k_3} O + O_2 && (k_3 = 6.120 \times 10^{-04} \cdot \sigma) \\ r4) \quad & O + O_3 \xrightarrow{k_4} 2\,O_2 && (k_4 = 1.576 \times 10^{-15}) \\ r5) \quad & O_3 + h\nu \xrightarrow{k_5} O^{1D} + O_2 && (k_5 = 1.070 \times 10^{-03} \cdot \sigma^2) \\ r6) \quad & O^{1D} + M \xrightarrow{k_6} O + M && (k_6 = 7.110 \times 10^{-11}) \\ r7) \quad & O^{1D} + O_3 \xrightarrow{k_7} 2\,O_2 && (k_7 = 1.200 \times 10^{-10}) \\ r8) \quad & NO + O_3 \xrightarrow{k_8} NO_2 + O_2 && (k_8 = 6.062 \times 10^{-15}) \\ r9) \quad & NO_2 + O \xrightarrow{k_9} NO + O_2 && (k_9 = 1.069 \times 10^{-11}) \\ r10) \quad & NO_2 + h\nu \xrightarrow{k_{10}} NO + O && (k_{10} = 1.289 \times 10^{-02} \cdot \sigma) \end{aligned}$$

Here $M = 8.120E + 16$ molec/cm^3 is the atmospheric number density; the rate coefficients are scaled for time t in seconds; and $\sigma(t)$ represents the normalized sunlight intensity,

$$T_L = \left(\frac{t}{3600}\right) \bmod 24 \; ;$$

$$T_R = 4.5 \text{ (SunRise)}; \quad T_S = 19.5 \text{ (SunSet)};$$

$$\sigma(t) = \begin{cases} \dfrac{1}{2} + \dfrac{1}{2}\cos\left(\pi \left|\dfrac{2\,T_L - T_R - T_S}{T_S - T_R}\right| \left[\dfrac{2\,T_L - T_R - T_S}{T_S - T_R}\right]\right) & \text{if } T_R \leq T_L \leq T_S, \\ 0 & \text{otherwise.} \end{cases}$$

It is easy to see that along any trajectory of the system (B.1) the number of oxygen and the number of nitrogen atoms are constant,

$$[O^{1D}] + [O] + 3[O_3] + 2[O_2] + [NO] + 2[NO_2] \;=\; \text{const} \; ,$$
$$[NO] + [NO_2] \;=\; \text{const} \; ,$$

therefore if we denote the concentration vector

$$y = \left[\, [O^{1D}], \; [O], \; [O_3], \; [O_2] \,, [NO] \,, [NO_2] \,\right]^T \, ,$$

the linear equality constraints have the form

$$A^T = \begin{bmatrix} 1 & 1 & 3 & 2 & 1 & 2 \\ 0 & 0 & 0 & 0 & 1 & 1 \end{bmatrix} \, , \qquad A^T y(t) = A^T y(t_0) = b \; .$$

The reduced stratospheric system (B.1) auto-corrects the negative values of O, O^{1D} and NO, which explains the good accuracy of the standard methods. For example, the atomic oxygen destruction term is $-[O]$ $(k_2[O_2] + k_4[O_3] + k_9[NO_2])$; since the parenthesis does not depend on any "possibly-negative" concentration, whenever $[O] < 0$ this destruction term is positive (produces O!) and the oxygen concentration increases toward positive values. Not all chemical systems have the auto-correction property. Appending the extra reaction

$$(\text{B.2}) \qquad r11) \quad NO \;+\; O \;\xrightarrow{k_{11}}\; NO_2 \qquad (k_{11} = 1.0E - 8)$$

leads to a non-correcting kinetic scheme as (B.2) will continue to destroy O and NO.

REFERENCES

[1] Ascher, U.M., Chin, H., and Reich, S.; Stabilization of DAEs and invariant manifolds. *Numerical Mathematics*, **67**: 131–149, 1994.

[2] Bolley, C. and Crouzeix, M.; Conservation de la positivite lors de la discretization des problemes d'evolution parabolique. *R.A.I.R.O. Numerical Analysis*, **12**(3): 237–245, 1978.

[3] Carmichael, G.R., Peters, L.K., and Kitada, T.; A second generation model for regional-scale transport/ chemistry/ deposition. *Atmospheric Environment*, **20**: 173–188, 1986.

[4] Damian-Iordache, V., Sandu, A., Damian-Iordache, M., Carmichael, G.R., and Potra, F.A.; KPP – A symbolic preprocessor for chemistry kinetics – User's guide. *Technical report*, The University of Iowa, Iowa City, IA 52246, 1995.

[5] Goldfarb, D. and Idnani, A., A numerically stable dual method for solving strictly convex quadratic programs. *Mathematical Programming*, **27**: 1–33, 1983.

[6] Hairer, E., Norsett, S.P., and Wanner, G.; *Solving Ordinary Differential Equations I. Nonstiff Problems.* Springer-Verlag, Berlin, 1993.

[7] Hairer, E. and Wanner, G.; *Solving Ordinary Differential Equations II. Stiff and Differential-Algebraic Problems.* Springer-Verlag, Berlin, 1991.

[8] Hundsdorfer, W.; Numerical solution of advection-diffusion-reaction equations. *Technical report* NM-N9603, Department of Numerical Mathematics, CWI, Amsterdam, 1996.

[9] Kinnison, D.E.; NASA HSRP/AESA stratospheric models intercomparison. *NASA ftp site*, contact `kinnison1@llnl.gov`.

[10] Owren, B. and Simonsen, H.H.; Alternative Integration Methods for Problems in Structural Dynamics. *Computer Methods In Applied Mechanics And Engineering*, **122**(1/2): 1–10, 1995.

[11] Sandu, A.; Numerical aspects of air quality modeling. *Ph.D. Thesis*, Applied Mathematical and Computational Sciences, The University of Iowa, 1997.

[12] Sandu, A., Blom, J.G., Spee, E., Verwer, J., Potra, F.A., and Carmichael, G.R.; Benchmarking stiff ODE solvers for atmospheric chemistry equations II - Rosenbrock Solvers. *Atmospheric Environment*, **31**: 3459–3472, 1997.

[13] Sandu, A.; Positive numerical integration methods for chemical kinetic systems. *Journal of Computational Physics.* **170**: 1–14, 2001.

[14] Shampine, L.F.; Conservation laws and the numerical solution of ODEs. *Computers and Mathematics with Applications*, **12**B(5/6): 1287–1296, 1986.

[15] Verwer, J.G., Hunsdorfer, W., and Blom, J.G.; Numerical time integration of air pollution models. *Modeling, Analysis and Simulations report*, MAS-R9825, CWI, Amsterdam, 1998.

[16] Verwer, J., Spee, E.J., Blom, J.G., and Hunsdorfer, W.; A second order Rosenbrock method applied to photochemical dispersion problems. *SIAM Journal of Scientific Computing*, **20**: 1456–1480, 1999.

[17] Blom, J.G. and Verwer, J.; A comparison of integration methods for atmospheric transport-chemistry problems. *Journal of Computational and Applied Mathematics*, to appear.

SOME ASPECTS OF MULTI-TIMESCALES ISSUES FOR THE NUMERICAL MODELING OF ATMOSPHERIC CHEMISTRY*

BRUNO SPORTISSE[†] AND RAFIK DJOUAD[‡]

Abstract. We review in this article some issues related to multi-timescales for the numerical modeling of atmospheric chemical kinetics. There is indeed a wide range of characteristic timescales for atmospheric chemistry and the time evolution is characterized by a slow-fast behaviour. We present here some related points: the use of appropriate stiff solvers (numerical point of view), the existence of an underlying reduced model (the counterpart for modeling) and some by-products for other aspects of numerical modeling (e.g. splitting methods and adjoint modeling).

1. Introduction. Atmospheric chemistry (chemical kinetics in general) is characterized by a wide range of timescales ([1, 2]). Many chemical species and reactions are associated with *fast* phenomena (that is: phenomena occuring at small timescales): such unstable chemical species are usually referred as *radicals* (e.g: atomic oxygen). On the other hand some species and reactions are related to a *slow* dynamical behaviour (that is: they only influence the long-term behaviour of the system): this is for instance the case of the stable species (such as methan CH_4).

This has a lot of consequences for the numerical modeling of atmospheric reactive flows. This slow-fast property of chemical kinetics induces the well-known *stiffness* of the resulting evolution equations to be integrated ([3–5]). If explicit solvers were used, the stability requirements would constrain the numerical timesteps to be similar to the smallest timescales. This is of course unaffordable and implicit solvers have to be used. This is a highly classical issue for the numerical integration of chemical kinetics (more generally of reactive flows).

There is however an interesting counterpart of stiffness as far as modeling is concerned: there exists a so-called *reduced model* that is a "good" approximation of the initial stiff model. This reduced model can be obtained by filtering the initial transient phase and by eliminating all the fast phenomena and the fast chemical species ([6, 7]). A similar point of view is much more powerful: in the phase space of chemical concentrations, all the trajectories (that is: whatever the initial conditions are) converge to a "unique" attracting manifold after a fast transient phase. This manifold is defined by some algebraic relations among the chemical concentrations. The reduced model is then given by projecting the trajectories along the low-dimensional manifold. In order to compute this reduced model one

---

*Presented at the IMA Workshop Atmospheric Modeling, March 17–24, 2000.

[†]Centre d'Enseignement et de Recherche sur l'Eau, la ville et l'environnement, École Nationale des Ponts et Chaussées (ENPC-CEREVE), rue Blaise Pascal, F-77455 Champs sur Marne la Vallée Cedex 2, France. E-mail: `sportiss@cereve.enpc.fr`

[‡]CRESS-York University, 4700 Keele Street, M3J 1P3 Toronto, Ontario, Canada. E-mail: `rafik@yorku.ca`

has then to determine this manifold (some algebraic constraints) and the projection along this manifold (the "rate of motion" along the manifold).

This point is closely related to classical approximations used in chemical kinetics, namely Quasi-Steady State Assumption or Partial Equilibrium Assumption ([8, 9]). The mathematical background is provided by the singular perturbation ([10]) and the center manifold theories ([11]). The practical implementation of such models and the efficient solving of the resulting differential-algebraic models are another issue. A first possibility is to use appropriate differential-algebraic solvers ([12]). A second possibility is to build look-up tables of the manifold in a preprocessed step ([7, 13, 14]).

Building a reduced model and solving it with appropriate tools is then an alternative to the use of stiff solvers.

There are other interesting consequences of the slow-fast behaviour of chemical kinetics. The first example is provided by the use of operator splitting methods ([15, 16]). Operator splitting methods are widely used in atmospheric modeling for many reasons ([17, 5]). The first one is the need for black boxes. The second reason is directly related to stiffness and to the large number of chemical species to take into account. By coupling linear spatial phenomena (such as advection and diffusion) and nonlinear local phenomena (chemical kinetics), the number of variables to handle with would be the product of the number of chemical species by the number of grid cells. This would lead to the inversion of large systems if implicit solvers are used and to an expensive computational run time. This justifies the use of operator splitting methods in this context.

Splitting methods induce however a loss of accuracy that is classically investigated through asymptotic expansion with respect to the splitting timestep. In the slow-fast case this analysis is unfortunately no more valid since the splitting timestep is in practice always larger than the smallest timescales. We refer to [18–20] for more details and a deeper understanding of some particular points.

In the same vein another by-product of this slow-fast behaviour for atmospheric chemistry concerns inverse modeling. In few words it seems intuitive that parameters associated with fast phenomena (initial conditions, kinetic rates) are not easy to recover.

The aim of this review paper is to illustrate these points by refering to other articles.

We present in Section 2 an example of stiff integration with the numerical modeling of aqueous-phase chemistry. We derive the homogeneized model and we use a nonautonomous second-order Rosenbrock solver for integrating the resulting set of ODEs ([21–23]).

We present some results for reduction in Section 3. We give the mathematical background ([24]) and we present an algorithm for generating the reduced model ([25]). An appropriate solver is used in order to integrate the differential-algebraic system efficiently ([26]).

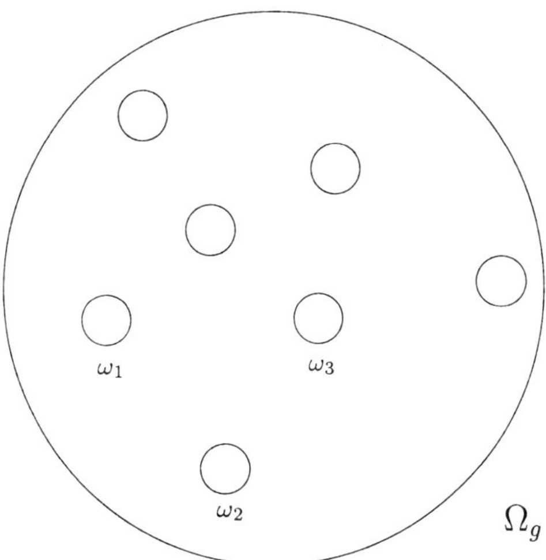

FIG. 1. *Microphysical level.*

In Section 4 we present some consequences of this analysis, mainly for operator splitting. We illustrate why the classical analysis may fail and we apply these ideas to a monodimensional case with gas-phase chemistry. The key point is that chemical kinetics has always to end the splitting sequence ([18, 20, 19]).

2. Numerical modeling of aqueous-phase chemistry. We illustrate in this section the stiff issue with the case of aqueous-phase chemistry.

2.1. Homogeneization of the microphysical model. We consider a cloud Ω_g with a set of n_d liquid droplets ω_i. Let c_a be the concentration of the dissolved species (with respect to the liquid phase) and c_g be the concentrations for the gas-phase species.

At the microphysical level the evolution of chemical concentrations is given by the following coupled systems of Partial Differential Equations:
- Inside the droplets ω_i:

$$(1) \qquad \frac{\partial c_a}{\partial t} = \frac{D_a}{\epsilon} \Delta c_a + \chi_a(c_a).$$

- In gas phase Ω_g:

$$(2) \qquad \frac{\partial c_g}{\partial t} = \frac{D_g}{\epsilon} \Delta c_g + \chi_g(c_g)$$

where $\frac{D_a}{\epsilon}$ (resp. $\frac{D_g}{\epsilon}$) stands for the molecular diffusive coefficients in aqueous phase (resp. in gas phase). χ_g (resp. χ_a) stands for gas-phase (resp. aqueous) chemical sources including loss and production terms.

The outer boundary condition for the gas phase is then typically given by:

$$(3) \qquad \frac{D_g}{\epsilon} \frac{\partial c_g}{\partial n} = 0$$

where n is the outward pointing normal on the boundary.

At the interface between droplets and gas-phase one has a Robin boundary condition ([27]):

$$(4) \qquad \frac{D_g}{\epsilon} \frac{\partial c_g}{\partial n} = -\frac{D_a}{\epsilon} \frac{\partial c_a}{\partial n} = \lambda(c_g - \frac{c_a}{HRT})$$

where n stands here for the unitary normal vector (with an orientation towards the droplets) and λ describes mass transfer through:

$$(5) \qquad \lambda = \frac{1}{4}\alpha\bar{v}$$

α is the accomodation coefficient and $\bar{v}$ is the mean molecular velocity. H stands for the Henry coefficient. R is the perfect gas constant. T is the temperature.

It is of course difficult to solve mass transfer at such scales (the radius of a cloud droplet is typically of magnitude 10 micrometers). We propose here a systematic technique in order to derive homogeneized models for instance with the case of large molecular diffusion.

By assuming that the molecular diffusive coefficients take large values (that is ϵ tends to 0 with fixed D_a and D_g) one can prove the validity of the so-called *lumped parameter assumption* ([28, 29]):

$$(6) \qquad c_g(x,t) \simeq \bar{c}_g(t) \;, \quad c_a(x,t) \simeq \bar{c}_a(t)$$

where $\bar{c}_g(t)$ and $\bar{c}_a(t)$ are approximations of the spatial means (respectively over gas phase and droplets). The convergence ($\simeq$) has to be specified in a certain sense, which is out of scope of this paper.

The time evolution of these averaged concentrations is then given by the following set of ODEs:

$$(7) \qquad \frac{d\bar{c}_g}{dt} = \chi_g(\bar{c}_g) - \frac{3\lambda}{a} \frac{L}{1-L}(\bar{c}_g - \frac{\bar{c}_a}{HRT})$$

$$(8) \qquad \frac{d(L\bar{c}_a)}{dt} = L\chi_a(\bar{c}_a) + \frac{3\lambda}{a}L(\bar{c}_g - \frac{\bar{c}_a}{HRT})$$

where $L(t)$ is the so-called liquid water content and is defined as the ratio of the liquid volume to the gas volume. Let us notice that $L(t)$ takes typically values from 10^{-8} to 10^{-6}. a stands for the droplet radius (supposed to have a constant value for more clarity).

Let us notice that this rigorous justification introduces a new source term for the evolution of $\bar{c}_a$:

$$(9) \qquad -\frac{dL/dt}{L}\bar{c}_a$$

which is usually omitted in classical derivations. This term accounts for
the time evolution of the liquid water content.

Up to this term we recover here the classical parameterization for mass
transfer ([2, 27]) in the case of large diffusion when $L \simeq 0$. Let us point out
the fact that this assumption is not met for species such as OH for which
another idem has to be used. The main point is that the mass transfer
coefficient has to take into account chemical kinetics as well ([2, 27, 30]).

2.2. Some numerical tests. The time integration of the homogeneized system (7)–(8) is particularly difficult since this system is stiff
and there is a strong dependence with respect to time through the liquid
water content. We present in the sequel a comparison of some numerical
solvers.

2.2.1. Modeling. We consider a multiphase kinetic scheme ([31])
with 30 chemical species (14 gaseous and 16 aqueous species), 39 chemical reactions (28 gaseous and 11 aqueous reactions) and 11 mass transfer
laws.

We refer to the initial article for a complete definition of chemical
kinetics and to appendix A. The values for some important parameters
are:

- Temperature $T = 293$ K,
- accomodation coefficient $\alpha = 0.1$ (constant in our tests),
- $pH = 4.16$,
- droplet diameter $D = 40\mu m$.

We have used the preprocessor SPACK ([32]) in order to generate the
FORTRAN code describing chemical kinetics. SPACK (Simplified Preprocessor for Atmospheric Chemical Kinetics) may handle gas-phase and
aqueous-phase kinetic schemes. Its main advantage is that it automatically computes lumping of species, which may be a difficult task if Henry's
equilibrium and dissociation equilibria are taken into account.

The time evolution of the cloud event is plotted in Fig. 2 (based
on [31]).

2.2.2. Numerical solvers. We have used three different solvers in
order to compute the numerical solution:

- the classical LSODE package based on a second-order BDF method
  ([33]),
- a second-order Rosenbrock method applied to autonomous systems
  (ROS2) ([34, 35, 21]),
- a second-order Rosenbrock method applied to non autonomous
  systems (ROS2-NA).

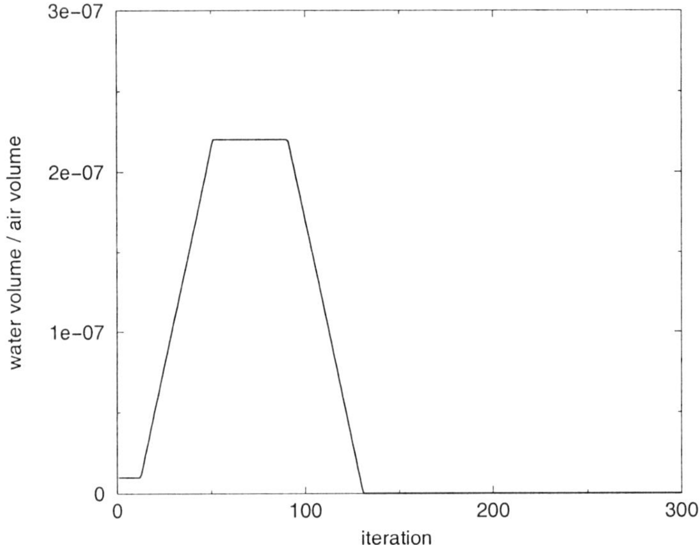

FIG. 2. *Cloud event. For one iteration $\Delta t = 60$ seconds.*

Let us recall the form of the ROS2 methods applied for the time integration of

$$(10) \qquad \frac{dc}{dt} = f(c, t).$$

The ROS2 solver computes a numerical approximation c_{n+1} for $c(t_{n+1})$ through the iterative scheme:

$$(11) \qquad c_{n+1} = c_n + b_1 k_1 + b_2 k_2$$

where

$$(12) \qquad k_i = \Delta t f\left(c_n + \sum_{j=1}^{i-1} \alpha_{ij} k_k, t_n + \alpha_i \Delta t\right) + \Delta t \frac{\partial f}{\partial c} \sum_{j=1}^{i} \gamma_{ij} k_j.$$

For ROS2-NA another term has to be added in Eq. (12):

$$(13) \qquad \gamma_i \Delta t^2 \frac{\partial f}{\partial t}.$$

The coefficients b_i, α_i, α_{ij}, γ_i and γ_{ij} take fixed values.

We have computed a reference solution given by LSODE with a strong accuracy requirement. A L_2 relative error (with respect to species) has been

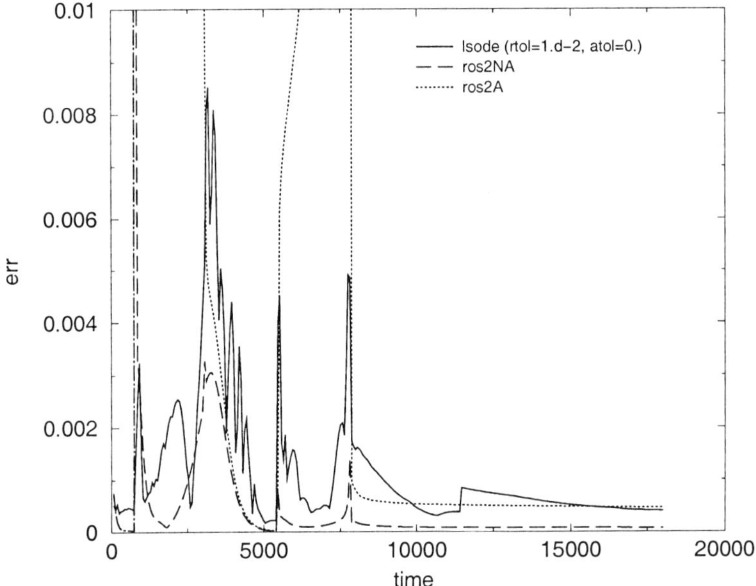

FIG. 3. *Time evolution of relative errors.*

computed by comparing the reference solution c_{ref} and the solution given by a given solver S (c_S):

$$(14) \qquad err = \frac{1}{N} \sum_{i=1}^{i=N} \left(\frac{c_{ref}^i - c_S^i}{c_{ref} + atol} \right)^2$$

where N is the number of chemical species and i is the index for chemical species. *atol* controls the small computed values. ROS2 and ROS2-NA have been used with $\Delta t = 60$ seconds. LSODE has been used with a desired relative accuracy of 0.01. The time evolution of *err* has been plotted for these three solvers in Fig. 3.

The performance of these numerical solvers is compared in Table 1. The CPU costs of these solvers are quite similar while $ROS2NA$ is much more accurate. These results prove the efficiency of ROS2-NA and the need for taking into account the non-autonomous term.

TABLE 1

Performance. CPU cost for one iteration.

Solver	LSODE	ROS2	ROS2NA
CPU	3.E-2	2.4E-2	2.8E-2

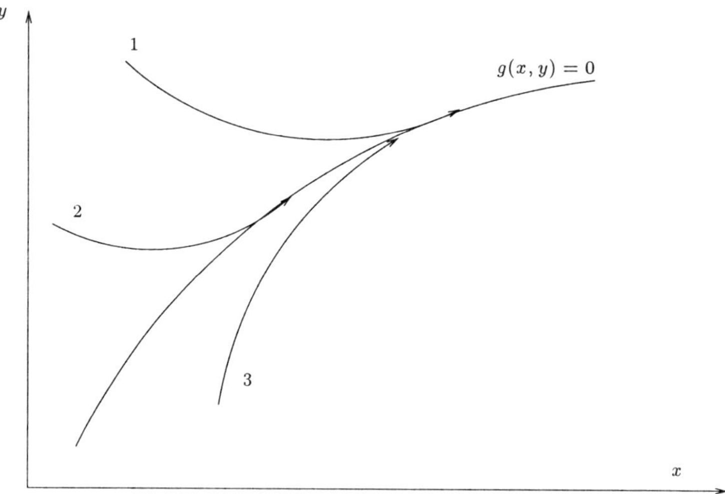

FIG. 4. *Convergence to the reduced model. 1, 2 and 3 stand for three sets of initial conditions.*

3. Reduction of chemical kinetics.

3.1. Background.

3.1.1. Theoretical aspects. We have illustrated in previous section some classical issues related to the existence of slow and fast timescales, namely *numerical stiffness*.

There exists however an alternative to the use of appropriate stiff solvers and we have to go back to modeling in order to illustrate this feature. The dynamical behaviour of such stiff systems is characterized by the existence of a fast transient phase followed by a slow long-term evolution. In other words, all the trajectories in the phase space of chemical concentrations (parameterized by time t) converge to an attracting manifold after a transient phase (see Fig. 4).

Such systems are mathematically described by the singular pertubation and the center manifold theories. One of the key tools is Tikhonov's theorem ([10]). Let us consider the following evolution equation:

$$(15) \qquad \frac{dx}{dt} = f(x,y) \,, \qquad \epsilon \frac{dy}{dt} = g(x,y)$$

x (resp. y) stands for *slow* (resp. *fast* species) associated to timescales of magnitude 1 (resp. ϵ). ϵ is once more a small positive parameter defined as the ratio of fast timescales to slow ones.

After a transient phase whose length is of magnitude ϵ, Eq. (15) can be approximated up to first-order in ϵ by the following reduced model:

$$(16) \qquad \frac{dx}{dt} = f(x,y) , \quad 0 = g(x,y).$$

Some assumptions have to be met and we refer to [10, 11] for more details. The key point is of course that $g(x,y) = 0$ has to provide a way for computing y as a function of x (let say $y = h(x)$).

3.1.2. Lumping species. Dynamical systems arising in practice do not have the nice form (15) and we have to perform a splitting of species into two sets: a set of *slow species* (x) and a set of *fast species* (y). A general technique is based on the linearization of the system: the search for the partitioning is then related to the search for the eigenvectors of the Jacobian matrix ([36, 37]).

We have proposed in [25] an algorithm based on the study of the stoichiometric matrix. Let us consider an evolution system

$$(17) \qquad \frac{dz}{dt} = S\,\omega(z) = P(z) - L(z)z$$

where S is the stoichiometric matrix and ω is the vector of reaction rates. $P(z)$ is the (positive) vector of production terms. $L(z)$ is the (positive) diagonal matrix of loss terms. We recall here the main steps of this algorithm:

1. Compute the QSSA index for each species i:

$$(18) \qquad I_i = \frac{|P_i(z) - L_i(z)\,z_i|}{P_i(z) + L_i(z)\,z_i}.$$

2. Select the fast species i such that $I_i < \epsilon$ (ϵ given by the user)
3. Compute for each fast species i, the contribution index J_i^r of the reaction r:

$$(19) \qquad J_i^r = \frac{|S_{ir}\,\omega_r|}{\displaystyle\sum_r |S_{ir}\,\omega_r|}.$$

4. Select the fast reactions r associated with the fast species i such that $J_i^r > \epsilon$.
5. Partition the stoichiometric matrix S in a slow part (S_s) and a fast part (S_f): $S = S_s + S_f$. The fast part is defined by step 4.

6. Compute a basis $(\vec{u}_1, \cdots, \vec{u}_l)$ of the left kernel of S_f:

$$(\vec{u})_i^t \cdot S_f = 0 \qquad\qquad 1 \leq i \leq l$$

7. There are then l lumped species $L_1, \cdots L_l$ given by:

$$L_i = \vec{u}_i^t \cdot \vec{z} \qquad\qquad 1 \leq i \leq l$$

This algorithm may be used in a preprocessed step in order to compute the partitioning of chemical species.

3.1.3. Coupling with other processes. The key point is of course to know how to couple a reduced kinetic scheme with other processes (supposed to be slower) such as diffusion, boundary conditions and advection.

The main result is that the low-dimensional manifold (that is to say the algebraic constraints) are uniquely determined by chemical kinetics. The other processes have however to be projected along the reduced model. The most efficient way in order to compute the reduced model is then to use lumped species. We refer to [25, 38] for more details and a deeper understanding.

3.2. Some numerical tests. We have used our algorithm in order to compute the lumped species for photochemical kinetic schemes ([25] for different conditions. In day-like conditions we recover the usual lumped species:

$$(20) \qquad NO_x = NO + NO_2 + O \ , \quad O_x = NO_2 + O_3 + O.$$

The accuracy of the resulting reduced model has been tested in a monodimensional case including diffusion, gas-phase chemistry, emissions and dry deposition. The resulting differential-algebraic systems have been integrated with LIMEX ([39]) with a strong accuracy requirement. The number of correct digits computed on the basis of a spatial relative error (L_2 norm) is always greater than 2 **if** slow processes are projected. This proves the acuracy of the reduced model and the need for projecting the slow processes.

The CPU performance of such reduced models is of course related to the solver used for solving the resulting differential-algebraic system. Look-up tables can of course be used and we refer for instance to [40]. An alternative is to use differential-algebraic solvers with the required accuracy. The accuracy of the numerical solution for the reduced model is of course the sum of the reduction error and of the numerical error. There is no need for using very accurate solvers in this context.

We have proposed in [26] an hybrid solver (DAN2) based on a second-order explicit method for the integration of slow species and on a fixed number of Newton's iterations for the algebraic constraints. This has been applied for solving the reduced CBM IV model ([41]) in a box case. The comparison has been made with an "optimal" version of the best available stiff solver applied to the exact model (that is: ROS2 with a low number of Jacobian updates).

This algorithm is reffered as DAN2 in Table 2. The results prove the speed-up with the same accuracy.

TABLE 2

Exact model versus reduced model (CBMIV). The number of correct digits is computed through a L_2 relative error with respect to species.

Solver	Model	CPU	Number of correct digits
ROS2	exact	1.4	2.23
DAN2	reduced	0.9	2.20

4. An example of by-product: operator splitting methods and multi timescales. We have seen above some *immediate consequences* of the wide range of chemical kinetics. In Section 2 this led to the use of stiff solvers. In Section 3 we have introduced the notion of reduced model and proposed some algorithms in order to compute it. The slow-fast behaviour of chemical kinetics has however many other numerical consequences. That is what we call "by-products". We illustrate that with one important example: the case of operator splitting techniques, that are a very popular tool in atmospheric modeling.

4.1. Background. We only consider atmospheric gas-phase. In this context, the time and space evolution of the concentration c_i of species labelled by i is given by the following PDE:

$$(21) \quad \frac{\partial c_i}{\partial t} + V(x,t) \cdot \nabla c_i = div(K(x,t)\nabla c_i) + \chi_i(c, T(x,t), t) + S_i(x,t)$$

where V is the advection wind field, K is the diffusion matrix (associated with a gradient-like parameterization of turbulence), T is the temperature, S is the source term and χ_i is the chemical production for species i. t designs time and x is the space coordinate. The key point is that the fields V, K and T are supposed to be known (computed off line by meteorological solvers). Some appropriate boundary conditions (including dry deposition and emissions) have to be added to this set of PDEs.

n_s is the number of chemical species and n_g is the number of grid cells (let us say in a Finite Differences framework).

The use of a coupled algorithm for the integration of such a PDE is quite difficult for at least two reasons:
- as far as modularity is concerned (a strong request in [42, 43]), this is better to use black boxes with appropriate solvers and then to solve independently the phenomena to be taken into account,

- as far as CPU time is concerned, if we use as usual the Method of Lines (first discretize with respect to space and then integrate the resulting Ordinary Differential Equations) the stiff ODEs to be integrated have a dimension $n_g \times n_s$. The complexity of matrix inversions (with a LU method) is then of magnitude $O((n_g n_s)^3)$.

Operator splitting methods are then commonly advocated in this context ([17]). We refer to [16, 15] for a general presentation of such methods. We present now some classical methods and the corresponding numerical analysis in the linear case.

In the linear case, we consider the following evolution equation:

$$(22) \qquad \frac{dc}{dt} = Ac + Bc \ , \quad c(0) = c_n$$

to be integrated. A and B are two linear operators. c_n is the current value of the solution at time $t_n = n\Delta t$ where Δt is the timestep of the splitting method and n is the index for iteration. We want to compute an estimation c_{n+1} of the solution $c(\Delta t)$ at time $t_{n+1} = t_n + \Delta t$ without solving (22) as explained before.

The simplest splitting method to be defined is of course:

- First step: integrate operator A

$$(23) \qquad \frac{dc^*}{dt} = Ac^* \quad \text{on} \quad [0, \Delta t] \ , \quad c^*(0) = c_n.$$

- Second step: integrate operator B

$$(24) \qquad \frac{dc^{**}}{dt} = Bc^{**} \quad \text{on} \quad [0, \Delta t] \ , \quad c^{**}(0) = c^*(\Delta t).$$

The approximation of $c(\Delta t)$ is then:

$$(25) \qquad c_{n+1} = c^{**}(\Delta t).$$

For an obvious reason we will name $(A - B)$ this algorithm.

The classical analysis for this method is based on asymptotic expansions with respect to the splitting timestep Δt supposed to be small. The exact solution (without splitting) is of course:

$$(26) \qquad c(\Delta t) = exp((A + B)\Delta t)c_n$$

while the solution computed with method $(A - B)$ is:

$$(27) \qquad c_{A-B}(\Delta t) = exp(B\Delta t)exp(A\Delta t)c_n.$$

The local error is then

$$(28) \qquad le = c_{A-B}(\Delta t) - c(\Delta t) = \frac{AB - BA}{2}\Delta t^2 c_n + O(\Delta t^3)$$

after having performed an asymptotic expansion with respect to Δt. We have then a first-order method unless A and B are commuting operators.

Second-order methods can be easily defined (the so-called Strang splitting: see [44]). We refer to [45] for the study of the nonlinear case. A completely different approach is to perform the splitting process at the level of linear algebra. This is sometimes referred as *internal splitting* (see [46, 5] for instance). We will not investigate this case here.

To conclude this review an alternative method has been proposed in [47] (*No Time Splitting*), [48] (*Source splitting*) and [49] (*Incremental splitting*). We will name this method as *Source splitting*.

The key idea is to avoid the change of initial conditions by adding artificial terms. Let us assume that B is a stiff operator. As we do not want to create a transient phase during the integration of B, Source splitting appears as a slight modification of the $(A - B)$ method:

- First step: integrate operator A

$$(29) \qquad \frac{dc^*}{dt} = Ac^* \quad \text{on} \quad [0, \Delta t] \,, \quad c^*(0) = c_n.$$

- Second step: integrate operator B with a complementary Source term

$$(30) \qquad \frac{dc^{**}}{dt} = Bc^{**} + \frac{c^*(\Delta t) - c_n}{\Delta t} \quad \text{on} \quad [0, \Delta t] \,, \quad c^{**}(0) = c_n.$$

Note that the initial condition for the second step has not been modified.

A classical analysis of this method proves that this a first-order method. The key point is that transient phases have been eliminated ([18]). A stability analysis of this algorithm coupled with Rosenbrock methods can be found in [21].

4.2. Splitting in the stiff case. As it has been seen before, the classical analysis for splitting methods is based on asymptotic expansions with respect to *small* timesteps Δt.

In practical cases (for instance in Air Pollution Modeling) it is quite difficult to ensure that Δt is actually *small* with respect to the physical timescales. The key point is indeed that chemical kinetics is characterized by a wide range of timescales.

At a mathematical level we will now study an evolution equation written under the form:

$$(31) \qquad \frac{dc}{dt} = Ac + \frac{B}{\epsilon}c \,, \quad c(0) = c_n$$

where ϵ is a small positive parameter defined as the ratio of fast timescales to slow ones. In this case, we assume that the operator B (typically chemical kinetics) is associated with the fastest phenomena as compared with A (typically vertical diffusion). In the previous subsection we have taken $\epsilon = 1$ which corresponds to the "classical case" (no stiffness).

Let us point out that one of the main motivations for using operator splitting is to use large timesteps with respect to the smallest (fastest) timescales:

$$(32) \qquad \Delta t \gg \epsilon.$$

If this is not met, both operators (A and B) have to be integrated with small timesteps. That is to say that the stability requirement associated with stiffness have to be taken into account: this is exactly what a careful modeler does not want by using implicit schemes!

The conclusion is that $\Delta t \gg \epsilon$ is actually met. If the classical analysis was valid, we would have for the local error (for instance for method ($A - B$)):

$$(33) \qquad le \simeq (\frac{B}{\epsilon}A - A\frac{B}{\epsilon})c_n\Delta t^2$$

while high-order terms could explode much faster with ϵ tending to 0.

Let us notice moreover that this analysis would be in contradiction with the intuitive feeling that a splitting method is much more appropriate if the timescales are well splitted (ϵ small). This is exactly the opposite of what the previous local error indicates with Eq. (33).

A last point concerns the sequential order for integration. The classical analysis does not give any influence for the choice of the sequence. The methods ($A - B$) and ($B - A$) (or ($A - B - A$) and ($B - A - B$)) are associated with errors of similar magnitudes. However, the choice of the sequence may play a role as noted in [18, 20].

We recall now two alternative studies of the splitting error in the so called *stiff* case. The key point is to perform the analysis with ϵ tending to 0 and then Δt tending to 0. This is a double limit approach, which is quite classical in numerical analysis (see [50] for Advection-Diffusion problems for instance). The idea behind this is that "a small physical parameter is supposed to tend faster to 0 than a small numerical parameter" (crudely speaking).

- The first approach is related with reduction of chemical kinetics. That is to say that there is an underlying reduced model associated with some algebraic constraints. In order to meet these constraints, it is better to end the splitting sequence with chemical kinetics.

  The key point is that the operator associated with chemical kinetics (operator B) acts as a projection onto this manifold on the contrary to the slow operator (A), unless some particular invariance properties are met (see Fig. 5).

  This gives some indications for the importance of the sequential order in splitting methods. By finishing the sequence with the stiff operator, one may be sure to be near the low-dimensional manifold even if the motion along the manifold is probably not completely exact (see cases C and D in Fig. 5).

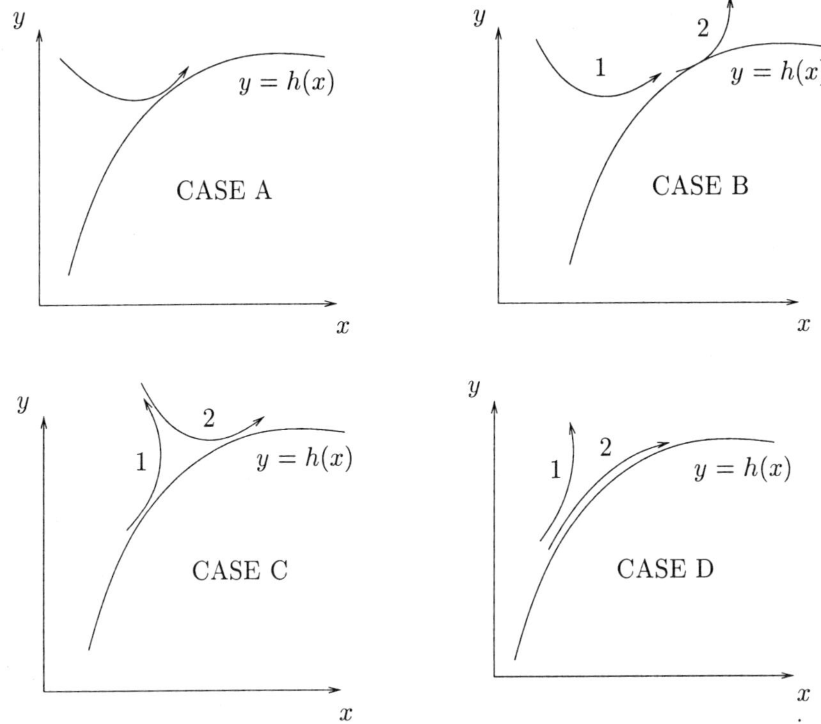

FIG. 5. *Dynamical behaviour of splitting schemes. Case A: exact solution. Case B: (B-A) splitting. Case C: (A-B) splitting. Case D: Source splitting.* $y = h(x)$ *defines the reduced model.*

- The second approach is related with numerical analysis and the concept of order reduction for stiff ODEs ([19]).

These approaches lead to the following recommendations:

- The usual orders of splitting methods are not kept in the stiff case. Strang splitting is only a first-order method.
- The orders are not the same ones for slow and fast species. The loss of accuracy only concerns fast species.
- Splitting sequences with chemical kinetics at the end have a better accuracy.
- It is better to add boundary conditions (dry deposition and emission terms) to chemical kinetics.

This has been tested in [18, 20] for monodimensional cases including gas-phase (see Appendix B), diffusion, dry deposition and emissions. Some numerical results can be found in Table 3. The relative error is computed over species with a spatial L_2 norm. D^*C (resp. DC^*) stands for instance for the first-order splitting method with diffusion in the first substep and chemistry in the second substep. Boundary Conditions are taken into account with diffusion (resp. with chemistry). CD^*C stands for Strang Splitting with an appropriate sequence. SS stands for Source Splitting. This confirms the analysis presented above.

TABLE 3
Final relative error(in %). Δt is the splitting timestep.

	D^*C	CD^*	DC^*	SS	CD^*C	D^*CD^*	C^*DC^*
$\Delta t = 1800$	8.5	48.4	12.8	3.3	0.9	17.6	4.8
$\Delta t = 900$	4.3	17.8	6.5	1.6	0.4	7.1	2.1
$\Delta t = 300$	1.4	4.9	2.1	0.5	0.2	2.0	0.9
$\Delta t = 100$	0.5	1.5	0.7	0.2	0.1	0.6	0.1

5. Conclusion. In this article we have reviewed some results related to the wide range of characteristic timescales in atmospheric chemistry:
- numerical stiffness (with an example provided by aqueous-phase chemistry),
- reduced models (with examples given by gas-phase chemistry with and without diffusion),
- and some interesting by-products (such as the careful use of operator splitting in this framework).

These results prove the need for joining physical modeling, numerical analysis and applied mathematics.

Other by-products of such an analysis are currently in progress. One application is for instance the study of inverse modeling for atmospheric chemistry: what is the influence of this slow-fast behaviour for inverse modeling? Does it give some recommendations for the numerical strategies and for the strategy of observations: what species have to be measured and when?

Acknowledgements. We thank Jan Verwer (CWI) for fruitful discussions concerning stiffness, operator splitting and numerical analysis in general. We thank our colleague Nicole Audiffren (LAMP) for continuing interactions concerning aqueous-phase chemistry.

Part of this work has been done in the framework of the A3UR Project carried on at Electricite de France (1995–1998) and of the INRIA Cooperative Research Action COMODE 2000–2002 ([51]).

APPENDIX

A. Aqueous-phase chemistry ([31]).

TABLE 4

Gas phase reactions. The kinetic rate is computed through $k = A\,10^{-n}\,exp\{B/T\}$ except for reactions [b] for which $k = A\,10^{-n}\,exp\{-B/C\}\,exp\{B/T\}$.

Nr	Reaction	Kinetic rate $A(n), B$
G1	$O_3 + H_2O + h\nu \rightarrow 2\,OH + O_2$	$SPEC^1$
G2	$OH + O_3 \rightarrow HO_2$	$1.6(12), -940$
G3	$O_3 + HO_2 \rightarrow OH$	$1.1(14), -500$
G4	$HO_2 + HO_2 \rightarrow H_2O_2$	$SPEC^2$
G5	$H_2O_2 \rightarrow 2.OH$	$4.6(6), 0$
G6	$H_2O_2 + OH \rightarrow HO_2$	$3.3(12), -200$
G7	$CH_4 + OH \rightarrow CH_3O_2$	$2.3(12), -1700$
G8	$CH_3O_2 + HO_2 \rightarrow CH_3O_2H$	$4.0(12), 0$
G9	$CH_3O_2H \rightarrow CH_2O + HO_2 + OH$	$4.6(6), 0$
G10	$CH_3O_2H + OH \rightarrow CH_3O_2$	$5.6(12), 0$
G11	$CH_3O_2H + OH \rightarrow CH_2O + OH$	$4.4(12), 0$
G12	$CH_2O \rightarrow CO + 2.HO_2$	$1.7(5), 0$
G13	$CH_2O \rightarrow CO$	$3.3(5), 0$
G14	$CH_2O + OH \rightarrow CO + HO_2$	$1.1(11), 0$
G15	$CO + OH \rightarrow HO_2$	$2.4(13), 0$
G16	$NO + O_3 \rightarrow NO_2$	$2.0(12), -1400$
G17	$NO_2 \rightarrow NO + O_3$	$5.6(3), 0$
G18	$NO + HO_2 \rightarrow NO_2 + OH$	$3.7(12), 240$
G19	$NO + CH_3O_2 \rightarrow NO_2 + CH_2O + HO_2$	$4.2(12), 180$
G20	$NO_2 + OH \rightarrow HNO_3$	$1.1(11), -1400, 298$ [b]
G21	$HNO_3 \rightarrow NO_2 + OH$	$3.2(7), 0$
G22	$CH_2O + HO_2 \rightarrow O_2CH_2OH$	$6.3(14), 620, 298$ [b]
G22 bis	$O_2CH_2OH \rightarrow CH_2O + HO_2$	$1.3(2), -7000, 298$ [b]
G23	$O_2CH_2OH + HO_2 \rightarrow HCO_2H$	$2.0(12), 0$
G24	$O_2CH_2OH + NO \rightarrow HCO_2H + HO_2 + NO_2$	$7.0(12), 0$
G25	$O_2CH_2OH + O_2CH_2OH \rightarrow 2.HCO_2H + 2.HO_2$	$1.2(13), 0, 298$ [b]
G26	$HCO_2H + OH \rightarrow HO_2$	$3.2(13), 0$
G27	$HNO_3 + OH \rightarrow 0.89 O_3 + 0.89 NO_2 + 0.11 NO$	$SPEC^3$

TABLE 5

Aqueous reactions. The kinetic rate is computed with $k = A\,10^{-n}\,exp\{-B/C\}$ $\times exp\{B/T\}$.

Nr	Reaction	Kinetic rate $A(n), B$
A1	$AH_2O_2 \rightarrow 2.AOH$	$SPEC^4$
A2	$AO_3 \rightarrow AH_2O_2$	$7.36(6), 0, 0$
A3	$ACH_2(OH)_2 + AOH \rightarrow AHCO_2H + AHO_2$	$2.00(-9), -1500, 298$
A4	$AHCO_2H + AOH \rightarrow AHO_2$	$1.60(-8), -1500, 298$
A5	$HCO_2^- + AOH \rightarrow AHO_2$	$2.50(-9), -1500, 298$
A6	$AO_3 + O^{2-} \rightarrow AOH$	$1.50(-9), -1500, 298$
A7	$AHO_2 + O^{2-} \rightarrow AH_2O_2$	$1.00(-8), -1500, 298$
A8	$AH_2O_2 + AOH \rightarrow AHO_2$	$2.70(-7), -1715, 298$
A9	$ACH_3O_2 + O^{2-} \rightarrow ACH_3O_2H$	$5.00(-7), -1610, 298$
A10	$ACH_3O_2H + AOH \rightarrow ACH_3O_2$	$2.70(-7), -1715, 298$
A11	$ACH_3O_2H + AOH \rightarrow ACH_2(OH)_2 + AOH$	$1.90(-7), -1860, 298$

TABLE 6

Henry's laws. The kinetic rate is computed with $k = A\,10^{-n}\,exp\{-B/C\}\,exp\{B/T\}$.

Nr	Reaction	Kinetic rate $A(n), B, C$
H1	$O_3 \leftrightarrow AO_3$	$1.1(2), 2300, 298$
H2	$H_2O_2 \leftrightarrow AH_2O_2$	$7.4(-4), 6615, 298$
H3	$CH_3O_2H \leftrightarrow ACH_3O_2H$	$2.2(-2), 5653, 298$
H4	$CH_2O \leftrightarrow ACH_2(OH)_2$	$6.3(-3), 6425, 298$
H5	$HNO_3 \leftrightarrow AHNO_3$	$2.1(-5), 8700, 298$
H6	$HO_2 \leftrightarrow AHO_2$	$2.0(-3), 6600, 298$
H7	$OH \leftrightarrow AOH$	$9.0(-3), 0, 298$
H8	$NO_2 \leftrightarrow ANO_2$	$6.4(3), 2500, 298$
H9	$NO \leftrightarrow ANO$	$1.9(3), 1480, 298$
H10	$CH_3O_2 \leftrightarrow ACH_3O_2$	$2.0(-3), 6600, 298$
H11	$HCO_2H \leftrightarrow AHCO_2H$	$3.7(-3), 5700, 298$

TABLE 7

Dissociation equilibria. The kinetic rate is computed with $k = A\,10^{-n}\,exp\{B\,(1/T -1/298)\}$.

Nr	Reaction	Kinetic rate $A(n), B$
AEQ1	$AHNO_3 = NO_3^- + H^+$	$15.4(0), 0$
AEQ2	$AH_2O_2 = HO_2^- + H^+$	$2.2(12), -3730$
AEQ3	$AHO_2 = O^{2-} + H^+$	$3.5(5), 0$
AEQ4	$AHCO_2H = HCO_2^- + H^+$	$1.8(4), -1510$

For specific reactions:

- $SPEC^1$:

  $x_{air} = N\,P/(8.314d6\,T)$

  N Avogadro number, T Temperature and P pressure.

  $x_{h_2o} = $ Water vapour content

  $x_{o_2} = 0.2 \times x_{air}$

  $x_{n_2} = 0.8 \times x_{air}$

  $v_{o_2} = x_{o_2} \times 4.0d - 11 * \exp\{67\,(1/T - 1/298)\}$

  $v_{n_2} = x_{n_2} \times 2.6d - 11 * \exp\{110\,(1/temp - 1/298)\}$

  $v_{h_2o} = x_{h_2o} \times 2.2d - 10$

  $k_{G1} = 1.6d - 05\,v_{h_2o}/(v_{h_2o} + v_{n_2} + v_{o_2})$

- $SPEC^2$:

  $rk_0 = 1.4d - 21 * \exp\{2200/T\}$

  $k_{G4} = (1 + rk_0 * x_{h_2o}) \times (1.5d - 12\,\exp\{600\,(1/T - 1/298)\} + 4.9d - 32$
  $\times \exp\{500\,(1/T - 1/298)\}\,x_{air})$

- $SPEC^3$:

  $rk_0 = 7.2d - 15 \times \exp\{785/T\}$

  $rk_2 = 4.1d - 16 \times \exp\{1440/T\}$

  $rk_3 = 1.9d - 33 \times \exp\{725/T\}$

  $k_{G27} = rk_0 + rk_3 \times x_{air}/(1 + rk_3 * x_{air}/rk_2)$

- $SPEC^4$:

  $k_{A1} = 1.6\,k_{G1}$

Aqueous reaction rates, Henry's equilibrium rates and dissociation constants are expressed in unit system (mol, l, atm, sec). Gas phase reaction rates are expressed in unit system ($molec, cm^3$).

B. Gas phase chemistry for Section 4.

TABLE 8
The kinetic rate is computed with $k = a\,10^b$.

Nr	Reaction	Kinetic rate $a(b)$
1	$NO_2 \rightarrow NO + O_3$	9.60 (-3)
2	$O_3 + NO \rightarrow NO_2$	4.30 (-1)
3	$O_3 \rightarrow 2OH$	2.95 (-6)
4	$OH + NO \rightarrow HONO$	1.63 (2)
5	$OH + NO_2 \rightarrow HNO_3$	2.82 (2)
6	$HONO \rightarrow OH + NO$	2.80 (-3)
7	$HO_2 + NO \rightarrow NO_2 + OH$	2.07 (2)
8	$HO_2 + NO_2 \rightarrow HO_2NO_2$	2.82 (1)
9	$HO_2NO_2 \rightarrow HO_2 + NO_2$	8.48 (-2)
10	$HO_2 + HO_2 \rightarrow H_2O_2$	1.20 (2)
11	$H_2O_2 \rightarrow 2OH$	8.32 (-6)
12	$OH + CO \rightarrow HO_2 + CO$	7.28 (0)
13	$O_3 + NO_2 \rightarrow NO_3$	7.96 (-4)
14	$NO + NO_3 \rightarrow 2NO_2$	4.70 (-2)
15	$NO_2 + NO_3 \rightarrow N_2O_5$	4.35 (1)
16	$N_2O_5 \rightarrow NO_2 + NO_3$	7.40 (-2)
17	$N_2O_5 + H_2O \rightarrow 2HNO_3 + H_2O$	7.43 (-8)
18	$NO_3 \rightarrow .3NO + .7NO_2 + .7O_3$	1.49 (-1)
19	$OH + O_3 \rightarrow HO_2$	1.67 (0)
20	$HO_2 + O_3 \rightarrow OH$	3.84 (-2)
21	$OH + SO_2 \rightarrow H_2SO_4 + HO_2$	3.17 (1)

REFERENCES

[1] B.J. Finlayson-Pitts and J.N. Pitts. *Atmospheric chemistry: fundamentals and experimental techniques.* John Wiley and Sons, 1986.

[2] J.H. Seinfeld. *Atmospheric physics and chemistry of Air Pollution.* Wiley, 1985.

[3] G.D. Byrne and A.C. Hindmarsh. Stiff ode solvers: a review of current and coming attractions. *J. Comp. Phys.*, **70**: 1–62, 1987.

[4] J.G. Verwer, J. Blom, M. Van Loon, and E.J. Spee. A comparison of stiff odes solvers for atmospheric chemistry problems. *Atmos. Environ.*, **30**: 49–58, 1996.

[5] J.G. Verwer, W.H. Hundsdorfer, and J.G. Blom. Numerical time integration for air pollution models. In *Proceedings of the Conference APMS'98*. ENPC-INRIA, October 26–29, 1998.

[6] S.H. Lam and D.A. Goussis. Understanding complex chemical kinetics with computational singular perturbation. Twenty-Second symposium on combustion, 1988.

[7] U. Maas and S.B. Pope. Simplifying chemical kinetics : intrinsic low-dimensional manifolds in composition space. *Combustion and Flame*, **88**: 239–264, 1992.

[8] M. Bodenstein and H. Lutkemeyer. Quasy-steady state assumption. *Z. Phys. Chem.* (114), 1924.

[9] N. Semenov. On the kinetics of complex reactions. *J. Chem. Phys*, **7**(8): 683–699, 1939.

[10] A.N. Tikhonov, A.B. Vasileva, and A.G. Sveshnikov. *Differential equations.* Springer Verlag, 1985.

[11] J. Carr. *Applications of Center manifold theory*. Springer Verlag, 1981.

[12] K.E. Brenan, S.L. Campbell, and L.R. Petzold. *Numerical solution of initial value problems in DAEs*. SIAM, 1989.

[13] T. Turanyi. Application of repro-modeling for the reduction of combustion mechanisms. *25th Symposium on Combustion*, pp. 949–955, 1994.

[14] R.M. Lowe and A.S. Tomlin. The application of repro-modeling to a tropospheric chemistry model. In *Proceedings APMS'98*, pp. 423–433. ENPC-INRIA, October 1998.

[15] N.N. Yanenko. *The method of fractional steps*. New-York, 1971.

[16] G.I. Marchuk. *Mathematical models in environmental problems*, Volume **16**. North Holland, 1986.

[17] G.J. McRae, W.R. Goodin, and J.H. Seinfeld. Numerical solution of the atmospheric diffusion equation for chemically reacting flows. *J. Comp. Phys.*, **45**: 1–42, 1982.

[18] B. Sportisse. An analysis of operator splitting techniques in the stiff case. *J. Comp. Phys.*, **161**: 140–168, 2000.

[19] J.G. Verwer and B. Sportisse. A note on operator splitting analysis in the stiff linear case. Technical Report MAS-R9830, CWI, 1998.

[20] B. Sportisse, G. Bencteux, and P. Plion. Method of lines versus operator splitting methods for reaction-diffusion systems in air pollution modelling. *Environmental Modeling and Software*, 15, 2000. Special Issue for Conference APMS'98, 1998, October 26–29.

[21] J.H. Verwer, E.J. Spee, J.G. Blom, and W.H. Hundsdorfer. A second order rosenbrock method applied to photochemical dispersion problem. Technical Report MAS-R917, CWI, 1997. to appear in SIAM J. Sci. Comput.

[22] R. Djouad, B. Sportisse, and N. Audiffren. Numerical Simulation of Aqueous-Phase Atmospheric Models: Use of a Nonautonomous Rosenbrock Method. Accepted for publication in *Atmos. Environ.*, 2001.

[23] R. Djouad. *Simulation numérique de la pollution atmosphérique. Aspects multiphasiques*. PhD thesis, ENPC et INSA Rouen., 2001.

[24] B. Larrouturou and B. Sportisse. Some mathematical and numerical aspects of reduction in chemical kinetics. In J. Wiley & Sons, editor, *Computational Science for the 21st century, conference in honor of Professor R. Glowinski*, 1997.

[25] B. Sportisse and R. Djouad. Reduction of chemical kinetics in air pollution modelling. *J. Comp. Phys.*, **164**: 1–23, 2000.

[26] R. Djouad and B. Sportisse. Solving differential-algebraic reduced systems arising in air pollution modelling. *Submitted to Applied Numerical Mathematics*, 2001.

[27] S.E. Schwartz. Mass-transport considerations pertinent to aqueous phase reactions of gases in liquid-water clouds. NATO ASI Series, Vol. **G6**, Springer Verlag, 1986. Chemistry of multiphase atmospheric systems.

[28] E. Conway, D. Hoff, and J. Smoller. Large-time behaviour of solutions of systems of nonlinear reaction-diffusion equations. *SIAM J. of Appl. Math.*, **35**(1): 1–16, July 1978.

[29] J. Smoller. *Shock wawes and reaction-diffusion equations*. Springer Verlag, 1982.

[30] D.J. Jacob. Chemistry of oh in remote clouds and its role in the production of formic acid and peroxymonosulfate. *J.Geophys.Res.*, **91**: 9807–9826, 1986.

[31] N. Audiffren, M. Renard, E. Buisson, and N. Chaumerliac. Deviations from the henry's law equilibrium during cloud events: a numerical approach of the mass transfer between phases and its specific numerical effects. *Atmospheric Research*, **49**: 139–161, 1998.

[32] Spack: a simplified preprocessor for atmospheric chemical kinetics. http:// binky.enpc.fr/~sportiss/pole/spack.html.

[33] A.C. Hindmarsh. *Scientific computing*, chapter ODEPACK: a systematized collection of ODE solvers, pp. 55–74. North Holland, 1983.

[34] C. Kessler, A. Griesel, and J.G. Verwer. A rosenbrock solver in chemistry-transport modelling: what about the speed in 3d? In *Proceedings EUROTRAC-2 Symposium*, 1998.

[35] A. Sandu, J.G. Verwer, J.G. Blom, E.J. Spee, and G.R. Carmichael. Benchmarking stiff odes solvers for atmospheric chemistry problems ii: Rosenbrock solvers. *Atm. Env.*, **31**: 3459–3472, 1997.

[36] S.H. Lam and D.A. Goussis. The csp method for simplifying kinetics. *Int. J. Chem. Kinet.*, **26**, 1994.

[37] U. Maas. *Automatische Reduktion von Reaktionsmechanismen zur Simulation reaktiver Stromungen*. PhD thesis, Stuttgart Univers., 1993.

[38] B. Sportisse and P. Rouchon. Reduction of slow-fast chemistry with slow processes. Technical Report 98-129, CERMICS, July 1999. Submitted to Chem. Eng. Sc.

[39] P. Deuflhard, E. Hairer, and J. Zugck. One step and extrapolation method for differential-algebraic systems. *Num. Math.*, **51**: 501–516, 1987.

[40] R. Lowe and A. Tomlin. Low-dimensional manifolds and reduced chemical models for tropospheric chemistry simulations. *Atmospheric Environment*, **34**: 2425–2436, 2000.

[41] M.W. Gery, G.Z. Whitten, J.K. Killus, and M.C. Dodge. A photochemical kinetics mechanism for urban and regional scale computer modeling. *J. Geophys. Research*, **94**(D10): 12925–12956, 1989.

[42] G.R. Carmichael, A. Sandu, F.A. Potra, V. Damian-Iordache, and M. Damian-Iordache. The current state and the future directions in air quality modeling. *SAMS*, 1996. In Modelling and simulation of complex environmental problems, Dagstuhl Seminar, Springer Verlag.

[43] L.K. Peters et al. The current state and future directions of eulerian models in simulating the tropospheric chemistry and transport of trace species: a review. *Atm. Env.*, **29**: 189–222, 1995.

[44] G. Strang. On the construction and comparison of difference schemes. *SIAM J. Numer. Anal.*, **5**: 506–517, 1968.

[45] L. Lanser and J.G. Verwer. Analysis of operator splitting for advection-diffusion-reaction problems from air pollution modelling. In *Proceedings 2nd Meeting on Numerical methods for differential equations*. Coimbra, Portugal, February 1998.

[46] J.G. Verwer, J.G. Blom, and W.H. Hundsdorfer. An implicit-explicit approach for atmospheric transport-chemistry problems. *Appl. Num. Math.*, **20**: 191–209, 1996.

[47] Pu Sun. A pseudo non-time splitting method in air quality modeling. *J. Comp. Phys.*, (127): 152–157, 1996.

[48] O. Knoth and R. Wolke. *Air Pollution Modelling and its applications X*, chapter A comparison of fast chemical kinetic solvers in a simple vertical diffusion model. Plenum Press, NY, 1994.

[49] B. Sportisse, A. Jaubertie, and P. Plion. Reducing mechanisms in chemical kinetics for the simulation of reactive transport; an application to air pollution modelling. In *Proceedings of the St Venant Symposium*, August 1997.

[50] H.G. Roos, M. Stynes, and L. Tobiska. *Numerical methods for singularly perturbed differential equations*. Springer, 1991.

[51] INRIA Cooperative Research Action COMODE. http://binky.enpc.fr/~sportiss/medee.html.

RESOLUTION OF POLLUTANT CONCENTRATIONS USING A FULLY 3D ADAPTIVE METHOD

SAKTIPADA GHORAI*, ALISON S. TOMLIN[†], AND MARTIN BERZINS[‡]

Abstract. The development of a high resolution model for air pollution dispersion and chemical transformation is described here. A cell-vertex finite volume scheme based on an adaptive tetrahedral mesh is used to solve the atmospheric diffusion equation. Preliminary studies of dispersion from an accidental release of pollutants into the atmosphere have been carried out. The results show the efficiency and accuracy of this adaptive method compared to fixed methods for mesh refinement. Some comments about the interpolation of input data such as wind fields onto unstructured meshes are also made.

Key words. Adaptive meshes, dispersion, accidental release, boundary layer, unstructured.

1. Introduction. The issue of mesh resolution in regional scale air pollution models is an important one since there are many examples of the detrimental effects of using course meshes on solution accuracy. Previous work has shown (cf. [10, 12, 4]) that coarse horizontal resolution can have the effect of increasing horizontal diffusion to values many times greater than that described by models, resulting in the smearing of pollutant concentrations and an underestimation of maximum concentration levels. For reactive pollutants the effects can be compounded by nonlinear chemical reactions and inaccurate predictions of secondary species budgets can occur. The effects of mesh resolution have been well noted by the atmospheric modelling community and attempts have been made to improve mesh resolution at the same time as trying to avoid excessive extra computational work. The usual approach is to use nested or telescopic grids as in [6], where the mesh is refined in certain regions of the horizontal domain which are considered of interest. This may include for example regions of high emissions such as urban areas, or close to regions where significant monitoring is taking place. Previous work (cf. [11]) has shown however that such telescopic grids often cannot resolve plume structures occurring outside of the nested regions and that adaptive refinement in the horizontal domain can provide higher accuracy without entailing large extra computational costs.

In the vertical domain usually a stretched mesh is used, placing more solution points close to the ground. As in the horizontal domain, the resolution of the mesh in the vertical direction affects the vertical mixing of

---

*Department of Mathematics, Indian Institute of Technology, Kanpur 208016, INDIA. This work was carried out while the first author was a research fellow at the University of Leeds.

†Dept. of Fuel and Energy, University of Leeds, Leeds LS2 9JT, UK. To whom all correspondence should be addressed.

‡School of Computer Studies, University of Leeds, Leeds LS2 9JT, UK.

pollutant species. The use of adaptive meshes in the vertical domain has so far received little attention. A recent study in [3] has shown the importance of vertical resolution by the adaptive strategy discussed in this paper. It has been shown that in order to resolve pollutant concentrations in unstable boundary layers, more mesh points need to be placed near the inversion layer rather than at ground level as is common. The work in [3] was based on a dispersion from a single source without reaction terms in a one-dimensional wind field. This paper describes preliminary results of the use of 3D adaptive method to a test problem representing the accidental release of reactive pollutants. Both high resolution and efficiency are crucial issues for such problems since they may represent hazardous releases where predictions have to made much faster than real time with high spatial accuracy. The CBM-LEEDS scheme of [5] has been adopted for the nonlinear chemistry terms. The solution method is a 3D finite volume cell-vertex approach based on a tetrahedral mesh. The test problems have been designed to determine the importance of mesh structure on both horizontal and vertical mixing for typical meteorological conditions. The paper therefore concentrates on simple scenarios and on the comparison between different mesh structures in order to highlight the solution accuracy of different modelling approaches. In particular we compare a high resolution fixed mesh method to a fully adaptive one. The advantages of the adaptive method will be shown in that it can produce a very accurate solution using a resolved mesh in regions of the domain where the pollution cloud reaches, whilst avoiding unnecessary computation by using a very course grid in other regions of the domain.

2. Method of solution. The atmospheric diffusion equation in three dimensions is given by

$$
\begin{aligned}
(2.1) \quad &\frac{\partial c}{\partial t} + \frac{\partial(u\,c)}{\partial x} + \frac{\partial(v\,c)}{\partial y} + \frac{\partial(w\,c)}{\partial z} \\
&= \frac{\partial}{\partial x}\left(K_x \frac{\partial c}{\partial x}\right) + \frac{\partial}{\partial y}\left(K_y \frac{\partial c}{\partial y}\right) + \frac{\partial}{\partial z}\left(K_z \frac{\partial c}{\partial z}\right) + S,
\end{aligned}
$$

where c is a column vector representing the concentrations with the dimension of c equal to the number of species. The scalers u, v, w are mean wind velocities and K_x, K_y, K_z turbulent diffusivity coefficients in the x, y and z directions with respect to a Cartesian coordinate system with the z axis pointing upwards. The vector S represent the source term containing reaction terms which are usually nonlinear, emission and deposition terms. A cell vertex finite volume scheme has been chosen so that the number of variables per partial differential equation is less than that of a cell centred scheme. The solution method consists of two major parts, a mesh adaption routine and a flow integration routine. The mesh adaption routine changes the connectivity in the data structure of the mesh in response to changes in the solutions. The flow integration routine advances the solution in time.

3. Mesh adaption. The basic mesh objects of *nodes, edges, faces* and *elements*, which together form the computational domain, map onto the data objects within the adaption algorithm tree data structure. The data objects contain all flow and connectivity information sufficient to adapt the mesh structure and flow solution by either local *refinement* or *derefinement* procedures. The mesh adaption strategy assumes that there exists a "good quality" initial unstructured mesh covering the computational domain. The refinement process adds nodes to this base level mesh by edge, face and element subdivision, with each change to the mesh being tracked within the code data structure by the construction of a data hierarchy. The derefinement is the inverse of refinement, where nodes, faces and elements are removed from the mesh by working back through the local mesh refinement hierarchy.

The main adaption routine is driven by refining and derefining element edges. If an edge is refined by the addition of a node along its length, then all the elements which share the (parent) edge under refinement must be refined. In the case of derefinement all the elements which share the node being removed must be derefined. Numerical criteria derived from the flow field will mark an edge to either refine, derefine or remain unchanged. It is necessary to make sure the edges targeted for refinement and derefinement pass various conditions prior to their adaptation. These conditions effectively decouple the regions of mesh refinement from those of derefinement meaning that, for example, an element is not both derefined and refined in the same adaption step.

For reasons of both tetrahedral quality and algorithm simplicity only two types of element subdivisions are used (cf. [9]). The first type of subdivision is called *regular subdivision* where a new node bisects each edge of the parent element resulting in eight new elements. The second type of dissection, *green subdivision*, introduces an extra node into the parent tetrahedron, which is subsequently connected to all the parent vertices and any additional nodes which bisect the parent edges. Green refinement removes inconsistently connected or "hanging" nodes without the introduction of additional edge refinement. The green elements may be of poorer quality in terms of aspect ratio and so the green element is not further refined. Figure 1 demonstrates regular and green refinement for a tetrahedron. The five possible refinement possibilities (if all the edges are refined then the parent element is regularly refined) give rise to between 6 and 14 child green elements.

The choice of the adaption criteria is very important since it can produce large number of nodes or small number of nodes depending on the condition used to flag an edge for adaption. Also, when there are a large number of species, the choice of a given criteria might result in high resolution for some species but low resolution for the other species. Let c_q be the qth species and 0 and i the nodes for a given edge $e_{0,i}$. We calculate tolg_q and tolc_q for the qth species by

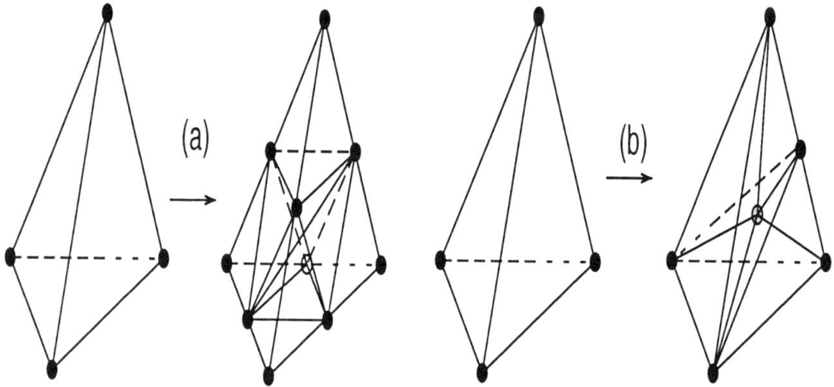

FIG. 1. *(a) Regular refinement based on the subdivision of tetrahedron by dissection of interior diagonal (1:8) and (b) "green" refinement by addition of an interior node (1:6).*

$$\text{tolg}_q = \frac{|(c_q)_0 - (c_q)_i|}{\text{length}[e_{0,i}]} \qquad \text{and} \qquad \text{tolc}_q = \frac{(c_q)_0 + (c_q)_i}{2}.$$

We refine the edge $e_{0,i}$ if tolg_q and tolc_q exceed some tolerances, otherwise it is derefined. Also a maximum refinement level is specified at the beginning so that if an edge is targeted for refinement but it is in the maximum level, then it is kept unchanged.

Suppose we have two edges with $\text{tolg}_q = 100$ and 200. If we take the tolerance parameter, Tg_q say, for tolg_q equal to 150, then only the second edge is refined to maximum level. On the other hand, if $\text{Tg}_q = 50$, then both edges are refined to maximum level. But we expect that the solution error for edge with $\text{tolg}_q = 200$ is greater than the error in the edge with $\text{tolg}_q = 100$. It might be advantageous to use two sets of $\text{Tg}_q = 50$ and 150. If $\text{tolg}_q > 150$, than we refine an edge to maximum level and if $50 < \text{tolg}_q < 150$, then we refine an edge to the level just lower than the maximum level. Thus the idea is to refine to the maximum level in the steepest gradient regions but to lower levels in the regions of less steep gradients.

4. Meteorological field. The test wind field (m s^{-1}) is given by

$$u = f(z)\cos(\theta), \quad v = f(z)\sin(\theta/2) \qquad \text{and} \qquad w = 0,$$

where $f(z)$ is a logarithmic function which is zero at the ground level and attains a maximum value of 6 at 800 m height. The wind direction varies according to the time of the day:

$$\theta = \pi\left(\frac{t}{9} - \frac{3}{2}\right),$$

where t is in hours.

The horizontal and vertical diffusivities are estimated as follows. We define a variable k_c such that

$$k_c = 1.0 + (t - 10), \qquad 10 \le t < 21$$
$$k_c = 0.0, \qquad\qquad\quad \text{otherwise},$$

where t is given in hours. The height (m) of the inversion layer is calculated by

$$h_i = 1200 - 200 \times |k_c - 6|.$$

The horizontal diffusivity is estimated by

$$K_x = K_y = (50 - 5 \times |k_c - 6|) \ \mathrm{m^2\,s^{-1}}.$$

The vertical diffusivity K_z is estimated using stable boundary layer profile when $h_i = 0$, otherwise it is estimated under a moderately unstable boundary condition with h_i as the inversion height. For example, for a stable boundary layer (i.e. $h_i = 0$), the vertical diffusivity at the height z is given by

$$K_z = u_* \kappa z \exp(-8 f z/u_*)/(0.74 + 4.7 z/L),$$

where L is the Monin-Obukhov length, u_* the friction velocity, u_*/f the Ekman layer height, and κ the von Karman constant. Similarly, when the boundary layer is moderately unstable with $z = h_i$ as the inversion layer, then K_z is calculated by the following formulae:

$$
\begin{aligned}
K_z = {}& w_* h_i \left[2.5\kappa(z/h_i)^{4/3}(1 - 15z/L)^{1/4} \right], & (z/h_i) < 0.05 \\
= {}& w_* h_i \left[0.021 + 0.408(z/h_i) + 1.35(z/h_i)^2 \right. \\
& \left. - 4.096(z/h_i)^3 + 2.56(z/h_i)^4 \right], & 0.05 \le (z/h_i) \le 0.06 \\
= {}& w_* h_i \left[0.2 \exp(6 - 10(z/h_i)) \right], & 0.06 < (z/h_i) \le 1.1 \\
= {}& 0.0013 \, w_* h_i, & (z/h_i) > 1.1.
\end{aligned}
$$

Here w_* is the convective velocity scale (cf. [8]). Using the above formulae, the vertical and horizontal diffusion coefficients can be estimated at any hour of the day.

5. Numerical scheme. Although sophisticated space-time error control techniques have been used in two space dimensional calculations (cf. [1], [11]), the computational cost and the need to evaluate spatial mesh adaption has led us to focus here on less costly methods. It is thus primarily for computational efficiency that, in common with many others, we have used an operator splitting technique. In this approach, the chemistry part is decoupled from the transport part. We use a node based scheme

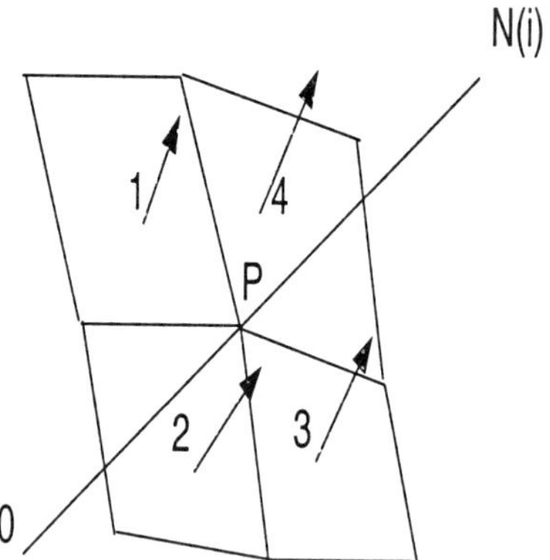

FIG. 2. *Dual mesh boundary faces attached to an edge.*

contrary to the cell-centred scheme, since the number of nodes is far less than the number of cells for a reasonable grid size. The diffusion equation is discretized over special volumes that form the dual mesh. The dual mesh is constructed by dividing each tetrahedron into four hexahedra of equal volumes, by connecting the mid-edge points, face centroids and the centroid of the tetrahedron. The control volume around a node 0 is thus formed by a polyhedral hull which is the union of all such hexahedra that share that node.

5.1. Flux evaluation using edge-based operation. The evaluation of flux around a dual cell can be cast into an edge-based operation. Let us discretize the divergence term

$$\frac{\partial f}{\partial x} + \frac{\partial g}{\partial y} + \frac{\partial h}{\partial z}$$

over the control volume Ω_0 enclosing the node 0. This divergence form is converted to flux form using Gauss divergence theorem:

$$\int_{\Omega_0} (\frac{\partial f}{\partial x} + \frac{\partial g}{\partial y} + \frac{\partial h}{\partial z})\, d\Omega = \int_{\partial \Omega_0} (f\, n_x + g\, n_y + h\, n_z)\, dS$$

$$(5.1) \qquad \qquad = \sum_k (f\, S_x + g\, S_y + h\, S_z),$$

where the summation is over all the dual mesh faces that form the boundary of the control volume around the node 0 and the areas S_x, S_y, S_z are projections of the dual quadrilateral face.

Consider edge **i**, formed by nodes 0 and N(i). The quadrilateral faces of the dual mesh that are connected to the edge at its mid-point P are shown in Figure 2. The number of such quadrilateral faces attached to an edge depends on the number of neighbouring tetrahedra. There are four tetrahedra sharing the edge **i** in Figure 2. The projected area, $\boldsymbol{A}_i$, associated with the edge **i** is calculated in terms of the quadrilateral face areas, $\boldsymbol{a}_1, \boldsymbol{a}_2, \boldsymbol{a}_3, \boldsymbol{a}_4$, as

$$(5.2) \qquad (A_i)_x = \sum_{j=1}^{4} (a_j)_x, \quad (A_i)_y = \sum_{j=1}^{4} (a_j)_y, \quad (A_i)_z = \sum_{j=1}^{4} (a_j)_z.$$

The projections are computed so that the area vector points outward from the control volume surface associated with a node. The boundary of the control volume around the node 0 is formed by the union of all such areas $\boldsymbol{A}_i$ associated with each edge **i** that share the node 0. The contribution of the edge **i** to the fluxes across the faces of the control volume surrounding the node 0 is given by

$$f_p \, (A_i)_x + g_p \, (A_i)_y + h_p \, (A_i)_z.$$

Hence equation (5.1) is replaced by

$$(5.3) \qquad \int_{\Omega_0} \left(\frac{\partial f}{\partial x} + \frac{\partial g}{\partial y} + \frac{\partial h}{\partial z} \right) d\Omega = \sum_{i} \left[f_p \, (A_i)_x + g_p \, (A_i)_y + h_p \, (A_i)_z \right],$$

where the sum is over the edges that share the node 0. The fluxes are thus calculated on an edge-wise basis and conservation is enforced by producing a positive flux contribution to one node and an equal but opposite contribution to the other node that forms the edge.

5.2. Advection scheme. Our aim is to discretize the term

$$\int_{\Omega_0} \left[\frac{\partial(u\,c)}{\partial x} + \frac{\partial(v\,c)}{\partial y} + \frac{\partial(w\,c)}{\partial z} \right] d\Omega \equiv \int_{\Omega_0} \nabla \cdot \boldsymbol{F} \, d\Omega,$$

where

$$\boldsymbol{F} = (u\hat{i} + v\hat{j} + w\hat{k}) \, c.$$

Using (5.3), the above equation can be written as

$$(5.4) \qquad \sum_{i} \left[u_p \, (A_i)_x + v_p \, (A_i)_y + w_p \, (A_i)_z \right] (c)_p = \sum_{i} (\boldsymbol{F}_p \cdot \boldsymbol{A}_i) \, (c)_p,$$

where $\boldsymbol{A}_i$ is called the edge-normal associated with the edge i and the sum is over all the edges sharing node 0 with control volume Ω_0. Let $(c)_p^-$ denote the value of c at P interpolated from node 0 and $(c)_p^+$ denote the value at P interpolated from node i (refer to Figure 2). Then an upwind version of (5.4) is given by

$$(5.5) \qquad \sum_{i} \left[\frac{(\boldsymbol{F}_p \cdot \boldsymbol{A}_i + |\boldsymbol{F}_p \cdot \boldsymbol{A}_i|)}{2} \, (c)_p^- + \frac{(\boldsymbol{F}_p \cdot \boldsymbol{A}_i - |\boldsymbol{F}_p \cdot \boldsymbol{A}_i|)}{2} \, (c)_p^+ \right].$$

Let r_0 and r_i denote the position vectors of the nodes 0 and i respectively. If r_p is the position vector of P, then

$$r_p = \frac{r_0 + r_i}{2}.$$

The extrapolation formulae are

$$(c)_p^- = (c)_0 + \Phi_0 \, (\nabla c)_0 \cdot (r_p - r_0)$$
$$(c)_p^+ = (c)_i + \Phi_i \, (\nabla c)_i \cdot (r_p - r_i),$$

where Φ_0 and Φ_i are the limiter functions at nodes 0 and i respectively. The limiter function at a node is chosen such that the reconstructed value is bounded by the values at the node and its neighbours (cf. [2]). Specifically let us consider node 0. Let N_0 denote the set of neighbouring nodes to the node 0. First we compute the maximum and minimum of all adjacent neighbours.

$$(c)_0^{min} = \min_{i \in N_0} \left((c)_0, (c)_i \right),$$

$$(c)_0^{max} = \max_{i \in N_0} \left((c)_0, (c)_i \right)$$

and require that

$$(c)_0^{min} \leq (c)_p^- \leq (c)_0^{max},$$

where $(c)_p^-$ is computed at each of the edges sharing the node 0. Considering the edge sharing the node 0 and i, we calculate ϕ_i according to

$$(5.6) \quad \phi_i = \begin{cases} \min \left(1, \dfrac{(c)_0^{max} - (c)_0}{(\nabla c)_0 \cdot (r_p - r_0)} \right), & \text{if} \quad (\nabla c)_0 \cdot (r_p - r_0) > 0 \\[3mm] \min \left(1, \dfrac{(c)_0 - (c)_0^{min}}{(\nabla c)_0 \cdot (r_p - r_0)} \right), & \text{if} \quad (\nabla c)_0 \cdot (r_p - r_0) < 0 \\[3mm] 1 & \text{if} \quad (\nabla c)_0 \cdot (r_p - r_0) = 0. \end{cases}$$

Now we calculate Φ_0 by

$$\Phi_0 = \min_i \{ \phi_i \},$$

where the minimum is taken over all the edges that share the node 0. The calculation of limiter functions and gradients at the nodes *is not on a node by node basis* which is CPU intensive. Instead they are calculated *in an edge-based operation*. The time step for the advection scheme is chosen so that it satisfies the CFL condition (cf. [14]). Again let us consider the node 0. We define S_0^1 and S_0^2 as follows.

$$S_0^1 = \sum_{\substack{i \in N_0 \\ (c_s)_p^+ \geq (c_s)_0}} -\frac{(\mathbf{F}_p \cdot \mathbf{A}_i - |\mathbf{F}_p \cdot \mathbf{A}_i|)}{2} + \sum_{\substack{i \in N_0 \\ (c_s)_p^- \leq (c_s)_0}} \frac{(\mathbf{F}_p \cdot \mathbf{A}_i + |\mathbf{F}_p \cdot \mathbf{A}_i|)}{2}$$

$$S_0^2 = \sum_{\substack{i \in N_0 \\ (c_s)_p^+ < (c_s)_0}} -\frac{(\mathbf{F}_p \cdot \mathbf{A}_i - |\mathbf{F}_p \cdot \mathbf{A}_i|)}{2} + \sum_{\substack{i \in N_0 \\ (c_s)_p^- > (c_s)_0}} \frac{(\mathbf{F}_p \cdot \mathbf{A}_i + |\mathbf{F}_p \cdot \mathbf{A}_i|)}{2}.$$

The time step at node 0 is Δt_0 and is given by

$$\Delta t_0 = \min\left(\frac{V_0}{S_0^1}, \frac{V_0}{S_0^2}\right)$$

and in a similar way, the time step at every node is calculated. The minimum of the time steps over all the vertices constitutes the time step for the advection scheme. Again this computation can be cast into an edge-based operation.

5.3. Diffusion scheme. The diffusion term is discretized as

$$\int_{\Omega_0} \left[\frac{\partial}{\partial x}(K_x \frac{\partial c}{\partial x}) + \frac{\partial}{\partial y}(K_y \frac{\partial c}{\partial y}) + \frac{\partial}{\partial z}(K_z \frac{\partial c}{\partial z}) \right] d\Omega \equiv \int_{\Omega_0} \nabla \cdot \mathbf{R} \, d\Omega,$$

where

$$\mathbf{R} = K_x \frac{\partial c}{\partial x}\hat{i} + K_y \frac{\partial c}{\partial y}\hat{j} + K_z \frac{\partial c}{\partial z}\hat{k}.$$

Here K_x, K_y, K_z are the turbulent diffusivity coefficients. Using (5.3), the above equation can be written as

$$(5.7) \qquad \sum_i \left[(K_x \frac{\partial c}{\partial x})_p (A_i)_x + (K_y \frac{\partial c}{\partial y})_p (A_i)_y + (K_z \frac{\partial c}{\partial z})_p (A_i)_z \right],$$

where $\mathbf{A}_i$ is the edge-normal associated with the edge i and the sum is over all the edges sharing the node 0 with control volume Ω_0. The first term in (5.7) can be expressed as

$$(K_x \frac{\partial c}{\partial x})_p = (K_x)_p \, (\frac{\partial c}{\partial x})_p$$

with similar expressions for the other two terms.

Thus we need to calculate the edge gradient term $(\nabla c)_p$ for the edge $e(0, i)$. For this we again apply the Gauss divergence theorem over a control volume. For edge $e(0, i)$, this control volume is the union of elements sharing the edge $e(0, i)$. Once we calculate $(\nabla c)_p$, its components are substituted in (5.7) and the same is done for all the edges that share the node 0. The discretized version of (5.7) can be expressed as

$$\tag{5.8} \sum_i a_i \, (c)_i - a_0 \, (c)_0,$$

where the sum is over the nodes that share the node 0 and

$$a_0 = \sum_i a_i.$$

Thus we need to store only the values of a_i and these values can be attached to the edge pointer. Importantly, the values of a_i depend only on the mesh geometry and turbulent diffusivity coefficients. This is very useful and results in the use of less CPU time since the values of a_i between two time steps remain the same as long as there is no mesh adaption between these steps.

5.4. Solution of the advection-diffusion equation. If c^n denotes the species concentration at time level n, then the species concentration at the next time step is given by

$$\tag{5.9} c^{n+1} = c^n + \tau \, (g(c) + f(c) + S),$$

where τ is the time step, $g(c)$ and $f(c)$ are due to the advection and diffusion operator respectively. Here S is the contribution due to the sources except the chemistry terms. In a fully explicit scheme, f and g are evaluated using values at the time level n. However, the time restriction for stability due to vertical diffusion is severe since the grid spacings along the vertical can be small. Hence we use an implicit-explicit formulation for (5.9), where the advection is evaluated explicitly and the diffusion is calculated implicitly. The time step τ is chosen as the time step due to advection only. It's value mainly depends on the wind speed and the vertical mesh spacings near the ground level. The time step is smaller for higher wind speed and vice versa. The resulting linear system of equations is solved using the Gauss-Seidel iteration technique with over-relaxation and the iteration is stopped when the relative error is less than some prescribed tolerance.

5.5. Chemistry scheme. The nonlinear chemistry part gives rise to stiff ordinary differential equations. The stiff chemistry part is solved by two different methods: the SPRINT2D method of [1] and the 2nd order Rosenbrock method (cf. [13]). The first method is based on error control techniques and is preferred when the splitting is carried out at the discretization level (cf. [11]). However, it is expensive when the splitting is done at the PDE level, since the start-up cost at every split interval is rather high. The Rosenbrock method is suitable when the splitting is done at the PDE level, since it is an explicit method with a constant time step. Both methods give the same results for the problem investigated in this work.

5.6. Solution procedure. The solution proceeds as follows. A split-time interval is fixed for the flow integration. The transport term is advanced over half the interval. The solution of this transport part is used as initial conditions for the chemistry part and integration is carried out for the full interval. The resulting solution is used as initial conditions for the transport part and the solution is computed for the other half interval. This procedure is repeated again for the next split-time interval.

6. Grid. The dimensions of the region used for the test case are 80×80 km^2 in the horizontal directions and 1.6 km along the vertical direction. The tetrahedral mesh is generated by dividing the whole region into cuboids and then subdividing a cuboid into 6 tetrahedral elements. The solution of the fully adaptive grid is compared to the solution on a reference grid.

6.1. Reference grid. The cuboids are 1 km and 1 km along the x and y axis respectively and 200 m in the vertical direction. This grid is kept constant throughout the calculations. The wind field is calculated at the mid-edge points. Since this grid is regular, the wind field remains mass conservative numerically.

6.2. Adaptive grid. The cuboids are 4 km and 4 km along the x and y axis respectively and 800 m in the vertical directions. We impose level 2 depth for the mesh adaption. Thus a horizontal edge will be 1 km in length and a vertical edge will be 200 m in length under level 2 refinement. Once the grid is refined, the wind data at the mid-edge points no longer remains mass conservative. This wind data is made mass conservative using a variational method due to Mathur and Peters (cf. [7]) and fully described in [3].

If the wind data is given only in coarse grid points, then we adopt the following method. We take a mid-edge point and determine the cuboid within which it lies. We determine the points of intersection of a vertical line through the mid-edge points with the lower and upper planes of the cuboid. The wind data at these two points is calculated by linear interpolation from the four points at each of these planes. Next we calculate the value at the mid-edge points by fitting a logarithmic profile

$$v = A \log(z) + B$$

between these two points.

7. Results. We compute the solution on the adaptive grid and compare the solution with the reference solution. The problem describes that of an accidental release of pollutants in the atmosphere and we discuss solutions with and without chemical reaction terms. A point source at $(12, 64)$ km and 400 m height releases pollutants from 9 to 11 am during the simulation.

When there are no chemistry terms, we consider a single species only with a background concentration of 5 ppb and the strength of the point source is 2.2×10^{24} molecules s^{-1}.

For the diffusion equation with chemical reaction terms we have taken the CBM-LEEDS scheme (cf. [5]) consisting of 59 reactions with 29 species and representing a reduced version of the Extended Carbon Bond mechanism for tropospheric chemistry. The species with non zero initial concentrations are given in Table 1. The other species and details of the chemical mechanism are described in [5]. The strength of the point source is 2.2×10^{24} molecules s^{-1} which consists of 90% of NO and 10% of NO$_2$.

TABLE 1

Initial conditions for the CBM-LEEDS mechanism. Concentrations of the other species are set to zero.

Species	Concentration (ppb)
NO$_2$	5
NO	25
O$_3$	20
H$_2$O	10^7
CO	2.14×10^3
FORM	0.1
ALD2	1
PAR	10
OLE	5.5
ETH	7.5
TOL	20
XYL	1

7.1. Solution without chemistry. The solution is started at 9 am and is run for 6 hours. The number of nodes in the reference grid is $59,049$. The initial number of nodes in the adaptive grid is $1,323$ and as a result of initial refinement around the source, it becomes $2,419$. We take the following adaptive criteria to flag an edge e_{0i} to refine. We calculate tolg_q, tolc_q (see Section 3) for the edge e_{0i} and let

$$\mathrm{Tg}_q = \max \mathrm{tolg}_q.$$

We refine the edge e_{0i} to level 2 if

$$(7.1) \qquad \mathrm{tolc}_q > 10^{11}, \ \mathrm{tolg}_q > 0.01 \times \mathrm{Tg}_q$$

and refine to level 1 if

$$(7.2) \qquad \mathrm{tolc}_q > 10^{11}, \ \mathrm{tolg}_q > 0.0005 \times \mathrm{Tg}_q.$$

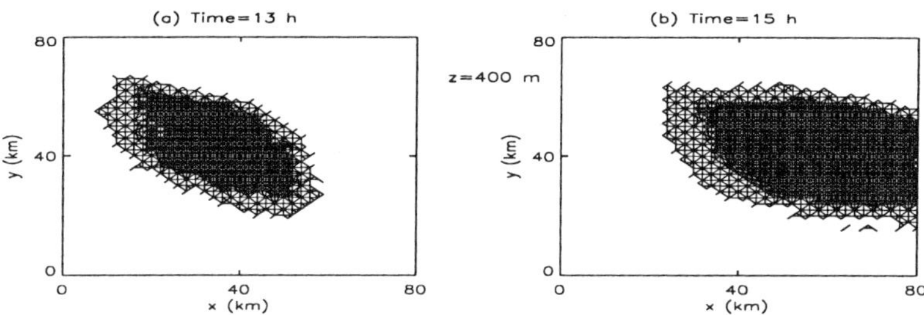

FIG. 3. *Grid refinement at 400 m height at 13 and 15 hours time.*

As the released pollutant spreads, the size of the region satisfying (7.1) and (7.2) increases. Thus the number of nodes in the adaptive mesh also increases. The numbers of nodes in the adaptive mesh are 10, 964 and 16, 919 at 13 and 15 hours respectively. The refinement at 13 and 15 hours time at 400 m height is shown in Figure 3. The edges at the outer region are refined to level 1 due to the small gradients there. Our initial mesh consists of three layers at 0, 800 and 1600 m height. Thus there are no nodes in the 400 m height outside of the adaptive region.

The comparison between the solution of the adaptive grid and reference grid is shown in Figure 4 and Table 2. The solutions of the dispersion problem at three different heights are shown in Figure 4. The values of the peak concentration at three different heights at 13 and 15 hours of time are given in Table 2. Table 2 and Figure 4 show that the solutions of the adaptive and reference grids are in good agreement.

TABLE 2

Peak concentrations at 13 and 15 hours for the reference and adaptive grid (level 2) solutions.

Height	Reference peak conc. (ppb)		Adaptive peak conc. (ppb)	
(m)	(13 h)	(15 h)	(13 h)	(15 h)
0	8.95	6.48	8.79	6.47
400	8.81	6.43	8.71	6.43
800	8.29	6.23	8.04	6.23

The program has been run serially on an Origin2000 computer. For the reference solution, the total CPU time is 1 hour 48 minutes. The corresponding time for the adaptive solution is 23 minutes. Thus the adaptive grid takes a much smaller time compared to the reference grid and achieves comparable accuracy.

We have refined the mesh to level 2. Thus the lengths of the horizontal and vertical edges become equal to that of the reference grid when refined to the maximum level. We might also take the cuboids of 8 km, 8 km along the x, y axis respectively and 1600 m in the vertical directions. Thus there

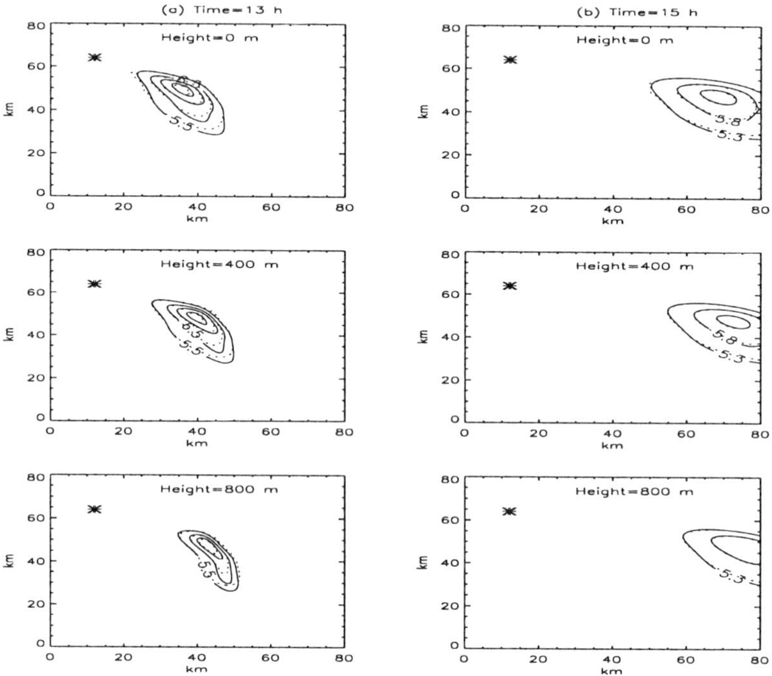

FIG. 4. *Comparison between the Solutions (equally spaced contour levels) of reference grid (solid line) and adaptive grid (dashed line) at 0, 400 and 800 m height at 13 and 15 hours time. The source is marked at the upper left corner.*

is only single layer at the top of the domain. We impose level 3 depth for the mesh adaption so that the horizontal and vertical edges become 1 km and 200 m respectively when refined to the maximum level. The initial number of nodes in the adaptive grid is 600 and as a result of initial refinement around the source, it becomes 1,510. We take the same adaptive criteria as in the case of level 2 adaption. The numbers of nodes in the adaptive mesh are 10,711 and 16,906 at 13 and 15 hours respectively. Thus the level 3 adaption produce similar refinement as the level 2 adaption. The CPU time is approximately 22 minutes which is almost equal to the level 2 adaption. The solution is also similar to the level 2 solution as can be seen from Table 3.

Thus it is clear that solution on the adaptive mesh is comparable with the solution on the reference grid. There are three important criteria to achieve this. First, the highest level of refinement must produce edges similar to the reference grid. Second, care must be taken to choose the adaptive criteria. The choice of very large tolerances leads to rapid numerical diffusion of the concentration which causes smearing in peak concentration profile. On the other hand, small tolerances lead to large number of nodes which increase the CPU time. Third, there is an extra cost with the adap-

TABLE 3

Peak concentrations at 13 and 15 hours for the reference and adaptive grid (level 3) solutions. See also Table 2 for the level 2 solution.

Height (m)	Reference peak conc. (ppb)		Adaptive peak conc. (ppb)	
	(13 h)	(15 h)	(13 h)	(15 h)
0	8.95	6.48	8.91	6.48
400	8.81	6.43	8.68	6.44
800	8.29	6.23	7.87	6.22

tive codes, that of periodically refining/coarsening the mesh. However, this cost is small if the mesh is not adapted every time-step. In the above problem, the mesh is refined/derefined every 15 minutes.

8. Solution with chemistry. The solution is started at 9 am and is run for 6 hours. The number of nodes in the reference grid is $59,049$. The initial number of nodes in the adaptive grid is $1,323$ and as a result of initial refinement around the source, it becomes $2,419$. We calculate adaptive criteria based on the species NO_2 and O_3 to flag an edge e_{0i} to refine. We calculate tolg_q, tolc_q (see Section 3) for the edge e_{0i} with respect to the species $q = NO_2, O_3$. Let

$$\text{tolg}_q = \frac{\text{tolg}_{no2} + \text{tolg}_{o3}}{2}, \ \text{tolc}_q = \frac{\text{tolc}_{no2} + \text{tolc}_{o3}}{2} \ \text{ and } \ \text{Tg}_q = \max \text{tolg}_q.$$

We refine the edge e_{0i} to level 2 if

$$(8.1) \qquad \qquad \text{tolc}_q > 10^{11}, \ \text{tolg}_q > 0.01 \times \text{Tg}_q$$

and refine to level 1 if

$$(8.2) \qquad \qquad \text{tolc}_q > 10^{11}, \ \text{tolg}_q > 0.0005 \times \text{Tg}_q.$$

The grid develops with time in a similar way to the dispersion only problem. The number of nodes under level 2 adaption is $11,593$ and $17,942$ at 13 and 15 hours time. The NO concentrations at 0, 400 and 800 m height are shown in Figure 5. The comparison of the peak NO concentration at these heights is given in Table 4. It is clear from Figure 5 and Table 4 that the NO solution of the adaptive grid is comparable to that of the reference grid.

Because of the photochemical reactions, the regions of the maximum NO concentration give rise to minimum O_3 concentration. The minimum O_3 concentration at 0, 400 and 800 m height is shown in Table 5. Again the agreement between O_3 solution of the reference and the adaptive solution is good.

The CPU times required for the adaptive and reference grids are 3 and 24 hours respectively. Thus the adaptive grid takes a much smaller time compared to the reference grid to achieve comparable accuracy.

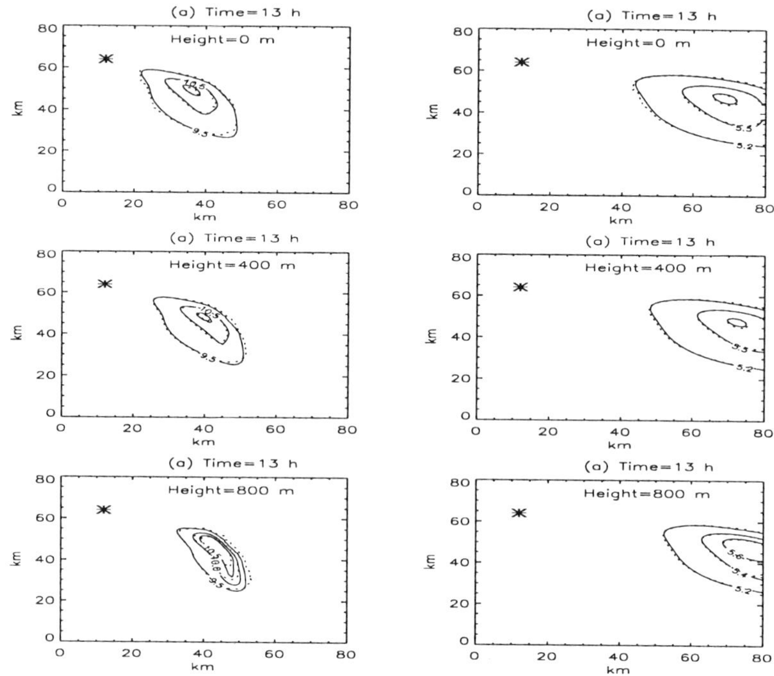

FIG. 5. *Comparison between the Solutions (equally spaced contour levels) of refer-ence grid (solid line) and adaptive grid (dashed line) at 0, 400 and 800 m height at 13 and 15 hours time. The source is marked at the upper left corner.*

TABLE 4

Peak NO concentrations at 13 and 15 hours for the reference and adaptive grid (level 2) solutions.

Height	Reference peak conc. (ppb)		Adaptive peak conc. (ppb)	
(m)	(13 h)	(15 h)	(13 h)	(15 h)
0	11.69	5.84	11.62	5.84
400	11.61	5.82	11.56	5.82
800	11.30	5.74	11.15	5.73

9. Conclusions. We have presented a solution method for the atmospheric diffusion equation based on an unstructured, 3D adaptive mesh. A splitting technique is employed to solve the transport and the chemistry part separately. In this preliminary study, we considered a test problem related to the accidental release of pollutant from a source. The test wind field is two-dimensional and varies its direction during the day. The inversion layer is parametrized by the hour of the day. The wind field

TABLE 5

Minimum O_3 concentrations at 13 and 15 hours for the reference and adaptive grid (level 2) solutions.

Height (m)	Reference min. conc. (ppb) (13 h)	(15 h)	Adaptive min. conc. (ppb) (13 h)	(15 h)
0	28.67	47.68	28.71	47.66
400	28.77	47.75	28.82	47.73
800	29.16	48.04	29.36	48.05

and the diffusion coefficients are chosen so that they are comparable to the existing values in the boundary layer. An interpolation technique has been suggested to estimate wind field for the fine grid from known values on the course grid in a mass conservative way.

The results of the problem with and without chemistry suggest that the fully adaptive method gives solutions comparable to those using a high resolution reference grid. We have not reported the result for the telescoping grid where the grid is refined only near the source. Here the solution retains accuracy when the pollutants remain near the source. The solution becomes totally unreliable when the wind field carries the peak concentration region away from the source. The ratios of the CPU times for adaptive and reference grids are approximately 1:8 and 1:5 for the diffusion equation with and without chemistry respectively. Thus we see that the performance of the adaptive method is excellent provided we choose a good quality initial mesh and appropriate adaptive criteria. It was also demonstrated that using a simple gradient criteria the mesh was able to follow the pollution cloud as it dispersed through the domain. Significant computational expense was saved by being able to use a coarse grid away from the pollution cloud.

Thus the preliminary results provide a potential high resolution numerical technique for atmospheric dispersion problems. Key questions for future research involve the comparison of the technique described here with structured mesh techniques for a realistic problem such as the ETEX test case or the Chernobyl release. Although the overhead for adaptivity has been shown here to be small in comparison with a fully resolved unstructured solution, the overhead with respect to the use of unstructured meshes has not been tested. Efficient parallelism using load balancing techniques would be necessary to reduce the overhead of using an unstructured approach. The use of unstructured meshes also raises issues with respect to mesh quality which may become important when used with lower mesh boundaries resulting from complex terrain. In the test case described here a good quality initial mesh was easily achievable but for more complex meshes the dependency of solution accuracy on the quality of tetrahedra needs to be examined. Comparisons with other adaption criteria such as techniques based on curvature would also be useful. Some issues of interpolation of input data have been discussed here. The variational method used for the mass conservative interpolation of wind data does not lead to significant errors on the unstructured mesh. The interpolation of emissions data (both point and area sources) onto the unstructured grid should also be examined in a comprehensive way. Point source data are likely to be better represented by the unstructured grid since it allows the possibility for specific refinement around emissions on small spatial scale. As emissions inventories become more sophisticated in terms of representing point, line and high resolution area sources within urban regions, the coarse resolution regional scale models currently in use may start to show a greater degree of resolution dependency. The use of adaptive methods may then become essential for certain applications.

Acknowledgements. Dr. Ghorai was supported throughout this work by a grant from NERC, UK. Computations were carried out on an Origin 2000 computer supported by EPSRC JREI funding.

REFERENCES

[1] M. BERZINS, R. FAIRLIE, S. PENNINGTON, J. WARE AND L. SCALES, *SPRINT2D: Adaptive Software for PDEs*, ACM Trans. Math. Software **24** (1999), pp. 475–499.

[2] T. Barth and D. Jesperson, *The Design and Application of Upwind Schemes on Unstructured Meshes*, AIAA-0366 (1989), pp. 9–12.

[3] S. GHORAI, A. TOMLIN AND M. BERZINS, *Resolution of pollutant concentration in the boundary layer using a fully 3D adaptive gridding technique*, Atmos. Env. **34** (2000), pp. 2851–2863.

[4] G. HART, A. TOMLIN, J. SMITH AND M. BERZINS, *Multi-scale atmospheric dispersion modelling by use of adaptive gridding techniques*, Environmental Monitoring and Assessment **52** (1998), pp. 225-238.

[5] A. HEARD, M. PILLING AND A. TOMLIN, *Mechanism reduction techniques applied to tropospheric chemistry*, Atmos. Env. **32** (1998), pp. 1059–1073.

[6] H. JACOBS, H. FELDMAN, H. KASS AND M. MESSESHEIMER, *The use of nested models for air pollution studies: an application of the EURAD model to a SANA episode*, J. Appl. Met. **34** (1995), pp. 1301–1319.

[7] R. MATHUR AND L. PETERS, *Adjustment of wind fields for application in air pollution modelling*, Atmos. Env. **24A** (1990), pp. 1095–1106.

[8] J. SEINFELD, *Air Pollution*, Wiley, NY, 1986.

[9] W. SPEARES AND M. BERZINS, *A 3D unstructured mesh adaption algorithm for time-dependent shock-dominated problems*, Int. j. num. methods fluids **25** (1997), pp. 81–104.

[10] O. TALAT, A QUANTITATIVE ANALYSIS OF NUMERICAL DIFFUSION INTRODUCED BY ADVECTION ALGORITHMS IN AIR QUALITY MODELS, Atmos. Env., **31**, (1997), pp. 1933-1940.

[11] A. TOMLIN, M. BERZINS, J. WARE, J. SMITH AND M. PILLING, *On the use of adaptive gridding methods for modelling chemical transport from multi-scale sources*, Atmos. Env. **31** (1997), pp. 2945–2959.

[12] A. TOMLIN, S. GHORAI, G. HART AND M. BERZINS, *The use of 3-D adaptive unstructured meshes in pollution modelling*, in Large-Scale Computations in Air Pollution Modelling, Z. Zlatev et al. (eds.), Kluwer Acad. Pub. (1999), pp. 339–348.

[13] J. VERWER, E. SPEE, J. BLOOM AND W. HUNDSDORFER, *A second-order Rosenbrock method applied to photochemical dispersions problem*, SIAM J. Sci. comput. **15** (1999), pp. 1243–1250.

[14] M. WIERSE, *A new theoretically motivated higher order upwind scheme on unstructured grids of simplices*, Adv. Comput. Math. **7** (1997), pp. 303–335.

EFFECT OF GRID RESOLUTION AND SUBGRID ASSUMPTIONS ON THE MODEL PREDICTION OF NON-HOMOGENEOUS ATMOSPHERIC CHEMISTRY

DAVID P. CHOCK*, SANDRA L. WINKLER*, AND PU SUN*

Abstract. We have carried out an elaborate study of the impact of grid resolution and subgrid chemistry assumption on the model prediction of grid-averaged species concentrations for a buoyant stack plume in a convective boundary layer. The Particle-Grid approach was used to describe both the particle (air parcel) turbulent transport and chemistry. This approach allows an identical transport process for all simulations. It also allows a description of subgrid chemistry. The ambient and plume particle transport follows the description of Luhar and Britter (1989, 1992). The chemistry follows that of the Carbon-Bond mechanism. Three different grid sizes were considered: fine, medium and coarse, together with three different subgrid chemistry assumptions: micro scale or individual particle, tagged-particle (plume versus ambient), and untagged-particle (plume and ambient particles treated indiscriminately). Reducing the subgrid information is not necessarily similar to increasing the model grid size. In our example, increasing the grid size leads to a reduction in the suppression of ozone in the presence of a high-NO_x stack plume, and a reduction in the effectiveness of the NO_x-inhibition effect. On the other hand, reducing the subgrid information (by using the untagged-particle assumption) leads to an increase in ozone reduction and an enhancement of the NO_x-inhibition effect insofar as the ozone extremum is concerned.

1. Introduction. An air quality grid model has a typical grid cell of several square kilometers by several tens or hundreds of meters. The typical time step for numerical integration is several minutes. These spatial and temporal resolutions are chosen out of practical necessity due to (1) a lack of high-resolution and reliable input information in chemical species emissions and removals, concentration and meteorological fields, (2) incomplete knowledge in atmospheric turbulence, especially in spatially non-homogeneous environments, and (3) limitation in computational resources. In a typical grid model, the flow and concentration fields are assumed to vary smoothly within a grid, and when chemistry is turned on, the species concentrations within the grid are assumed to be spatially homogeneous. However, the physical and chemical processes in the atmosphere cover a wide range of temporal and spatial scales. In particular, the photochemical reaction time scales typically cover a range of more than ten orders of magnitude. In other words, some reactions may occur so fast that physical transport cannot quickly smooth out the local variation of the species concentrations within the grid. As a result, non-uniformity or non-homogeneity of the species concentrations will result within a model grid. The lack of homogeneity in species concentrations leads to a departure of the chemical reaction rates from the case of assumed homogeneity in the model, and subsequently, errors in the model predictions. The impact of

---

*Ford Research Laboratory, Ford Motor Company, P.O. Box 2053, Mail Drop 3083, Dearborn, MI 48121-2053.

grid resolution on ozone chemistry and on the development of emission control strategy for ozone has been raised before (See, for example, Jang, *et al.* (1995a, b), Liang and Jacobson (2000).) because the degree of NO_x-inhibition appears to be reduced in model predictions when the model grid is large. This model artifact ought to be addressed for the development of a valid emission control plan.

The general approach in studying the effect of grid resolution is to conduct model runs, each run with a different grid size in the modeling domain, and determine the differences in model results. Of course, all input data for the finest grid resolution should first be available, and grid averaging of the input data is performed as grid size is increased in the separate model runs. There is a subtle numerical problem in this approach. Namely, the Eulerian-type advection scheme does not transport all chemical species in an air parcel at the same velocity. This is because, as a result of grid discretization, different Fourier modes of a concentration profile travel at different phase velocities, so that, depending on the spatial distributions of the species concentrations, the species can move at somewhat different velocities even though in principle, species in the same air parcels should all move together. So, as we change the grid size, the amplitudes of the Fourier modes also change, so that there is no guarantee that the concentrations of a given species will be transported identically for different grid size runs even if we could assume that the finest spatial resolution of the velocity field could be maintained for all grid sizes. The fact that the spatial resolution of the velocity field must be modified to conform to the grid size requirement will also modify the flow field. In fact, it may lead to non-conservation of mass, assuming that for one particular grid size of choice, the velocity field is mass conserving.

Chock and Winkler (1994a, b) presented a Particle-Grid approach for air quality modeling. In this approach, mini-air parcels (also referred to as particles) containing the masses of multiple chemical species undergo random walk and participate in chemical reactions with other mini-parcels within the same model grid. The species concentrations for the reactions at a given time step are initially the sums of the species masses from all mini-parcels within the grid divided by the grid volume. The resulting species concentrations after the reactions are redistributed as species masses back to the mini-parcels for continued transport at the next time step. The approach retains the information of the position of each parcel as it is transported, but simulates the chemical reactions at the spatial resolution of the model grid. The approach is ideal for the study of the effect of grid resolution on model predictions of non-homogeneous chemistry because (1) the advective transport of species concentrations is strictly determined by the flow field alone and is not influenced by the species concentration profiles due to numerical artifacts as would be the case in an Eulerian scheme; (2) the space-time dependence of the flow field can be made identical for all model runs while the chemistry can be run for different grid resolutions so that one can truly identify the effect of grid resolution on model prediction of non-homogeneous chemistry; (3) subgrid modeling can be accommodated to allow for distinct chemical processes for parcels or particles of different characteristics within the same grid cell. In principle, reducing the grid size should reduce the role of subgrid non-homogeneity. However, in practice, subgrid non-homogeneity may persist even as the grid size is reduced considerably. Subgrid chemistry is generally very difficult to handle and in an Eulerian scheme, requires a high-order closure procedure or a treatment based on probabilistic considerations.

The success of the Particle-Grid approach critically depends on how species masses are redistributed back to the mini-parcels participating in the chemical reactions in the same grid cells. The detail of species mass redistribution will be presented below. In what follows, we illustrate the application of this approach to an atmospheric chemical system containing a buoyant stack plume in a convective boundary layer. Because of the intensive computational demand of the study, we assume a 2D flow field and up to 20 minutes of photochemical reactions. The issue of the grid-size dependence of the NO_x-inhibition effect will be addressed.

2. Methodology. The 2D modeling domain is 1600 m in the horizontal or x direction and 400 m in the vertical or z direction. Three grid resolutions are considered: fine (10 m × 10 m), medium (20 m × 20 m) and coarse (40 m × 40 m). Therefore, the numbers of grid cells in the modeling domain are 160 × 40 (or 6400), 80 × 20 (or 1600), and 40 × 10 (or 400) for the fine, medium and coarse grid, respectively. The random walk simulations are all described in the fine-grid resolution. Other grid resolutions become important when chemistry is considered. We used a time step, Δt, of 1 s, and a uniform mean horizontal velocity of 1 m/s. Vertically uniformly spaced ambient particles were introduced at the upwind boundary at a rate of 10 per time step per fine grid cell so that, if there had been no random walk, then the particle number density would have been exactly 100 per cell. We shall first describe the transport simulation of a buoyant plume in a convective boundary layer, then the chemistry simulations, and the protocol for combining transport and chemistry. We then present the simulation results and our conclusions. Identical particle trajectories were preserved and used for all chemistry simulations under different grid resolutions. In so doing, we removed the potential artifacts inherent in Eulerian numerical methods that allow the results of chemistry under different grid resolutions to influence the advective transport of species concentrations.

2.1. Planetary boundary layer flow simulation. For the purpose of our study, it would be helpful to prescribe as realistic a flow field as possible even though this is not a requirement for the validity of the study outcome. The flow field used in the present study was generally based on the random walk model for buoyant-plume dispersion in inhomogeneous turbulence in a convective boundary layer described by Luhar and Britter

(1989, 1992). The inhomogeneous turbulence due to convection is assumed to occur under a sufficiently strong horizontal mean wind u such that horizontal diffusion is relatively unimportant to ensure the validity of Taylor's eddy translation hypothesis, but not sufficiently strong so that buoyancy rather than shear stress dominates the turbulent energy production. Thus the random walk is prescribed in the vertical direction where convection takes place and translated in the horizontal direction by the horizontal wind, which is assumed to have a homogeneous horizontal velocity fluctuation (Chock and Winkler, 1994a). The stochastic differential equations for both the fluid element and the mini-parcel or particle for the vertical motion are given by

$$
\begin{aligned}
dw &= a(z,w)dt + b(z,w)d\xi, \\
dz &= wdt.
\end{aligned}
$$

(1)

The random velocity increment $d\xi$ is an independent normal variate with mean zero and variance dt; z and w are the position and velocity of the fluid element or particle, respectively. Assuming that the evolution of w follows a Markov process, then the well-mixed condition (Thomson, 1987) requires that the probability density functions of both the fluid elements and the particles be equal $(=P)$ and satisfy the Fokker-Planck equation:

$$
w\frac{\partial P}{\partial z} + \frac{\partial}{\partial w}(aP) = \frac{1}{2}\frac{\partial^2}{\partial w^2}(b^2 P).
$$

(2)

In homogeneous, stationary turbulence, the basic Langevin equation with $a = -w/\tau_w$ and $b = (2\overline{w^2}/\tau_w)^{1/2}$ satisfies (2), with τ_w equal to the Lagrangian time scale and $\overline{w^2}$ being the variance of vertical turbulent velocity. These relations are assumed to be valid for the horizontal transport, with w and $\overline{w^2}$ replaced by u and $\overline{u^2}$, and τ_w replaced by τ_u defined for the Lagrangian time scale in the horizontal velocity. In inhomogeneous turbulence that occurs in a convective boundary layer, the third moment $\overline{w^3}$ of the vertical velocity fluctuations exists, even though the random forcing of (1) should be Gaussian. To satisfy the requirements of a non-vanishing third moment and Gaussian random forcing, Luhar and Britter (1989) assumed that P in (2), which should now be skewed, was the sum of two weighted normal distributions, P_A and P_B, with means $\bar{w}_A$ and $\bar{w}_B$, and standard deviations σ_A and σ_B, respectively. In fact, the designations A and B conveniently represent the updrafts and downdrafts, respectively. By explicitly writing down the equations for the zeroth to the third moment, and assuming $\sigma_A = \bar{w}_A$ and $\sigma_B = \bar{w}_B$ (Baerentsen and Berkowicz, 1984) for closure, (2) can be solved, and the random walk equation can be expressed as

$$
dw = \left(\frac{\overline{w^2}}{\tau_w}\frac{\partial \ln P}{\partial w} + \frac{\phi}{P}\right)dt + \left(\frac{2\overline{w^2}}{\tau_w}dt\right)^{1/2}\mu,
$$

(3)

where P is the probability density function of (2), ϕ/P is a complex but well-defined function of w and the downdraft and updraft velocity parameters (See Luhar and Britter, 1989), and $\mu = d\xi/\sqrt{dt} = N(0,1)$ is a standardized normal variate. To link (3) to realistic convective conditions, both $\overline{w^2}$ and τ_w are expressed as observed functions of height (z), boundary layer height (z_i), and the convective velocity $(w_* = (z_i H_0 g/T)^{1/3})$, which in turn is a function of the boundary-layer height, surface heat flux and temperature.

2.2. Buoyant plume simulation in convective boundary layer. In the case of a buoyant plume in a convective boundary layer, the description of Luhar and Britter (1991) applies only to the plume particles that move in the vertical direction. Luhar and Britter consider the effect of entrainment by ambient air on the plume particles but do not consider the effect of entrainment on ambient particles. The horizontal displacement of the plume is actually displacement in time (normalized by a mean horizontal wind velocity). Such a displacement is different from an actual horizontal displacement that also contains velocity fluctuation and entrainment of the ambient particles into the plume. In the present study, we used the same treatment as that of Luhar and Britter (1991) for the plume particles, and we also allowed the ambient particles entering the stack exhaust region to be converted to plume particles, and adjusted the "weights" (to be defined later) of the ambient and plume particles to conserve mass without explicitly simulating the entrainment of ambient particles into the plume. While this approach is not physically rigorous, it is physically sensible. Furthermore, for our study of grid resolution, whether the fine-grained description of dispersion is rigorous or not is of secondary importance compared to the requirement that the subgrid dispersion process is unchanged for the different grid sizes considered.

The dispersion of the buoyant plume follows the description of Slawson and Csanady (1967, 1971) and consists of three phases: the initial phase, which is dominated by self-generated turbulence and entrainment, the intermediate phase, which is dominated by ambient turbulence in the inertial subrange, and the final phase, which is dominated by the energy-containing eddies of the ambient turbulence. The locations where one phase crosses over to another are determined by requiring the equations for the neighboring phases to be equal at the points of transition. An added simplifying assumption is that the plume-rise equations for a convective boundary layer are the same as those for a neutral boundary layer. In addition, the acceleration of a buoyant plume particle is assumed to be the sum of the acceleration due to ambient turbulence, dw/dt, and that due to buoyancy, dw_b/dt described by the three phases. To account for additional dispersion due to self-generated turbulence induced by buoyancy, Luhar and Britter (1991) added another stochastic component, $\chi\, dw_b/dt$, where $\chi = N(0,0.2)$ is a normal variate with mean zero and standard deviation 0.2. The resulting stochastic differential equations in the vertical direction are

$$(4) \qquad \begin{aligned} dW &= dw + dw_b + \chi dw_b, \\ dz &= W dt. \end{aligned}$$

where χ changes with different plume particles while μ of (3), included as the first term on the right-hand side of (4), changes with different plume particles and every time step. The expression for dw_b is rather messy and we choose not to present it here. Interested readers can read Luhar and Britter (1991) for more detailed information. The dispersion in the horizontal direction is assumed to be the same for both ambient and plume particles and is described by the following stochastic differential equations:

$$(5) \qquad \begin{aligned} du &= -\frac{u}{\tau_u}dt + \left(\frac{2\overline{u^2}}{\tau_u}dt\right)^{1/2} \nu, \\ dx &= (\bar{u} + u)dt, \end{aligned}$$

where $\bar{u}$ and u are, respectively, the mean and random-walk horizontal velocities, and ν is again the standardized normal variate $N(0,1)$.

In our 2D simulation, ambient particles containing the ambient species masses are introduced uniformly with height from the left boundary upwind of the stack. The rate of introduction is such that, each upwind boundary cell should contain X number ($= 100$) of ambient particles at each time step on average. The time step is 1 s. Emissions are added to ambient particles that enter an emission volume, which is one-tenth the size of the grid cell. The emission volume (2×5 m^2) covers 1 m to the east and to the west and 5 m above the stack height, which is 50 m tall and 15 m from the upwind (west) boundary (See Figure 1). To simulate the stack plume, each ambient particle entering the emission volume not only absorbs the emission and buoyancy from the stack, but is also converted and split into 100 plume particles, each containing 1/100 of the mass and weight (see below) of the converted but pre-split particle. The new plume particles also have the same location and mean horizontal velocity of the original particle but have also acquired an upward vertical velocity due to buoyancy.

Because of their lower density, the plume particles move upward, and consequently, the number of ambient particles immediately downwind of the stack becomes depleted. This void should have been filled by other ambient particles participating in the vortices generated in the entrainment process. However, Luhar and Britter (1991) did not simulate the entrainment process for the ambient particles. Accordingly, we devised a weighting scheme that is physically sensible to remove the artificial void described above. We assigned a weight of one for each ambient particle as it enters the left boundary and require that the total number, X, of weighted particles for each of the fine grid cells be constant at all time. If at a given time step an ambient particle is converted and split into 100 plume particles within the finite volume above the stack, then the weight of each new plume particle will be 0.01 and the species masses of the particle

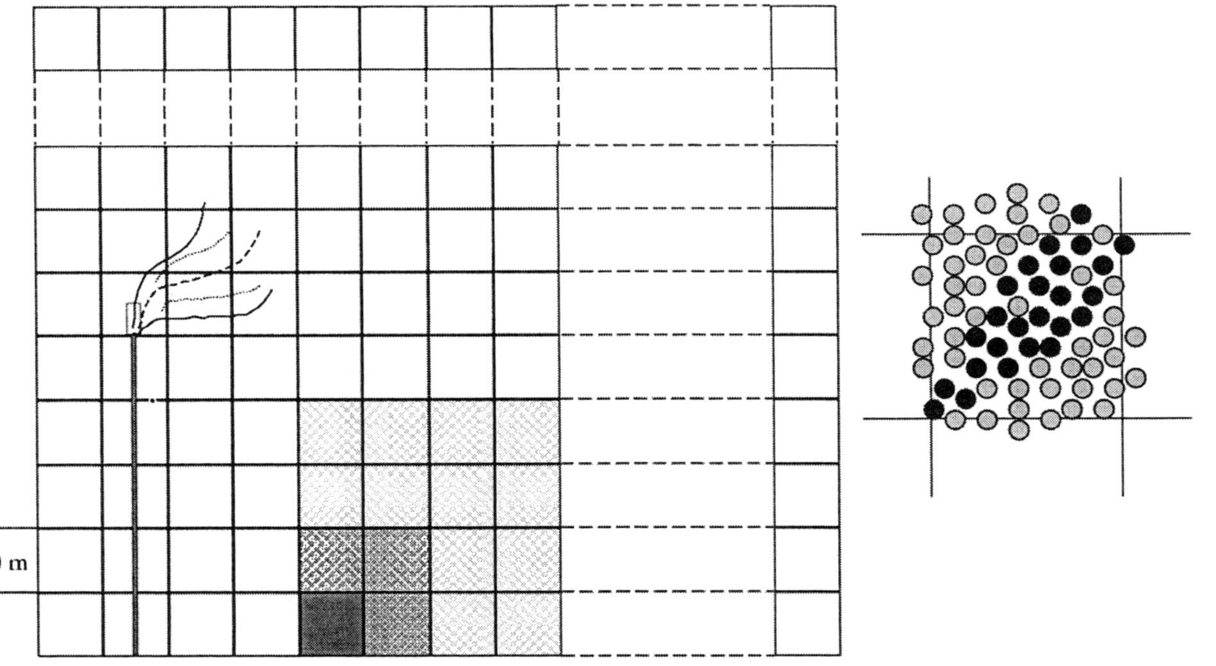

FIG. 1. *A schematic of the modeling domain (1600 m horizontal, 400 m vertical). The emission volume is 2×5 m^2 in cell (2, 6). The grid sizes used are 10×10 m^2 (fine, dark shade in Figure), 20×20 m^2 (medium, medium shade), and 40×40 m^2 (coarse, light shade). The right side of the figure shows the mixture of plume (solid) and ambient (shaded) particles in a grid cell.*

will also be 0.01 times those of the pre-split particle. Similarly, when particles merge (see below), their weights and corresponding species masses are added and assigned to the resulting particle. The weight, b, of the ambient particles in a given grid cell at a given time step is determined by

$$(6) \qquad b = \frac{X - N_p}{N_a},$$

where, N_p and N_a are the total weights of the plume and ambient particles, respectively, in the grid cell. It is highly unlikely that $X < N_p$ in our simulation. The weights of the ambient particles are constantly being adjusted at every time step to eliminate artificial voids and to assure the uniformity of the background concentrations, whereas the weights of the plume particles are adjusted, if necessary, but only to preserve mass consistency.

2.3. Simulation of ozone chemistry. We use the Carbon-Bond IV chemical mechanism (U.S. EPA 1990) with corrections for the $XO_2N + XO_2N$ reaction in the simulation. We generated species concentration profiles based on different spatial grid resolutions: fine, medium, and coarse. Furthermore, subgrid-scale simulations of the chemistry may also have a significant impact on the simulation results. We considered three subgrid chemistry assumptions: the micro scale or individual-particle (IP) chemistry, in which chemistry takes place within each individual particle independently of other particles in the same cell, and the tagged- and untagged-particle chemistry, in which plume particles and ambient particles are either separated (tagged) or combined together (untagged) in chemical reactions within a grid cell. The distinction in chemical reactions between separating or combining the plume and ambient particles points to the use of additional subgrid information that would otherwise be lost in the study of the effect of grid resolution on non-homogeneous chemistry. The capability of distinguishing tagged and untagged particles within the same grid is an advantage of the Particle-Grid method and is rather difficult to implement in a conventional Eulerian approach without invoking assumptions like higher-order closure or probability distributions of the species concentrations.

Since the individual-particle chemistry is closest to reality, its result, expressed in the fine grid resolution, is considered the "ground truth" against which results from other grid resolutions and subgrid chemistry assumptions are compared. Here, the weighted species masses of each particle were converted to concentrations using a fractional cell volume (area in our case), the fraction being the weight of the particle divided by the total weight ($X = 100$) of particles in the cell. Clearly, the IP chemistry is the most computationally expensive.

In the tagged- and untagged-particle chemistry cases, the species concentrations in the cell of a given grid resolution were determined by dividing the total mass of same-tag particles (in the tagged-particle case) or all particles (in the untagged-particle case) by their respective total fractional cell volume. Two chemistry calculations were performed per time step for tagged-particle chemistry, or one chemistry calculation for the untagged-particle chemistry. The time step for the chemical reaction is 1 s. The kinetic equations were solved with the Implicit-Explicit Hybrid stiff differential equation solver (IEH, see Sun *et al.*, 1994; Chock *et al.*, 1994). The resulting species concentrations were converted back to masses using the same total fractional volume. The grid-resolved plume and ambient species masses are then redistributed back to their respective particles. The redistribution procedure, which is of course not needed in the real world and in the micro scale case, is not unique especially when the chemistry is nonlinear and the species masses for each species in a particle may be different among the ambient particles and among the plume particles in the same grid cell. Redistribution of species concentrations is also implicitly present in the Eulerian approach but is simply ignored because only one set of species concentrations is assigned to each grid cell. It is implicitly present because of the implicit assumption on the concentration profiles in locations other than the cell centers, say, dictated by the flux-correction steps or by the order of accuracy in the transport numerical scheme.

To describe the distribution of species masses back to the particles in the cell, we shall use the tagged-particle case for illustration. The untagged-particle case simply corresponds to that of removing the distinction between plume and ambient particles in the chemistry and mass redistribution. For the most accurate possible redistribution of the cell mass of a given species after the plume or ambient chemistry back to the plume or ambient particles in the cell, we imposed the following rule: the cell species mass change should be distributed over all particles of the same type in the cell as proportionally as possible to the species concentration rate of change determined by the species concentrations of the each individual particle. This rule is subject to two constraints. First, the total cell mass change for a given species, as determined by the plume or ambient chemistry for all particles of the same type in the cell, must equal the sum of the species mass changes over all same-type particles in the cell. Second, no species mass can become negative as a result of mass redistribution. The following steps were taken for a given species for, say, the plume particles:

(1) The species-i concentration change ($\Delta C_{i,N}^{p}$) for, say, plume (p) particle N for a given time step Δt was calculated by performing chemistry on the individual particles. Using IP chemistry gives the most accurate information for redistributing the mass, but is also the most computationally expensive. When execution time is an issue, the chemical kinetic equations can be used to calculate an approximation to the rate of change without having to solve for the actual concentration change. To ensure that the species concentration does not go negative, we required that if ΔC was negative, then $\Delta C_{i,N}^{p} = \max(\Delta C_{i,N}^{p}, -C_{i,N}^{p}/\Delta t)$.

(2) The total mass change after the reaction step for species i for all plume particles in the cell is designated DM_{i}^{p}. Even though DM_{i}^{p} may

be, say, positive, the concentration change for species i in particle N (i.e., $\Delta C^p_{i,N}$) need not be positive. We separately summed the positive and negative concentration changes for a given species over all particles of the same type (plume in this case) in the cell. The resulting positive (or negative) concentration change sum $\Delta C^p_i(+)$(or $\Delta C^p_i(-)$), divided by the total concentration change sum $\Delta C^p_i(T) = \Delta C^p_i(+) + \Delta C^p_i(-)$ was used to scale the total mass change for the species to yield the positive (negative) mass change for all same-type particles in the cell. If $\Delta C^p_i(T)$ and DM^p_i have the same sign, say, negative, then for a plume particle N, the mass change for species i at the end of the reaction time step Δt, is determined by

$$\Delta M^p_{i,N}(-) = DM^p_i \frac{\Delta C^p_{i,N}}{\Delta C^p_i(-)} \frac{\Delta C^p_i(-)}{\Delta C^p_i(T)} = \Delta C^p_{i,N} \frac{DM^p_i}{\Delta C^p_i(T)}.$$

If the mass reduction for species i is greater than what is available in particle N, then the mass for species i in the particle was reduced to zero and the leftover of the mass change was added together for a normalized uniform reduction from particles of nonzero mass for species i. Similar equations are used when $\Delta C^p_i(T)$ and DM^p_i are all positive.

(3) When the cell mass change DM^p_i and the particle-based cell concentration change $\Delta C^p_i(T)$ have opposite signs, then more arbitrariness exists in mass redistribution. We used the following rule: If the species existed before the chemistry, the change in species mass was applied to the particles according to the original distribution of the species mass in the particles. If the species was new, the change in mass was applied equally to all particles. In both cases, all resulting masses must be non-negative.

The elaborate steps above are not unique. They are designed in an attempt to preserve the realistic subgrid structure as much as possible, and become trivial when subgrid non-homogeneity is minimal. They are applied separately to both plume and ambient particles in the cell in the case of tagged-particle chemistry or to all particles in the cell in the untagged-particle case.

2.4. Sequence of tasks. The sequence of tasks taken at each time step or every fixed number of time steps during the simulation is given below.

1. Read in new meteorological parameters (unchanged in our case).
2. Read in new plume parameters (unchanged in our case).
3. Split particles (every n^{th} step) in fine grid when necessary.
4. Merge particles (every m^{th} step) in fine grid when necessary.
5. Add plume emissions to particles in emission region.
6. Update particle counts and cell concentrations based on grid resolution of interest.
7. Do chemistry for grid resolution and subgrid assumption of interest.

8. Redistribute the mass back to the particles in a given grid resolution and subgrid assumption.
9. Determine species concentrations in each grid cell.
10. Do transport in fine grid.
11. Update the boundary.
12. Rescale the background particles' weights and masses.

In essence, at each time step, we did some particle management, then incorporated emissions, turned on the chemistry for a given grid resolution and subgrid assumption, output the species concentrations, and moved the particles in fine grid (regardless of grid resolution). The sequence repeated itself for each grid resolution. Because of the high computation demand, the simulation was performed for a total time period of 1200 seconds in real time. Accordingly, we assumed both the meteorology and the plume parameters to be unchanged throughout the simulation.

Particle management is one way to control the execution time, to assure sufficient homogeneity in particle number density, and to allow for air parcel mixing. This was carried out in the subgrid scale using the fine grid resolution. The identical spatial subgrid processes took place regardless of the grid resolution for the chemistry. Therefore, we could assure that identical transport processes took place for all grid resolution and that we could truly look into the effect of grid resolution and subgrid assumption on non-homogeneous chemistry. Before describing the particle management (splitting and merging), we should reiterate the importance of particle weighting described earlier to remove potential unphysical void in ambient particles due to the lack of treatment of entrainment for ambient particles in the presence of a buoyant plume (Luhar and Britter, 1992). Given the weighting, ambient particles were never allowed to split outside the emission region. To make sure that there were a sufficient number of particles (plume plus ambient) per cell (even though the particle weights remained 100 per fine grid), the plume particles in a cell were allowed to split when the cell contained less than one-half (50) the original number of particles. ("One-half" is an input parameter.) This criterion was checked every n steps. In our runs, $n = 9999$, so splitting never took place. In the splitting process, two particles of identical mass and weight but half that of the original particle would have been created. They would have retained the location and plume characteristics, etc. of the original particle.

In addition to the reduction of the execution time, particle merging also represents one form of subgrid-scale mixing. Merges among ambient particles are not meaningful because there is no change in species concentrations after the mixing. Therefore, we imposed a rule that merges must involve at least one plume particle. Merged particles always formed a plume particle. While the rule appears arbitrary albeit reasonable, one should realize that the rule becomes part of the subgrid mixing process that identically occurs in all grid resolutions of interest. Merging occurs when a cell contains more than a specified number of combined plume and ambient particles. The cells were checked for an excessive number of particles every m steps, where $m=1$ in our runs. The total number of particles in a fine-grid cell must exceed $p \times 100$ to trigger merging, where 100 is the original number of particles in each cell. The parameter p is input and depends on the distance of the cell from the point source. If the cell was close to the point source we wanted to retain detailed plume information, so we used $p=4$ to cause merging only when there were more than 400 particles in the cell. Far from the source, we could conserve computing resources by allowing particle merging when the cells contained as few as 200 particles ($p=2$). The distance from the source is also an input parameter; we used ≥ 15 cells (≥ 150 m) as the delineation point to trigger particle merging far from the plume source. Once the need for merging is established, the cell was divided into $k \times k$ subcells, where k is the square root of $p \times 100$—the number of particles that triggered the merge. When more than one particle was found in a subcell, and at least one was a plume particle, all the particles in the subcell were merged to form a new plume particle. The new particle is located at the center of mass of the merging particles and has their average velocity and plume characteristics. The mass and the weight of the new particle are the sum of the mass and weight of the merged particles. The original merged particles are simply deleted.

The incorporation of the buoyant plume emission was described in Sec. 2.2 above and will not be elaborated here. The chemistry step together with the mass redistribution protocol, was described earlier, in Sec. 2.3.

The transport step follows the random walk description of Secs. 2.1 and 2.2. The boundary condition was updated only at the inflow or western boundary. At every time step (1s), all particles in the first-column cells were deleted and replaced by evenly distributed 100 particles per cell with species masses equal to those consistent with continuously updated ambient chemistry. Because the mean ambient velocity is 1 m/s, our procedure allows 10 particles per cell on average with updated chemistry to enter the second-column cells per time step.

At the end of the time step the ambient particle weights were rescaled in order to eliminate unphysical voids as described earlier. The total weight for all particle types in the cell must sum to 100 per fine grid.

2.5. Input information. Table 1 shows the grid information. For all grid resolutions, $\Delta t=1$ s, the mean horizontal and vertical velocities are $\bar{u}=1$ m/s, $\bar{w}=0$ m/s. The planetary boundary height, z_i, is 400 m, the convective velocity, w_*, is 1.5 m/s. The horizontal Lagrangian time scale, τ_u, is 100 s, and the horizontal velocity fluctuation, $(\overline{u^2})^{1/2}$, is 0.822 m/s. Other parameters, like the local vertical velocity variance $\overline{w^2}$ and time scale τ_w for the convective boundary layer are identical to those given in Luhar and Britter (1989). The total number of particles was around 8,000 initially and was 487,779 at the end of the simulation. The total simulation time is 1200 s, or 20 minutes.

TABLE 1
Grid information.

	FINE GRID	MEDIUM GRID	COARSE GRID
ΔX	10 m	20 m	40 m
ΔZ	10 m	20 m	40 m
# cells in X dir.	160	80	40
# cells in Z dir.	40	20	10
# parts/cell	100	400	1600

TABLE 2
Initial species mass assigned to each particle [mass = ppm*(dx*dy*dz)/(#parts/cell)].

FAST SPECIES			SLOW SPECIES		
1	NO	8.194E−04	17	HONO	3.162E−05
2	NO2	9.220E−03	18	HNO3	1.462E−02
3	O3	1.703E−01	19	H2O2	2.191E−03
4	PNA	4.284E−05	20	FORM	1.039E−02
5	CRO	9.368E−16	21	ALD2	7.468E−03
6	O	2.927E−09	22	PAN	5.750E−03
7	NO3	1.686E−06	23	MGLY	2.195E−04
8	N2O5	2.429E−06	24	OPEN	5.456E−05
9	HO2	4.815E−05	25	CO	9.541E−01
10	OH	4.306E−07	26	PAR	1.797E−01
11	ROR	1.488E−10	27	OLE	4.764E−04
12	C2O3	6.413E−06	28	ETH	3.865E−03
13	XO2	3.339E−05	29	TOL	2.752E−03
14	TO2	2.800E−08	30	XYL	3.149E−04
15	XO2N	1.462E−06	31	CRES	1.874E−04
16	O1D	6.639E−08	32	ISOP	7.512E−05
			33	NTR	6.141E−04
			34	SO2	0.000E+00
			35	SULF	0.000E+00
			36	MEOH	0.000E+00
			37	ETOH	0.000E+00
			38	MTBE	0.000E+00

Table 2 shows the initial species masses (in units of ppm·m^3) of the ambient particles. They were obtained by multiplying the species ambient concentrations in ppm by the cell volume (100 m^3, assuming a Δy of 1 m) and dividing the product by the number of particles per cell (= 100). The species concentrations correspond to those predicted by the Urban Airshed Model (UAM) for Glendora, CA at 14:06 during an ozone episode day.

Because only 23 of the 38 species concentrations were explicitly provided in the model output, the 38 species masses in Table 2, including the free radical species, were obtained by solving the chemical kinetic equations for 6 minutes starting at 14:00 and using 10^{-16} ppm as the initial concentrations of the free radicals in the IEH solver (Sun, *et al.*, 1994). In our simulations, the free radical species concentrations were not set to zero at each time step but were carried along by the particles. The chemistry is assumed to take place under the solar radiation corresponding to 14:00 in the California South Coast Air Basin.

Table 3 shows the point source emission profile (in g-mol/hr), which is proportional to a point source obtained from a UAM dataset for the South Coast. The identity of the source is not known to us and such knowledge is unimportant for our purpose. We converted the emission of each species to mass (in units of ppm·m^3) per particle weight per second for all weighted particles in the emission volume by multiplying the emission value by 6.78 divided by the total weight of the particles in the emission volume. Note also that the plume contains a substantial amount of NO emissions. These emissions will titrate the ambient ozone to form NO_2 and cause the ozone concentration to reach a minimum at or near the grid containing the point source. The buoyancy flux of the plume was assumed to be 29.25 m^{4}s^{-3}. All other parameters are identical to those used in Luhar and Britter (1992).

TABLE 3

Point source emission profile (in g-mole/h) and the total emission species mass (in ppm-m^3) per time step (1 s) for particles in emission volume.

Species	Emission rate (g-mole/h)	Total emission mass (ppm-m^3) per time step
NO	4.939300×10^{-1}	3.3488×10^{0}
NO2	2.598000×10^{-2}	1.7614×10^{-1}
OLE	7.800001×10^{-4}	5.2884×10^{-3}
PAR	2.954000×10^{-2}	2.0028×10^{-1}
TOL	2.000000×10^{-4}	1.3555×10^{-3}
XYL	6.000001×10^{-5}	4.0666×10^{-4}
FORM	1.520000×10^{-3}	1.0302×10^{-2}
ALD2	0.000000×10^{0}	0.0000×10^{0}
ETH	1.520000×10^{-3}	1.0302×10^{-2}
MEOH	0.000000×10^{-0}	0.0000×10^{0}
ETOH	0.000000×10^{-0}	0.0000×10^{0}
ISOP	0.000000×10^{-0}	0.0000×10^{0}
SO2	1.409500×10^{-1}	9.5532×10^{-1}
CO	0.000000×10^{-0}	0.0000×10^{0}
AERO	2.000000×10^{-0}	1.3555×10^{1}

3. Results. We used the Particle-Grid approach to analyze the impact of non-homogeneous chemistry on air quality modeling of a stack plume in the convective boundary layer and to understand the effect of different chemistry time steps, grid sizes, and subgrid particle source type classifications on predicted concentrations of air pollutant species. We also looked at the impact of grid size and subgrid chemistry assumption on NO_x emissions control. The simulation with micro scale or individual-particle (IP) chemistry interpolated to a fine grid concentration field represents the base case or "ground truth" against which other results are to be compared. We present only the results at the end of the 1200 s simulations. In the figures, the far downwind portions of the images are not interesting and are cropped.

3.1. Effect of chemistry time step. In these simulations, the transport of particles was carried out at the time step of 1 s. However, the IP chemistry step was turned on every n seconds and lasted n seconds. It turned out that the results were similar for n equal to 1, 5 and 60. Table 4 lists the maximum concentrations of NO, NO_2 and the minimum O_3 concentration at the end of the simulation (1200 s) for each of the three time steps, all concentrations having been translated to the fine grid. The Table also contains the ranges of the same-cell concentration differences within the model domain between the 5 s or 60 s time step and the 1 s time step. Figure 2 shows the difference between the 1 s and 60 s chemistry time step results for NO_2 and O_3. The NO and O_3 extrema decrease slightly and the NO_2 maximum increases slightly as the time step increases. This may be because at the larger chemistry time step, more merging of plume and ambient particles (and therefore, more spreading of NO) could take place before the chemistry was turned on so that a localized excess of NO was less likely and a more efficient titration of ozone by NO within a particle would take place. For all three species the 60 s time step case changes the extreme concentrations 2.5 times more than the 5s case. As expected, the largest difference in concentration is located very close to the source, as shown in Figure 2. The shape and extent of the plume is essentially unchanged for the different chemistry time step.

TABLE 4

Extremum concentrations (in ppb) at the end of simulation for three different chemistry time steps and the ranges of the differences over the model domain between the larger time steps and the smallest time step.

	$\Delta t=1$ s	$\Delta t=5$ s		$\Delta t=60$ s	
	Min/max	Min/max	Range (5 s–1 s)	Min/max	Range (60 s–1 s)
NO	133.01 (max)	131.04	$(-8.13, 0.02)$	125.04	$(-21.22, 0.04)$
NO_2	54.735 (max)	56.716	$(-0.01, 8.130)$	62.689	$(-0.06, 21.20)$
O_3	149.79 (min)	147.81	$(-8.115, -0.010)$	141.94	$(-21.14, -0.07)$

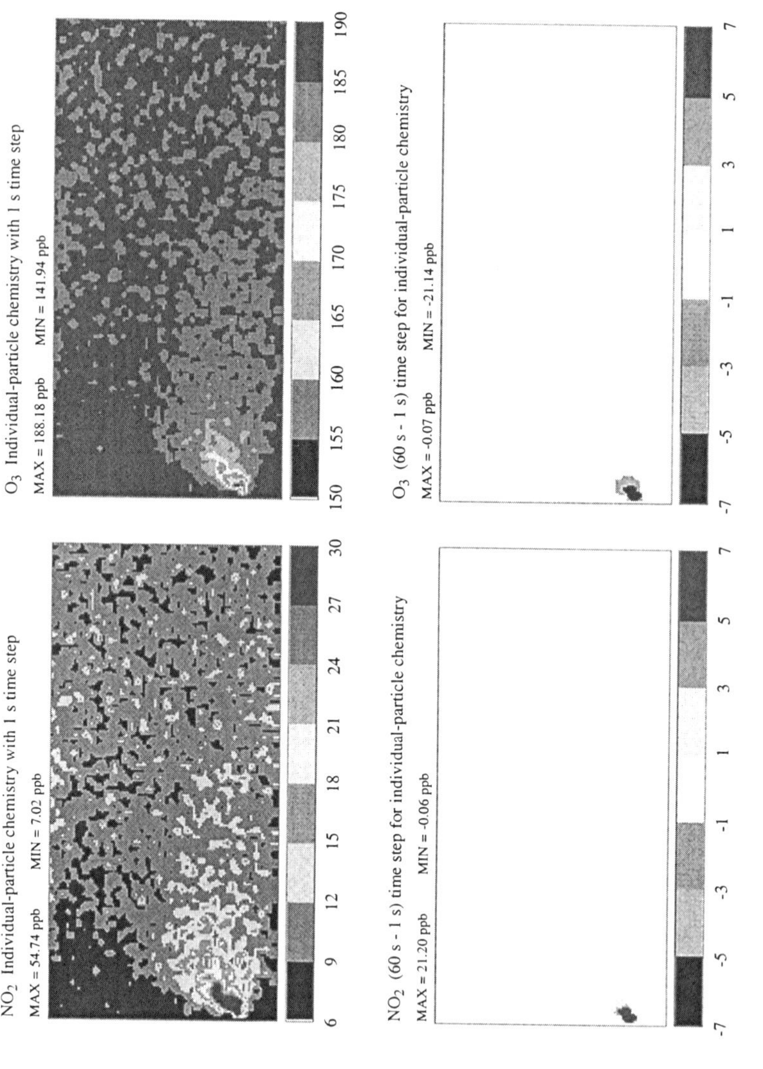

FIG. 2. *The concentration distribution of IP chemistry with 1 s time step (top), and the difference between the 1 s and 60 s chemistry time step results (bottom) for NO_2 (left) and O_3 (right).*

3.2. Effect of grid size and subgrid chemistry assumptions. Grid size has a much greater effect on the plume concentration than the time step. Henceforth, only the time step of 1 s is used for the chemistry step. We tested three grid resolutions to determine the effect of grid size on non-homogeneous chemistry. For a given grid resolution, we also compared the impact of three subgrid chemistry assumptions on the model outcome. The three subgrid chemistry assumptions are the micro scale or individual-particle (IP) chemistry, and the tagged- and untagged-particle chemistry. The expressed cell concentration for a species is the sum of the species masses of all particles in the cell divided by the cell volume, fine, medium or coarse.

Table 5 lists the extreme concentrations of NO, NO_2 and O_3 for the IP, tagged and untagged chemistry cases at three grid resolutions in columns two, three and five. Columns four and six show the maximum concentration differences for the tagged- and untagged-particle chemistry cases, respectively, relative to the IP chemistry case across the model domain. Obviously, the maximum concentration differences need not occur at the grid with the extremum concentrations. We shall not discuss columns four and six further. All subgrid chemistry results show reduced peak concentrations as the grid size increases from fine to coarse resolution for maximum NO (a decrease of 76% with IP chemistry, 78% with tagged chemistry, and 82% with untagged chemistry) and maximum NO_2 (a decrease of 56% with IP chemistry, 57% with tagged chemistry, and 50% with untagged chemistry).

TABLE 5

The extreme concentrations of NO, NO_2 and O_3 for the IP, tagged and untagged chemistry cases at three grid resolutions.

	IP	Tagged	Max. diff. (tagged-IP)	Untagged	Max diff. (untagged-IP)
NO max					
Fine	133.01	121.97	−12.30	94.53	−46.21
Medium	61.67	51.70	−9.97	35.96	−32.06
Coarse	31.38	26.87	−4.83	17.09	−24.28
NO_2 max					
Fine	54.74	65.49	12.51	94.43	62.30
Medium	41.67	51.04	11.35	72.07	34.24
Coarse	24.11	27.87	4.86	46.93	20.84
O_3 min					
Fine	149.79	139.04	−12.50	112.73	−39.69
Medium	158.66	146.42	−12.23	127.69	−30.97
Coarse	173.71	171.02	−4.05	162.74	−16.97

For O_3, the minimum concentration increases, and there are relatively large differences among the results for different subgrid assumptions. From fine to coarse grid, the minimum O_3 increases by just 16% with IP chemistry, but by 23% with tagged-particle chemistry and by 44% with untagged-particle chemistry. For a given grid resolution, different subgrid chemistry assumptions also lead to significantly different simulation results. In particular, the NO maximum decreases while the NO_2 maximum increases as the subgrid assumptions move from the most detailed case (micro scale or individual-particle case) to the least detailed case (untagged-particle case). The maximum NO decrease ranges from 29% for fine grid, 42% for medium grid, to 46% for coarse grid. The maximum NO_2 increase ranges from 72% for fine grid, 73% for medium grid, to 95% for coarse grid. The ozone minimum decreases as the subgrid chemistry assumption changes from the IP case to the untagged-particle case. The decrease is 25% for fine grid, 20% for medium grid, and 6% for coarse grid. Since the ozone minimum occurs near the source, the excess amount of NO tends to be only capable of depleting the local ozone but not the surrounding ozone in the IP chemistry, and this diffusion-restricted condition tends to be less sensitive to grid resolution in the IP chemistry than in other subgrid chemistries. Conversely, untagged-particle chemistry has the largest inherent artificial diffusion bounded by the grid resolution and the likelihood of a local excess of NO is minimal so that the largest possible amount of ozone is removed in this case. Figures 3 and 4 show the concentration distribution for NO and O_3, respectively, with IP chemistry (left) and tagged chemistry (right) for the fine (top) and coarse (bottom) grids. The coarse-grid results show considerably less detail than the fine grid, but the NO plume shape and downwind extent is clearly discernable. The major differences are near the emission source; the downwind concentrations are comparable despite the different resolutions. Ozone, a product rather than an emitted species, loses nearly all of its detail near the emission source in the coarse grid case, but again the downwind concentrations are comparable.

Figures 5, 6 and 7 compare the tagged-particle chemistry result with that of the untagged-particle chemistry in fine grid resolution for NO, NO_2 and O_3, respectively. The top panel in each figure shows the tagged chemistry result, the middle panel shows the untagged result and the bottom panel shows the difference between the two. These two subgrid chemistry assumptions lead to much greater differences than between the IP and tagged chemistry assumptions. For NO (Figure 5), the untagged-particle chemistry shows no downwind extent, but for NO_2 and O_3 (Figures 6 and 7, respectively), the untagged-particle chemistry case has greater downwind extent than the tagged-particle case. The high NO_2 and low O_3 concentration areas near the source are also larger in the untagged-particle case than in the tagged case.

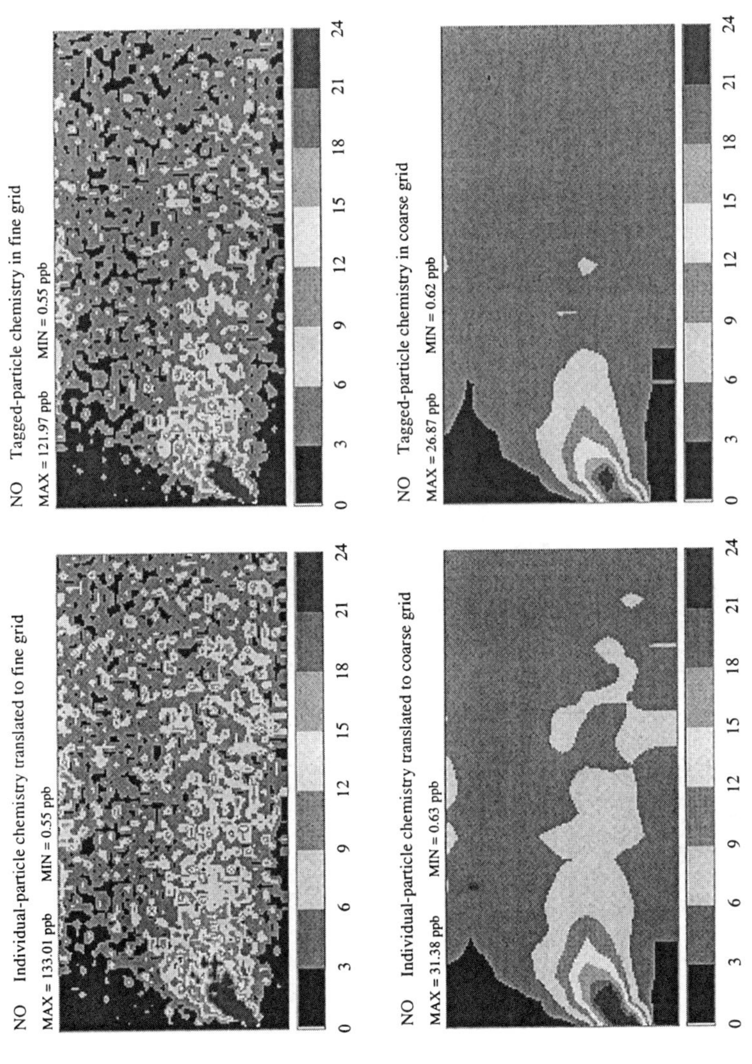

FIG. 3. *The concentration distribution for NO, with IP chemistry (left) and tagged chemistry (right) for the fine (top) and coarse (bottom) grids.*

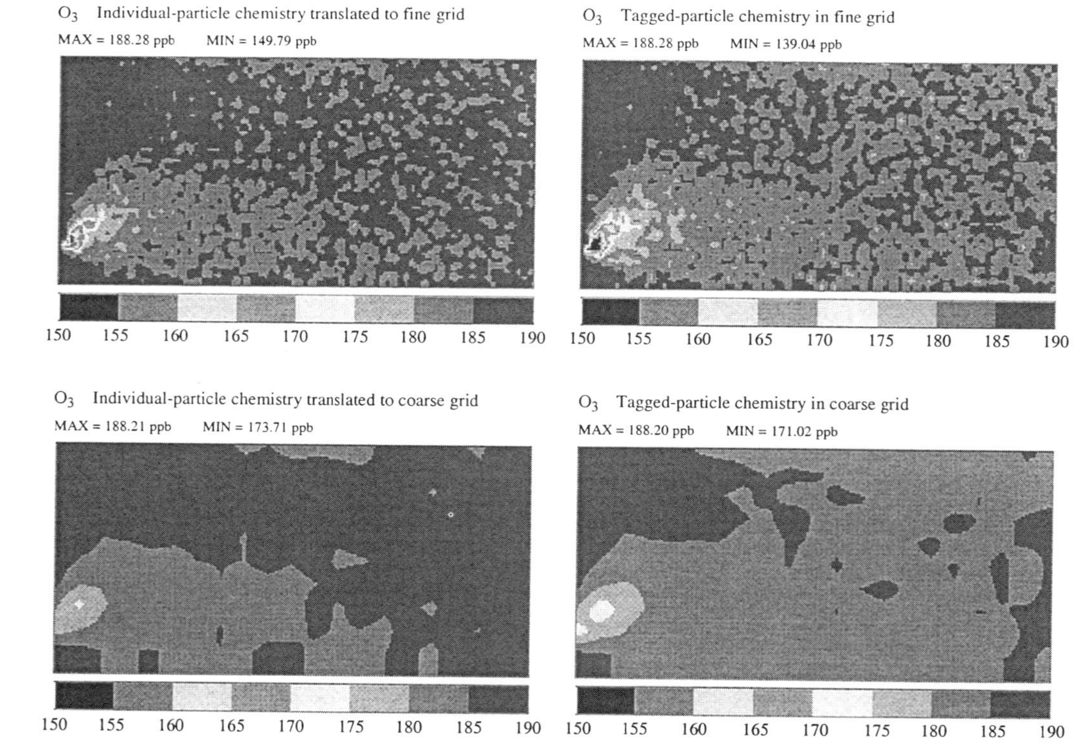

FIG. 4. *The concentration distribution for O_3, with IP chemistry (left) and tagged chemistry (right) for the fine (top) and coarse (bottom) grids.*

NO Tagged-particle chemistry in fine grid
MAX = 121.97 ppb MIN = 0.55 ppb

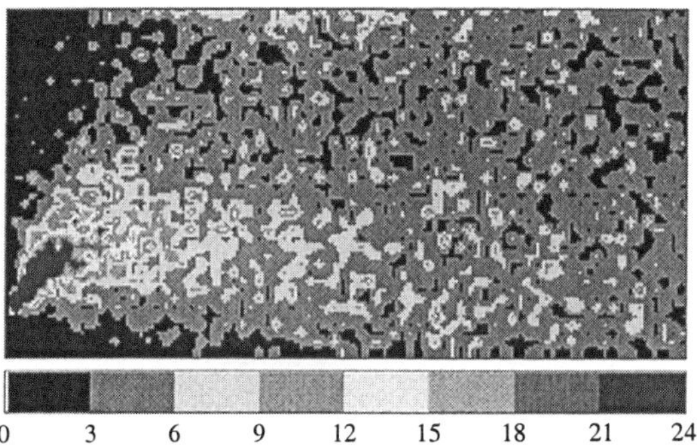

NO Untagged-particle chemistry in fine grid
MAX = 94.53 ppb MIN = 0.55 ppb

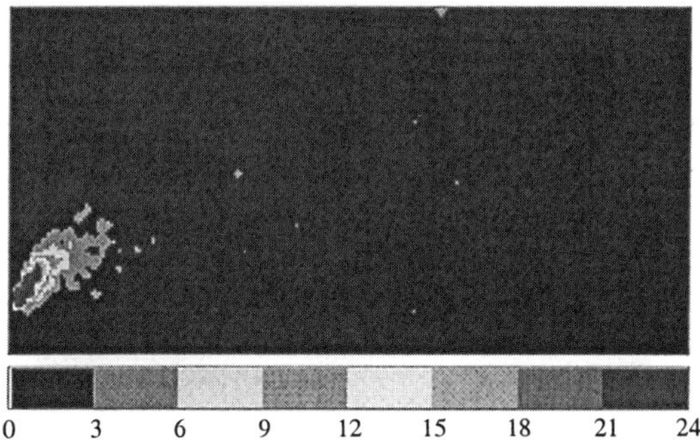

NO (Untagged-tagged)-particle chemistry in fine grid
MAX = 2.80 ppb   MIN = -37.93 ppb

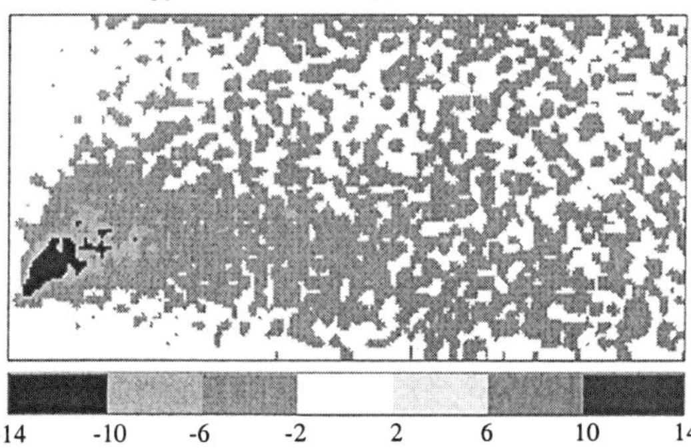

FIG. 5. *The NO concentration distribution in fine grid resolution. The top panel shows the tagged-particle chemistry result, the middle panel shows the untagged result and the bottom panel shows the difference between the two.*

NO$_2$ Tagged-particle chemistry in fine grid
MAX = 65.49 ppb   MIN = 7.02 ppb

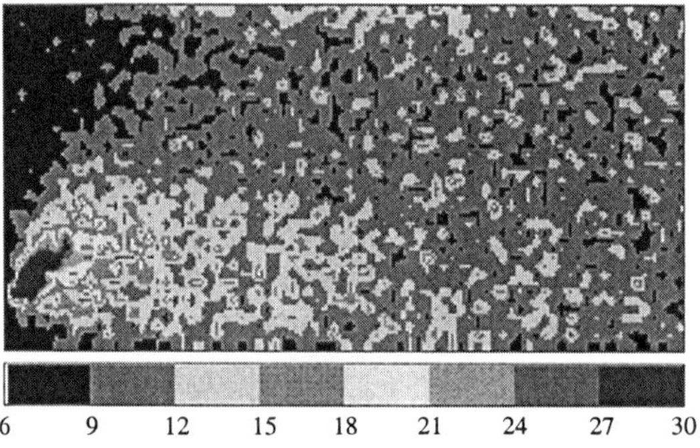

NO$_2$ Untagged-particle chemistry in fine grid
MAX = 94.43 ppb   MIN = 7.02 ppb

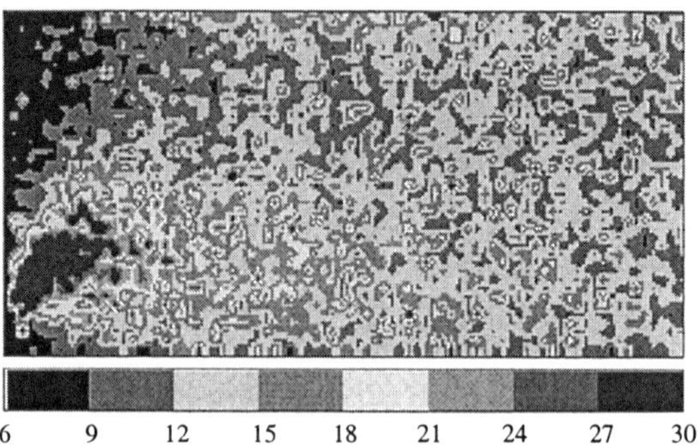

NO$_2$ (Untagged-tagged)-particle chemistry in fine grid
MAX = 58.84 ppb   MIN = -1.92 ppb

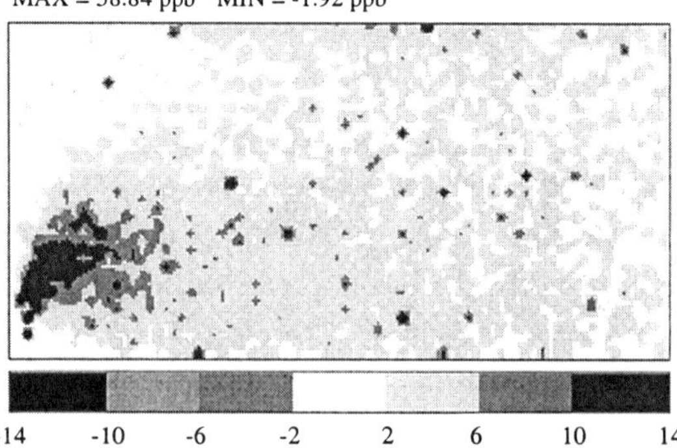

FIG. 6. *The NO$_2$ concentration distribution in fine grid resolution. The top panel shows the tagged-particle chemistry result, the middle panel shows the untagged result and the bottom panel shows the difference between the two.*

O$_3$ Tagged-particle chemistry in fine grid
MAX = 188.28 ppb MIN = 139.03 ppb

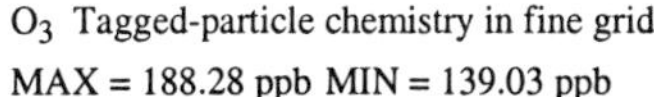
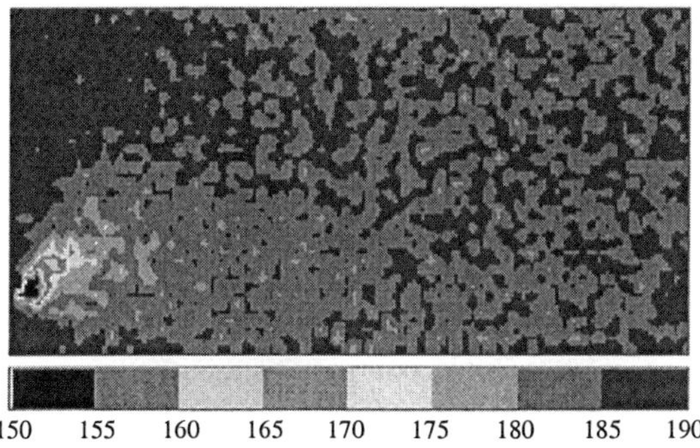

O$_3$ Untagged-particle chemistry in fine grid
MAX = 190.55 ppb MIN = 112.73 ppb

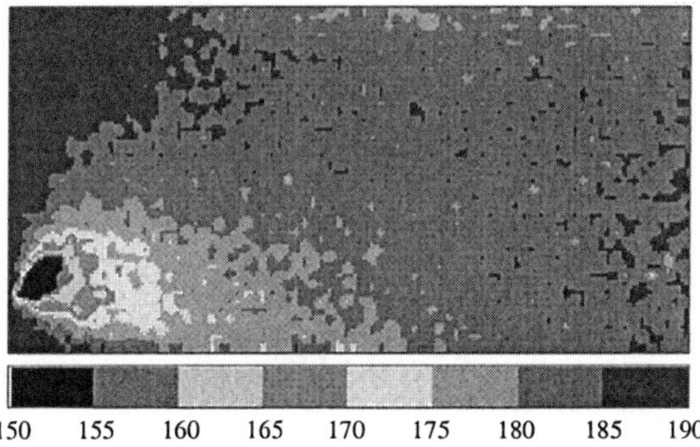

O$_3$ (Untagged-tagged)-particle chemistry in fine grid
MAX = 22.58 ppb   MIN = -29.71 ppb

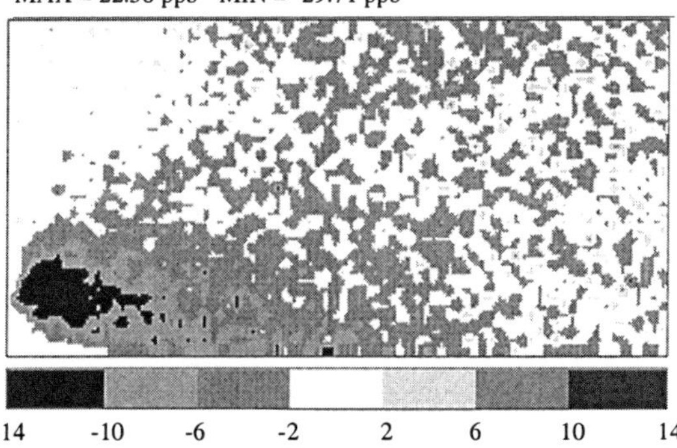

FIG. 7. *The O$_3$ concentration distribution in fine grid resolution. The top panel shows the tagged-particle chemistry result, the middle panel shows the untagged result and the bottom panel shows the difference between the two.*

3.3. Implication of grid size and subgrid chemistry on NO_x emission control. We tested an emission control scenario in which the NO_x ($NO + NO_2$) emissions were reduced by half, thus decreasing the NO_x inhibition effect on O_3. Table 6 shows the results for the minimum ozone concentrations and their changes relative to their respective original values for different subgrid assumptions and grid resolutions. Even though our simulation period is short, our results agree with those of more conventional models that show that use of a large grid size tends to lead to an outcome that reduces the effectiveness of NO_x-inhibition in ozone chemistry. This conclusion should also be coupled with the observation that, independent of the NO_x emission control, suppression of the ozone minimum is also the least with the large grid size. This is expected because large grid size tends to suppress the amplitude of concentration variations. The subgrid chemistry appears to have a large influence on the ozone impact of NO_x emission reduction. The reduction of NO_x increases the ozone minimum much more with the untagged-particle assumption than with the tagged-particle assumption, meaning that the NO_x-inhibition effect is the greatest for the untagged-particle case. In hindsight, this is again expected because the full force of the NO_x inhibition effect is operative within the whole grid in the untagged-particle case, and this observation cannot be readily appreciated in the Eulerian framework. Figure 8 shows the resulting change in O_3 concentrations for the IP and tagged-particle chemistry cases (left and right side of the figure, respectively) on the fine (top of the figure) and coarse (bottom) grids. The tagged chemistry cases predict a greater increase in O_3 near the source than the IP cases. This is again an indication that in the IP case, O_3 is locally exhausted in the NO_x-rich pockets so that reduction of NO_x would not make as great a difference in increasing ozone as in the tagged-particle chemistry case where the condition for the presence of NO_x-rich pockets is substantially removed. The coarse grid cases show a much smaller difference due to NO_x control, but the ranges of concentration variations in the coarse-grid case are considerably lower to start with (see Table 5).

TABLE 6

The minimum concentration of O_3 for the IP, tagged and untagged chemistry cases at three grid resolutions when the NO_x emissions are reduced by half.

	IP Chemistry		Tagged Chemistry		Untagged Chemistry	
	Min. O_3 for half NO_x emission	Percent diff. from original O_3	Min. O_3 for half NO_x emission	Percent diff. from original O_3	Min. O_3 for half NO_x emission	Percent diff. from original O_3
Fine	160.78	7	151.27	9	142.28	26
Medium	169.03	6	161.80	10	156.91	23
Coarse	177.39	2	175.37	2	173.04	6

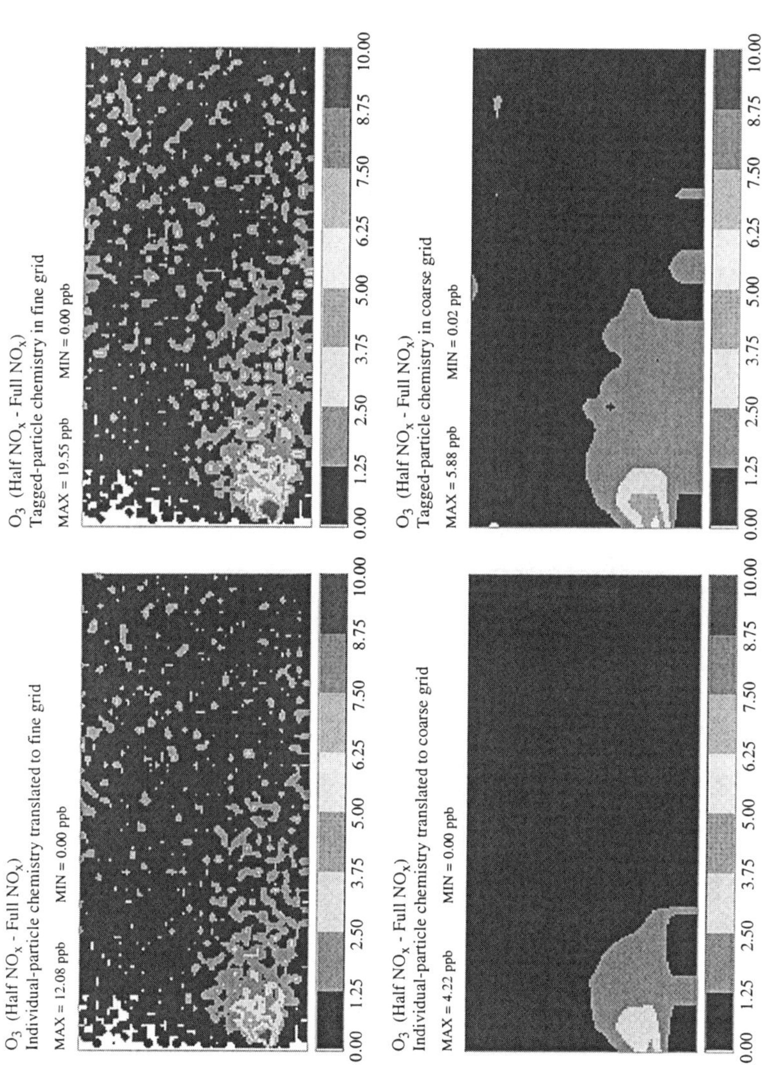

FIG. 8. *The resulting change in O_3 concentrations when the NO_x emissions are halved, for the IP and tagged-particle chemistry cases (left and right side of the figure, respectively) on the fine (top of the figure) and coarse (bottom) grids.*

4. Conclusions. We have carried out an elaborate study of the impact of grid resolution and subgrid chemistry assumption on the model prediction of grid-averaged species concentrations. The study takes advantage of the unique characteristics of the Particle-Grid approach that allows one to look at the effect of grid resolution and subgrid chemistry separately. The case we chose to study had highly non-homogeneous species concentrations such that the accuracy of the chemistry-transport model in describing the system was quite sensitive to the choice of the model's grid resolution, and interestingly, to the subgrid chemistry assumption of the model. As can be seen in our study, use of the untagged-particle assumption leads to a significant departure from the base-case result, indicating that reducing the grid resolution alone in a typical chemistry-transport model may not be sufficient to accurately describe a system with highly non-homogeneous species concentration distributions that also contain fast chemistry.

It is worth noting that reducing the subgrid information is not necessarily similar to increasing the model grid size. In our example, increasing the grid size leads to a reduction in the suppression of the ozone concentration and a reduction in the effectiveness of the NO_x-inhibition effect. On the other hand, reducing the subgrid information (by using the untagged-particle assumption) leads to an increase in ozone reduction and an enhancement of the NO_x-inhibition effect insofar as the ozone extremum is concerned. We believe that this is a result of not being able to allow for a possible localized complete depletion of ozone due to an excess of NO in the case of the untagged-particle assumption.

Our simulation is for a very short period. The question is how these short-term results influence the longer-term model predictions. It appears that if no new sources are encountered and no species removal processes are in place as the air parcel travels downwind, it is possible that the subgrid properties and even the fine grid resolution may not be highly critical as the fast chemistry quickly adjusts the species concentrations to the condition dictated by the transport process. Of course, one cannot tell for sure without an actual simulation study.

REFERENCES

BAERENTSEN J.H. AND BERKOWICZ R. (1984). Monte Carlo simulation of plume dispersion in the convective boundary layer. *Atmos. Environ.* **18,** 701–712.

CHOCK D.P. AND WINKLER S.L. (1994a). A Particle-Grid air quality modeling approach – 1. The dispersion aspect. *J. Geophys. Res.* **99**D, 1019–1032.

CHOCK D.P. AND WINKLER S.L. (1994b). A Particle-Grid air quality modeling approach – 2. Coupling with Chemistry. *J. Geophys. Res.* **99**D, 1033–1042.

CHOCK D.P., WINKLER S.L., AND SUN P. (1994). Comparison of Stiff Chemistry Solvers for Air Quality Modeling. *Environ. Sci, & Technol.* **28,** 1882–1892.

JANG J.C.C., JEFFRIES H.E., BYUN D., AND PLEIM J.E. (1995a). Sensitivity of ozone to model grid resolution – I. Application of high-resolution regional acid deposition model. *Atmos. Environ.* **29,** 3085–3100.

JANG J.C.C., JEFFRIES H.E., AND TONNESEN S. (1995b). Sensitivity of ozone to model grid resolution – II. Detailed process analysis for ozone chemistry. *Atmos. Environ.* **29**, 3101–3114.

LIANG J. AND JACOBSON M.Z. (2000). Effects of subgrid segregation on ozone production efficiency in a chemical model. *Atmos. Environ.* **34**, 2975–2982.

LUHAR A.K. AND BRITTER R.E. (1989). A random walk model for dispersion in inhomogeneous turbulence in a convective boundary layer. *Atmos. Environ.* **23**, 1911–1924.

LUHAR A.K. AND BRITTER R.E. (1992). Random-walk modeling of buoyant-plume dispersion in the convective boundary layer. *Atmos. Environ.* **26**A, 1283–1298.

SLAWSON P.R. AND CSANADY G.T. (1967). On the mean path of buoyant, bent-over chimney plumes. *J. Fluid Mech.* **28**, 311–322.

SLAWSON P.R. AND CSANADY G.T. (1971). The effect of atmospheric conditions on plume rise. *J. Fluid Mech.* **47**, 33–49.

SUN P., CHOCK D.P., AND WINKLER S.L. (1994). An Implicit-explicit hybrid solver for a system of stiff kinetic equations. *J. Computational Phys.* **115**, 515–523.

THOMSON D.J. (1987). Criteria for the selection of the stochastic models of particle trajectories in turbulent flows. *J. Fluid Mech.* **180**, 529–556.

U.S. EPA (1990). *User's guide for the urban airshed model*, Vol. **1**: *User's manual for UAM(CB-IV)*. EPA-450/4-90-007A, Research Triangle Park, NC.

NUMERICAL SOLUTION OF TRACE SPECIES ADVECTION UNDER NON-UNIFORM DENSITY DISTRIBUTION: EXPERIMENT WITH TWO-DIMENSIONAL LINEAR FLOWS

DAEWON W. BYUN* AND SANG-MI LEE[†]

Abstract. It has been known that certain Eulerian numerical advection schemes for trace species suffer serious mass conservation problems under mass-inconsistent meteorological conditions (i.e., density and wind fields do not exactly satisfy the continuity equation). Because this problem can grow out of bound for the long-term simulation of trace species transport, several mass adjustment schemes have been applied to photochemical air quality models. This paper compares different mass adjustment schemes for their numerical characteristics, such as preservation of constant field, monotonicity, and linearity. Two-dimensional steady-state linear flows are used for the test because analytical solutions to the advection equation are available for different combinations of the elementary flows. We investigate the mass conservation and other numerical transport characteristics for various types of mass-inconsistent flows.

1. Introduction. For the last two decades, steady progress has been made in numerical advection schemes to allow realistic and accurate simulation of atmospheric transport processes in air quality and weather forecasting models. There are many studies of this kind in the literature. Most of the research on the advection algorithms has focused on improving desired numerical characteristics such as mass conservation, numerical diffusion, monotonicity, and positive definiteness. The properties have been tested with idealized one- and two-dimensional wind fields such as simple translational and rotational flows. Occasionally, somewhat complex deforming wind fields are used with a particular signal for which analytical solution is available (Staniforth et al. 1987). With the limited number of signal and wind conditions, they studied characteristics of numerical solvers in flux form and advective form (i.e., semi-Lagrangian transport). Both types of the schemes can be diffusive and dispersive. A well-known shortcoming of most of the flux-form advection algorithms is that they fail to maintain the constancy of mixing ratio under certain flow conditions. Even when they can maintain the constancy of mixing ratio in one-dimension, they often fail to do so when applied to multi-dimensional problems with a time-splitting approach. On the other hand, most of the semi-Lagrangian transport (SLT) schemes can preserve the constancy of mixing ratio by design, but have problems with mass conservation. These SLT schemes are inherently multi-dimensional and therefore, do not incur additional time-splitting error, except for the interpolation error.

*Department of Geosciences, Joint Appointment in Department of Chemistry, Air Quality Modeling, University of Houston, 312 Science and Research Building 1, Houston, Texas 77204-5007 (`dwbyun@math.uh.edu`).

†Environmental Fluid Dynamics Program, Department of Mechanical and Aerospace Engineering, Arizona State University, Tempe, AZ 85287-9809 (`smlee@asu.edu`).

Many of the advection tests in the literature have implicitly assumed uniform air density without providing sufficient characterization of the numerical schemes under the non-uniform density conditions. Because of the practical utility of air quality models in assessing and controlling air quality problems, the evaluation of numerical transport algorithms must be performed under realistic weather conditions. Related research would simulate tracer transports with actual meteorological conditions to compare with atmospheric tracer experiments. Although these simulations generally match with tracer experiment results qualitatively, they could not quantify the key numerical characteristics in detail because of the nature of observations, which cannot separate effects of advection alone from other transport and transformation processes. The numerical advection may not conserve the mass of a trace species under divergent flow conditions. Having an accurate transport algorithm is important for atmospheric models. In the past, it has been assumed that the total mass of tracer species can be conserved during the advection process as long as wind fields are nondivergent. Therefore, many of the atmospheric models have relied on the incompressible flows for maintaining the conservation of mass. Diagnostic models based on non-divergent flow assumption are often used to provide wind fields neglecting the effects of density variation in time and space.

Recently, Byun (1999a and b) has highlighted the importance of dynamic consistency in meteorological and air quality modeling for multiscale atmospheric applications. Mass consistency quantifies how well the density and wind fields satisfy the continuity equation for air. One of the fundamental requirements of the numerical transport algorithms used in air quality models is the conservation of trace species in the domain. In ideal conditions, the input meteorological data for air quality simulations should be mass consistent. The need to bringing in realism and to characterize the actual atmospheric conditions using either numerical models with highly parameterized physical and cloud algorithms and four-dimensional data assimilation procedures, the input meteorological data could be less than perfectly mass consistent. In this situation, even precisely mass conserving numerical algorithms may fail to conserve tracer species mass in the domain. Byun (1999b) has shown that a corrective step is needed to use the realistic meteorological data, which are not perfectly mass consistent. He has reported several adjustment schemes used in current air quality models and proposed a two-step time splitting numerical algorithm that satisfies the mixing ratio conservation equation. The adjustment scheme is expected to preserve constancy of mixing ratio during the advection process. The constancy-preserving character for mixing ratios is a necessary condition for a numerical algorithm to conserve total mass of trace materials.

The present paper intends to study the effects of mass inconsistency on the numerical integration of trace species continuity equation using idealized wind and density distributions. One objective of the present study is to compare the performance of correction algorithms under different density fields and flow patterns. In addition, the linearity of advection process is studied because of its importance in maintaining the ratio of pollutant mix that affects photochemistry. The linearity property of the original advection equation must be closely maintained in the numerical advection process to attribute contributions of precursor emissions from selected sources to air quality and deposition at receptor locations. Note that we are not intending to compare numerical characteristics among different advection algorithms in the present research. The contention is that the mass continuity equation can be solved much more accurately with the existing advection algorithms when a proper adjustment scheme is used also. Our tests focus on two-dimensional, steady-state, non-divergent linear flows and non-steady divergent flows.

2. Characterization of trace species advection process.

2.1. Species continuity equation and mass conservation. The governing conservation equations for air and trace species can be written in a generalized coordinate form (e.g., Byun, 1999a)

$$(1) \qquad \frac{\partial(\rho J_s)}{\partial t} + m^2 \nabla_s \bullet \left(\frac{\rho J_s \hat{\mathbf{V}}_s}{m^2} \right) + \frac{\partial(\rho J_s \hat{\nu}^3)}{\partial s} = 0$$

$$(2) \qquad \frac{\partial(c_i J_s)}{\partial t} + m^2 \nabla_s \bullet \left(\frac{c_i J_s \hat{\mathbf{V}}_s}{m^2} \right) + \frac{\partial(c_i J_s \hat{\nu}^3)}{\partial s} = J_s Q_{c_i}$$

where ρ is density of air, $J_s = |\partial s/\partial z|^{-1}$ is the component of the metric tensor associated with the vertical coordinate transformation (vertical component of the Jacobian matrix, hereafter Jacobian for simplicity), m is the map scale factor of the conformal projection, $s = \hat{x}^3$ is the generalized vertical coordinate, $\hat{\mathbf{V}}_s = i\hat{u} + j\hat{\nu}$ and $\hat{\nu}^3$ are contravariant horizontal and vertical wind components, respectively, and Q_{c_i} is the source of pollutant c_i. Species mixing ratio (q_i) is related with the concentration and air density as $c_i = \rho q_i$. Quality meteorological models are expected to satisfy the air density continuity equation, so the right hand side of Eq. (1) should vanish. The components of the Jacobian tensor (m and J_s) play the important role of defining the volume of air treated in a grid cell for numerical simulations. Because many numerical advection algorithms have been developed and tested for Cartesian computational coordinate system, inclusion of the Jacobian components allows direct use of these algorithms for applications with various different coordinates.

2.2. Correction methods for the mass conservation. Atmospheric modeling for a limited domain, as in urban or a regional scale simulation, is subjected to the inflow, outflow, and top boundary conditions imposed by the large-scale weather system. Also, numerical algorithms used in the models may not be able to maintain the continuity relation perfectly, rendering inconsistent air density and wind distributions over time (mass inconsistency). Then, the effective continuity equation is given as;

$$(1') \qquad \frac{\partial(\rho J_s)}{\partial t} + m^2 \nabla_s \bullet \left(\frac{\rho J_s \hat{\mathbf{V}}_s}{m^2} \right) + \frac{\partial(\rho J_s \hat{v}^3)}{\partial s} = J_s Q_\rho$$

where Q_ρ is the mass conservation error. In the presence of this error, atmospheric models, in particular air quality models in which nonlinear chemistry among the trace species magnifies the mass conservation problem, have sought various ways to minimize the effects of nonzero mass error term. Some air quality models attempt to modify vertical velocity component to remove the mass inconsistency (i.e., modify wind to make $J_s Q_\rho$ vanish). However, the results are not always satisfactory because the modification of vertical wind based on a non-divergent form diagnostic equation cannot take into account the mass consistency errors originated by the tendency term in Eq. (1), i.e., $\frac{\partial(\rho J_s)}{\partial t}$.

An alternative method is to force the conservation of species mixing ratio by modifying Eq.(2) accordingly (Byun, 1999b):

$$(3) \qquad \frac{\partial(c_i J_s)}{\partial t} + m^2 \nabla_s \bullet \left(\frac{c_i J_s \hat{\mathbf{V}}_s}{m^2} \right) + \frac{\partial(c_i J_s \hat{v}^3)}{\partial s} = J_s c_i \frac{Q_\rho}{\rho}.$$

With this method, the effects of mass inconsistency can be corrected with

$$(4) \qquad (c_i J_s)^{cor} = (c_i J_s)^T \exp\left(\int \frac{Q_\rho}{\rho} dt \right),$$

where superscripts 'T' and 'cor' represent values after transport (advection) and after correction, respectively. Many of the mass correction schemes used in air quality models are just incomplete expressions of Eq. (4) that might be valid for certain dynamic conditions, such as for uniform air density and/or non-divergent wind field. Table 1 summarizes such adjustment methods as reported in Byun (1999b). **A1** is the correction scheme used in Urban Airshed Model (UAM: Scheffe and Morris 1993) and **A2** is used in SARMAP Air Quality Model (SAQM: Chang et al. 1997). The scheme based on the anelastic approximation is assigned as **A3**, a scheme with the tendency term approximated by a finite-difference method as **A4**, and a scheme in which Eq. (3) is solved with a two-step time splitting integration as **A5**. **A1** and **A2** ignore effects of coordinate/grid structures during the advection process. For Cartesian coordinates where the Jacobian (J_s) is uniform and constant with time and, therefore, **A2** becomes equivalent to **A5**. For a steady-state flow, schemes **A3**, **A4**, and **A5** are identical. Note that all the adjustment algorithms in Table 1 become the same for the steady state flow with uniform density distribution in a Cartesian co-ordinate system. **A5** is expected to maintain conservation of mixing ratio

TABLE 1

Mixing ratio correction schemes tested.

Correction Method	Symbol	Mass correction algorithm
No correction	**A0**	
Advection of unity	**A1**	$c_i^{cor} = \dfrac{c_i^T}{c_r^T} c_r, \quad c_r = 1.0$
Advection of air density	**A2**	$c_i^{cor} = \dfrac{c_i^T}{\rho^T} \rho^{int}$
Advection of J_s with anelastic approximation	**A3**	$(c_i J_s)^{cor} = (c_i J_s)^T \exp\left[\dfrac{(\rho J_s)_n - (\rho J_s)^T}{(\rho J_s)_n}\right]$
Advection of J_s with finite differencing of tendency term	**A4**	$(c_i J_s)^{cor} = (c_i J_s)^T \exp\left[\dfrac{(\rho J_s)^{int} - (\rho J_s)^T}{(\rho J_s)^{int}}\right]$
Advection of J_s with two-step time splitting	**A5**	$(c_i J_s)^{cor} = \dfrac{(c_i J_s)^T}{(\rho J_s)^T}(\rho J_s)^{int}$

Note: Superscripts *cor*, *int*, and T represent corrected, interpolated, and transported (advected) quantities, respectively. Subscript n represents current time step.

to the machine precision all the time. Because the principal objective of these mass adjustment schemes is to conserve trace species mixing ratio, we will term a method is correct when such character can be demonstrated under given dynamic conditions. Under mass consistent flow and density distributions, a scheme that conserves trace mixing ratio will conserve total trace species mass.

2.3. Linearity of advection process. The physical advection process is linear because Eq. (2) satisfies the linearity condition

$$(5) \qquad f(ac_A + bc_B) = af(c_A) + bf(c_B),$$

where a and b are arbitrary constants, and c_A and c_B are concentrations of the same trace species from two different sources. On the other hand, the numerical implementation of Eq. (2) may or may not satisfy the linearity condition. For example, first-order upstream schemes maintain the linearity. But because these schemes are very diffusive, they are not used in operational air quality models. Many higher order schemes have focused on improving the peak accuracy and mass conservation while ignoring maintenance of the linearity. We will investigate the effects of mass correction schemes on the non-linearity of numerical advection process.

3. Description of wind and density fields for two-dimensional linear flows. Before investigating trace species advection with more realistic time-dependent wind fields from hydrostatic or non-hydrostatic atmospheric models using different terrain-following coordinate systems, it is instructive to study the effects of mass adjustment with simpler two-dimensional steady flows. For a two dimensional atmospheric flow, Eq. (1) can be rewritten as

$$
(6) \qquad \frac{\partial \rho^*}{\partial t} + \nabla_s \bullet \left(\rho^* \hat{\mathbf{V}}_s \right) = 0 \, ,
$$

where $\rho^* = \frac{J_s}{m^2}\rho$ is the grid-volume adjusted air density in the generalized coordinate system. Some of the mass conserving characteristics of non-divergent flows can still be studied with the simplified steady state continuity equation,

$$
(7) \qquad \nabla_s \bullet \left(\rho^* \hat{\mathbf{V}}_s \right) = \rho^* \nabla_s \bullet \hat{\mathbf{V}}_s + \hat{\mathbf{V}}_s \bullet \nabla_s \rho^* = 0 \, .
$$

For a two-dimensional mass conserving non-divergent flow, one can expect the mass flux term $\hat{\mathbf{V}}_s \bullet \nabla_s \rho^*$ in Eq. (7) to vanish. This relationship between the wind and density distributions can be used for classifying the idealized flow patterns.

Many numerical advection algorithms have been tested with certain two dimensional incompressible flows. Most of the tests have been subject to rather simple uniform (translational) or rotational flows. There are a few papers testing these algorithms with a rather complex flows and signals where analytical solutions can be found (e.g., Staniforth et al. 1987). But, characterization of the algorithms can be limited by the availability of these analytical solutions. The analytical solutions of the advection process for two-dimensional linear flow fields can always be found and therefore allow testing of many different flow situations and signal shapes. The components of a horizontal wind field centered at a reference point (x_c, y_c) in space can be expressed by the Taylor expansion in terms of the horizontal space variables:

$$
(8a) \qquad u = u_c + \left(\frac{\partial u}{\partial x} \right)_c (x - x_c) + \left(\frac{\partial u}{\partial y} \right)_c (y - y_c) + O(\varepsilon^2)
$$

$$
(8b) \qquad \nu = \nu_c + \left(\frac{\partial \nu}{\partial x} \right)_c (x - x_c) + \left(\frac{\partial \nu}{\partial y} \right)_c (y - y_c) + O(\varepsilon^2)
$$

where the subscript "c" denotes the reference point. Because the "hatted" variables in the generalized coordinate system are equivalent to those represented by plain symbols in the Cartesian coordinate system (i.e., u, ν, and ρ instead of $\hat{u}$, $\hat{\nu}$, and ρ^*), we have dropped the "hats" for the variables in Eqs. (8a and b) for simplicity. We are justified in neglecting higher-order derivatives here because we consider only the linear distribution of velocity in the vicinity of the reference point. The fact that the flow field may actually be nonlinear does not invalidate the use of this expression as long as the flow field is analyzed locally around the position of the interest. Furthermore, applicability of this expansion is not limited to the steady-state flow fields.

The elementary streamline patterns, i.e., divergence, vorticity, stretching, and shearing flows, are defined, respectively;

$$\text{(9a)} \qquad \frac{\partial u}{\partial x} + \frac{\partial v}{\partial y} = 2S_\Delta$$

$$\text{(9b)} \qquad \frac{\partial v}{\partial x} - \frac{\partial u}{\partial y} = 2S_\Omega$$

$$\text{(9c)} \qquad \frac{\partial u}{\partial x} - \frac{\partial v}{\partial y} = 2S_\Gamma$$

$$\text{(9d)} \qquad \frac{\partial v}{\partial y} + \frac{\partial u}{\partial y} = 2S_\Lambda \ .$$

In terms of these flows, we can rewrite Eqs. (8a, b), neglecting second or higher order terms, as:

$$\text{(10a)} \qquad u = u_c + (S_\Delta + S_\Gamma)(x - x_c) + (S_\Lambda - S_\Omega)(y - y_c)$$

$$\text{(10b)} \qquad v = v_c + (S_\Lambda + S_\Omega)(x - x_c) + (S_\Delta - S_\Gamma)(y - y_c) \ .$$

One advantage of decomposing the flow field into the four elementary patterns is that we can construct a pseudo-realistic wind field while controlling the kinematic characteristics of the flow exactly. An additional benefit of using a linear wind field is that the solution at a later time can be found analytically. For simple flows such as pure translation, rotation, or deformation, one can readily derive the characteristic streamline equations. For example, with uniform wind (i.e., $u = u_c$ and $v = v_c$), the streamline function is given as:

$$\text{(11)} \qquad v_c x - u_c y = \text{const.}$$

Now, we can express the mass consistent density field that satisfy the condition, $\mathbf{V}_s \bullet \nabla_s \rho = 0$ with the stream function:

$$\text{(12)} \qquad \rho - \rho_c = -\rho_c \kappa^{-1} \left[v_c(x - x_c) - u_c(y - y_c) \right] \ ,$$

where κ is an arbitrary constant with the dimensionality of diffusivity [m^2/sec]. From Eqs. (11) and (12), we see that the isopicnic lines that satisfy the mass continuity for a non-divergent flow are always parallel to the streamlines.

A derivation of the streamline formula for a general linear flow is provided in Appendix A.1. For non-divergent flow, which is the subject of discussion in this section, the streamline formula is given as:

$$\text{(13)} \qquad \begin{aligned} (x - x_c - x_p)^2(S_\Lambda + S_\Omega) &- 2(x - x_c - x_p)(y - y_c - y_q)S_\Gamma \\ &+ (y - y_c - y_q)^2(S_\Omega - S_\Lambda) = \text{cont.} \end{aligned}$$

Refer to Appendix A.1 for the coordinates of the intersecting reference point (x_p, y_q). Note that the reference point in a non-divergent two-dimensional flow can be characterized as a hyperbolic node, elliptic node, or a combination of the two. In order to maintain mass consistency in the steady-state two-dimensional non-divergent flow, the air density field should either be uniform in space, which is trivial, or should satisfy $\mathbf{V}_s \bullet \nabla_s \rho = 0$, which suggests that the density field should have a distribution defined with the following equation,

$$(14) \quad \begin{aligned} \rho - \rho_c = -\rho_c \kappa^{-1} \big[& (x - x_c - x_p)^2 (S_\Lambda + S_\Omega) \\ & -2(x - x_c - x_p)(y - y_c - y_q) S_\Gamma + (y - y_c - y_q)^2 (S_\Omega - S_\Lambda) \big] . \end{aligned}$$

Eq. (14) is a general solution of the mass-consistent density distribution for a given two-dimensional non-divergent linear wind field. The density field depends on the value of the coefficient κ, which describes the thermodynamic gradient (baroclinicity) of the computational domain. If the density and flow fields are not mass consistent, the term $\mathbf{V}_s \bullet \nabla_s \rho = 0$ becomes a source of error in Eq. (1), and Eq. (3) should be used to guarantee the conservation of species mass. In other words, not only the divergence, as discussed in Kitada (1987), but also the flux contribution due to the winds crossing the isopicnic surface should be considered to conserve air mass correctly. Note that for a divergent flow, the flow is not expected to be solenoidal and the original form of the mass continuity equation, which couples the wind and density fields, should be satisfied.

In Section 5, we present several numerical test results for various combinations of elementary flows and density distributions to demonstrate the effects of adjustment process given mass inconsistent conditions. A steady state wind field can be constructed by assigning magnitudes of the elementary flow patterns and the corresponding mass consistent density field can be provided with Eq. (14).

4. Numerical experiment design. To establish effectiveness and limitations of the of mass correction methods, we have studied mass conservation characteristics of the numerical advection for a set of idealized flows and signals for which we have analytical solutions. Numerical tests have been performed with the driver and advection modules of the Models-3 Community Multiscale Air Quality (CMAQ) modeling system (Byun and Ching 1999).

4.1. Flow fields. We have set up several idealized flow patterns to test a numerical algorithm's capability of correctly solving the advection equation. They include the elementary stream patterns, such as divergence (**dv**), vorticity (**vo**), stretching (**st**), and shearing (**sh**) flows, and combinations of the elementary flow patterns, such as **stvo, stsh, shvo, stshvo, stvodv, stshdv, shvodv,** and **stshvodv** flows. For example, the flow pattern **stshvo** was established by adding three elementary flow patterns **st**, **sh**, and **vo**. The strength of the pure rotational flow (S_Ω) was set to make a full circle for every 24 hour of simulation (i.e., $S_\Omega = 7.272205 \times 10^{-5}$ sec^{-1}). For other composite linear flows, the same flow strength of $S_\Gamma = S_\Lambda = S_\Omega = S_\Delta = 1.818051 \times 10^{-5}$ sec^{-1} was used unless specified otherwise. These values produce reasonable ranges of wind speeds that are comparable to real atmospheric conditions.

In addition, four different density distribution types were studied to demonstrate their influence on numerical integration of the continuity equation for trace species. **Type-0** represents a uniform density and the transformation metric field and **Type-1** is for a field with uniform $S_\Gamma = S_\Lambda = S_\Omega = S_\Delta$, in which density and the Jacobian are individually non-uniform. For example, a weather forecasting model with hydrostatic pressure as the vertical coordinate generates **Type-1** wind flows. **Type-2** was defined to be a mass consistent non-uniform density field, which was determined by Eq. (14) with $\kappa = 1.0 \times 10^7$ [m^2/sec] and a uniform $J_s = 8127.8$ m. Finally **Type-3** represents a field with an incremental density slope added to **Type-2**, producing a mass inconsistent flow. Note that the total mass of a trace species must be computed with the prescribed air density distribution even for the **Type-3** flows because only steady state flows are considered here. This means that the effective modification of density brought by the adjustment scheme is discarded at each integration time step.

Numerical test domain with 90×90 cells at 4 km grid resolution was used. These physical scales were chosen just for the convenience of simulation. To remove possible complications resulting from the Earth's curvature, the map scale factors were assigned to be 1.0 everywhere. The specified mass inconsistency, i.e., the added incremental density slope $\Delta\rho$ [kg m^{-3}] was defined to be

$$(15) \qquad \Delta\rho = 1.2 \times 10^{-6}\big[(x - x_c) + (y - y_c)\big] ,$$

where the distance should be measured in meters. For the tested domain, this provided about 10–20 % density variation depending on the base flow patterns. The simulation period was 24 hour with the integration time step set at 60 seconds. All the computations were done with IEEE single precision (32 bits) on a workstation.

Figure 1 shows composite flow patterns with density distribution of **Type-2** for the four flow patterns: **stvo**, **stsh**, **shvo**, **stshvo**. With this flow, the density gradients are distributed parallel to the streamlines. Figure 2 presents flow patterns **sh**, **vo**, **st**, **stsh** for density distribution **Type-3**. These flows produce nonzero mass fluxes violating the mass continuity with the steady state assumption. In real meteorological conditions, these mass fluxes will be responsible for the redistribution of momentum and density fields, so the atmosphere cannot be in the steady state. These

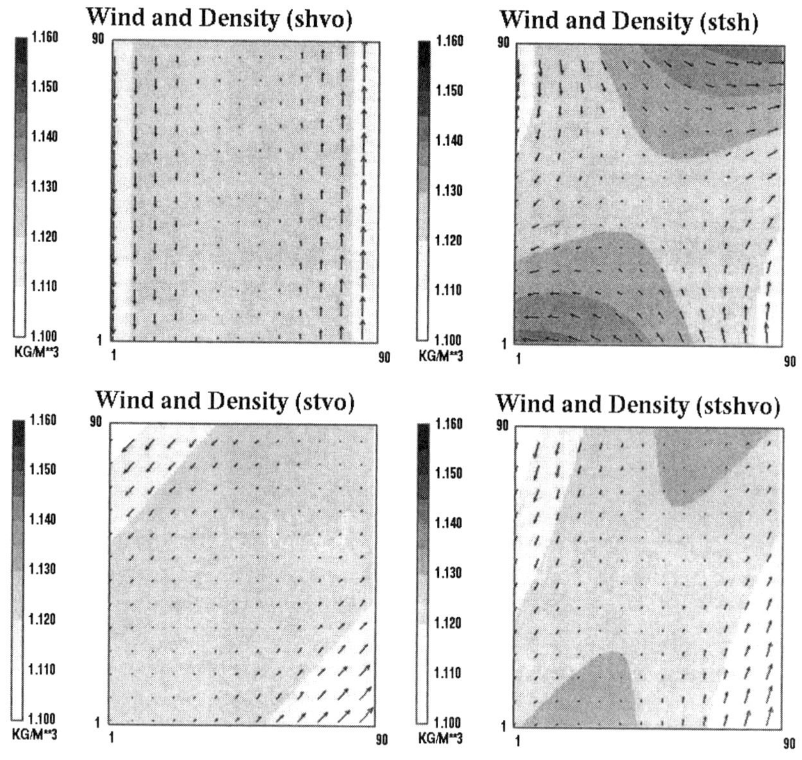

FIG. 1. *Composite flows with the mass-consistent density distribution of* **Type-2** *(with constant Jacobian) for the four flow patterns:* **stvo, stsh, shvo,** *and* **stshvo.** *Density fields are distributed parallel to the streamlines.*

linear flows are useful for testing numerical advection characteristics because the analytical solutions are available (see Appendix A). Once initial position is known, the new advected field can be computed with the prescribed wind and the necessary travel time with a back trajectory concept. Figure 3, for example, shows the travel times of the signals subject different linear flows.

In order to be mass consistent, a divergent flow must be accompanied by a corresponding change in the density field. Therefore, a steady state divergence flow condition cannot be achieved. When the density fields change with time corresponding to the mass fluxes, the flows and density fields can be mass consistent. Figures 4a and 4b provide flow and density distributions for a composite **stshvodv** flow with a weak divergence ($S_\Gamma = S_\Lambda = S_\Omega = 1.818051 \times 10^{-5}$ sec^{-1} and $S_\Delta = 3 \times 10^{-7}$ sec^{-1}) at the initial and final (24 hr later) times.

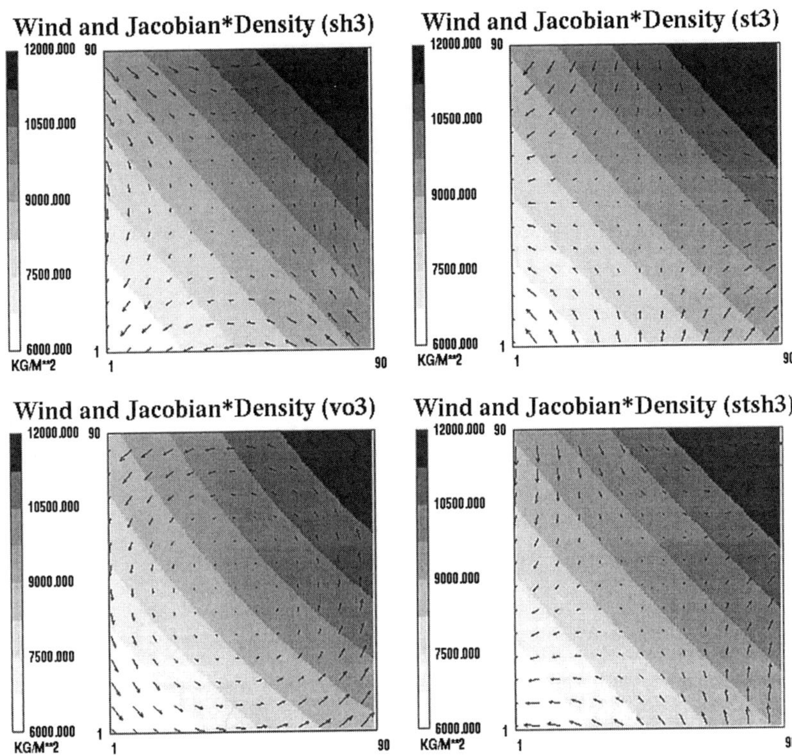

FIG. 2. *Flow patterns* sh, vo, st, *and* stsh *for mass-inconsistent density distribution,* **Type-3**.

4.2. Trace signals. To characterize the advection algorithms and mass adjustment schemes, we selected several base signals for the numerical experiment. The first signal tested was a tracer species with constant and uniform initial and boundary conditions with mixing ratio 1.0 ppm (i.e., IC=1.0 and BC=1.0 ppm). The unit of the mixing ratio was arbitrary in this test. This trace species behaves like air and the signal is useful to test whether an advection scheme can successfully maintain a uniform mixing ratio field. Resulting mixing ratio values different from 1.0 signify a lack of proper skill for the numerical advection algorithm. In addition, two associated signals (one with IC=1.0, BC=0.0; the other with IC=0.0, BC=1.0) were advected at same time to study how well the numerical algorithms satisfy the linearity condition, Eq. (5).

Another signal tested was a hill distribution. Because one of the key objectives of the present study is to assess effects of mass correction on the linearity property of a numerical advection scheme, three hill signals

DAEWON W. BYUN AND SANG-MI LEE

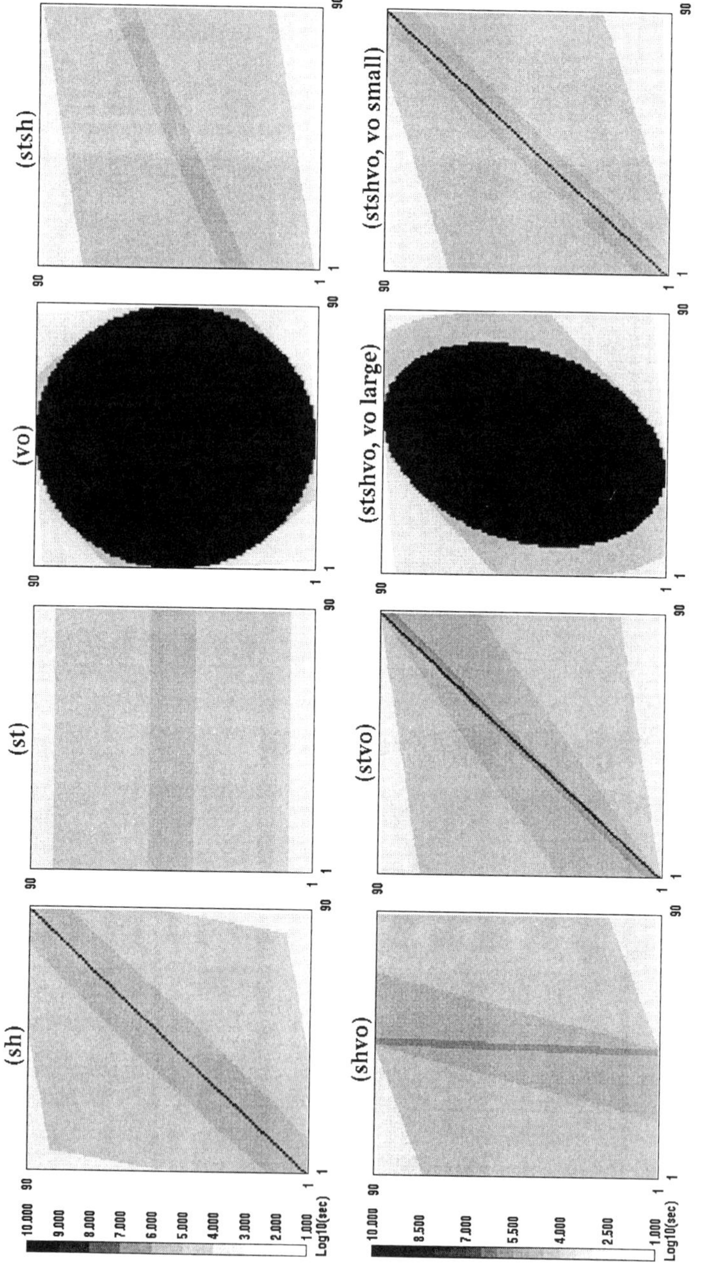

FIG. 3. *The minimum travel times of a signal from boundary subject to different linear flows. Time scale is in log base 10 and the number 10^{10} sec is used to represent infinite time.*

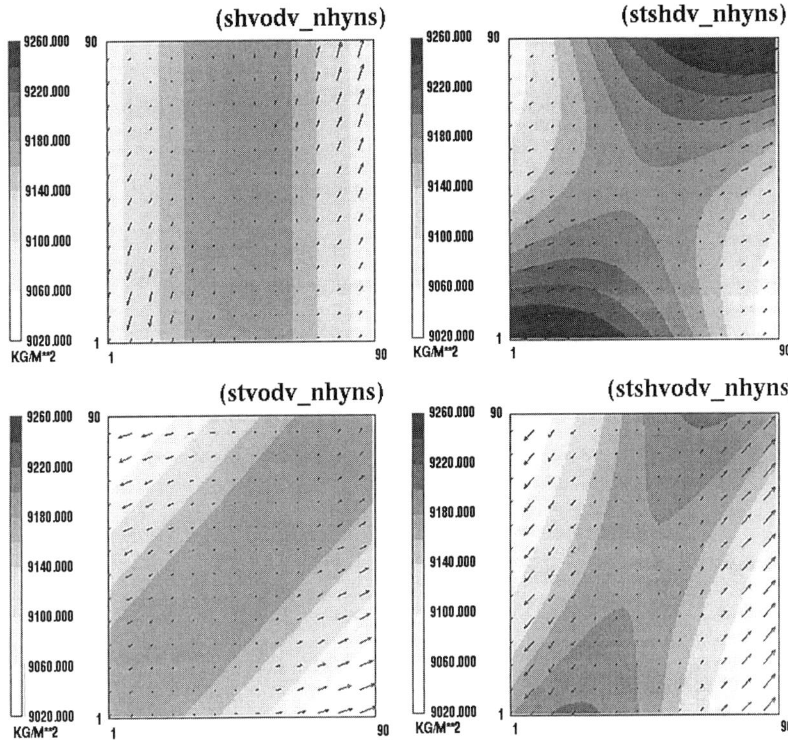

FIG. 4a. *Various* **Type-2** *flow patterns with weak divergence at the start of simulation time.*

were advected simultaneously. They satisfy the condition that the first two signals superpose to form the third one under ideal conditions. The hill signal was described by the 'Witch of Agnesi Mountain'

$$(16) \qquad h_s = \frac{a^2}{a^2 + (x - x_r)^2 + (y - y_r)^2}$$

where a is the radius of the mountain, and (x_r, y_r) is the position of the peak. The three signals are expressed as

$$(17a) \qquad h_A = q_{max}(1 + h_s) + q_{min}$$

$$(17b) \qquad h_B = 2q_{max}(1 + h_s) - q_{min}$$

$$(17c) \qquad h_C = -q_{max}(1 + h_s) + 2q_{min}$$

where mixing ratios q_{max} and q_{min} determine the amplitude and background values of the signals. When the hill h_B and the crater h_C are added, the original hill h_A is formed. The difference (from zero) provides a measure

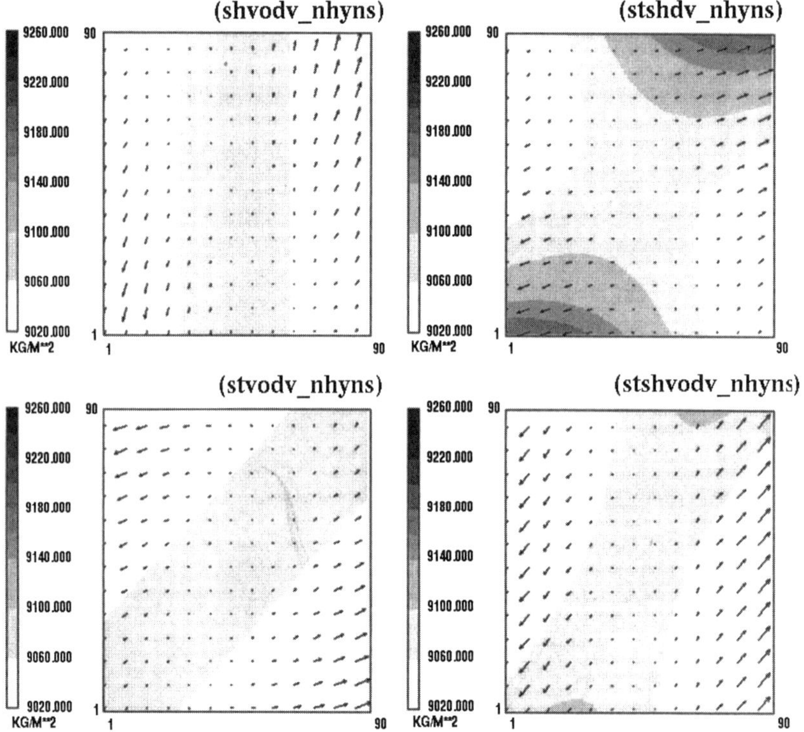

FIG. 4b. *The same as Figure 4a, but at the end of simulation time (24 hours later).*

the non-linearity of the advection process. In this experiment, $a = 3\Delta x$ and $(x_r, y_r) = (30\Delta x, 30\Delta y)$, $q_{min} = q_{max} = 50$ ppm were used. Boundary conditions of the signals h_A, h_B and h_C are described with the background values 100, 50, and 50 ppm, respectively. Note that these are continuous an extension of the hills, but fixed with time. Consequently, depending on the distance of the initial hills from the boundary, the shapes of the hills are changing continuously even under the pure rotational flow, unlike the case of cosine hill simulations often used. However, this does not pose any serious problem in the current tests because we can account for all the mass exchanges across the boundaries under the linear flow conditions. Because certain numerical advection methods clip negative values during the computation, 10 ppm (instead of zero) was used as the background value.

The third signal type tested is a set of checkerboard patterns with several different sizes, i.e., $n - \Delta x$ signals (for n=2, 4, 6, 8, 10, 12, 16, and 18). These signals were selected because most of numerical advection schemes have poor skill in maintaining the $2\Delta x$ wave but with improving results for longer wavelength signals. Here, the zero boundary condition was used for simplicity. Advection tests can show deformation characteristics of the $n - \Delta x$ signals during the advection. They provide some insight on how precursor pollutants injected into individual cells will propagate with time. Other signals tested included a set of three evenly spaced cosine cones, a sing ring, a slotted cylinder. Not all the results of these tests are presented here.

4.3. Numerical algorithms. The trace species advection equation for a two-dimensional flow is given as

$$
(18) \qquad \left.\frac{\partial c_i^*}{\partial t}\right|_{hadv} + \frac{\partial c_i^* \hat{\nu}^1}{\partial \hat{x}^1} + \frac{\partial c_i^* \hat{\nu}^2}{\partial \hat{x}^2} = 0 \,,
$$

where $c_i^* = c_i J_s$. For the convenience of notation, the superscript "*" is dropped when describing numerical algorithms. The equation can be solved as two consecutive one-dimensional equations using the solution of one as the initial condition of the other. This additional splitting can degrade the accuracy of a numerical algorithm as shown in Appendix B. It will be shown later that this error behaves as a source of mass inconsistency, which must be corrected by a mass adjustment scheme.

The one-dimensional advection equation was solved with a finite volume scheme, the piecewise parabolic method (PPM) of Colella and Woodward (1984). The algorithm is monotonic and shows relatively good peak preserving characteristics. Although some of the conclusions of the present study are transferable to other numerical schemes, certain characteristics such as peak conservation and linearity must be assessed for different numerical schemes used in the target air quality models. One of the main objectives of this paper is to compare effectiveness of different mass adjustment schemes. Although the choice of the numerical advection scheme clearly influences the final results, we can characterize different adjustment schemes within the limits of the accuracy of PPM. The computational time step was set to satisfy the Courant number less than 0.75. Wind components and density variables were on the staggered grid system.

Characteristics of the numerical advection algorithms are affected by the boundary conditions used. In this study, a positive definite no-flux outflow boundary condition with an appropriate flow divergence restriction is used. The no flux divergence at the boundary (flux out of the boundary cell set equal to the flux into the cell) is expressed as:

$$
(19) \qquad c_0 = \max\left\{0,\ c_1 - \frac{\nu_2}{\nu_1}(c_2 - c_1)\right\}
$$

where $c_0, c_1,$ and c_2 represent concentrations (precisely, they should be weighted by the Jacobian) outside the computational domain (c_0), at the boundary cell (c_1), and first cell of the inner domain (c_2), respectively. ν_1 is wind at the outer boundary flux point and ν_2 is wind at the inner boundary flux point. To prevent spurious flux situations at the boundary cell (that is often associated with the no-boundary flux divergence scheme), we applied the following constraints: when Δx was sufficiently small (for example smaller than say 10^{-3} m/s), or when the wind was divergent at the boundary cell (i.e., $\nu_1 \bullet \nu_2 < 0$), the no concentration gradient at the outflow boundary ($c_0 = c_1$) was imposed.

4.4. Performance measures. A few performance measures used to assess effects of the correction schemes on the mass conservation and other numerical characteristics are summarized in Table 2. A similar list, except for the linearity measure, has been used by many other researchers (e.g., Odman, 1998). The performance characteristics of a numerical algorithm should be judged on how well it maintains the signals of interest. One important benefit of using a steady linear wind field is that the analytical solution is available.

For example, Figure 5a compares behavior of the numerical advection of checkerboard patterns with the analytical solution for the rotational flow. At hour 1 (after 60 time steps), the short wave signals were smoothed in the numerical solutions while long wave signals were well maintained. The analytical solution projected on the Eulerian grid presented a discretization problem. At hours 6, 12, 18, and 24, the analytical solution maintained the checkerboard pattern perfectly (not presented here) while the numerical solutions continuously deteriorated as the integration continues. Because the Courant number increased (proportional to the distance from the center and the strength of S_Ω), short wave signals were damped more rapidly from the edges (See Figure 5b).

The analytical solution of the advection equation propagates along the characteristic curves (trajectories). When the solution is projected on an Eulerian grid, the total mass apparently fails to conserve depending on the relative locations of the grid points and origins of the trajectories. Figure 6 shows a cyclic fluctuation of domain average mixing ratios when a signal with three cones was advected by a rotational flow with the angular speed of one revolution per day. The cyclic fluctuation of sum of mixing ratios, although small in magnitude, is caused by the sampling process of the mixing ratios, which may or may not fall exactly on the grid points, from the initial condition field. The relative grid positions to the signal coincide at every six hours and the time series repeats after 24 hours, when the signal returns to the original position. The symptom is analogous to the mass conservation problem of a semi-Lagrangian transport (SLT) algorithm with interpolation. The difference is that the analytical solution can reconstruct the initial field when the positions in the trajectory coincide with the grid points. Unlike the analytical example, an actual SLT scheme does not remember the original field and thus the recovery is not possible. Because

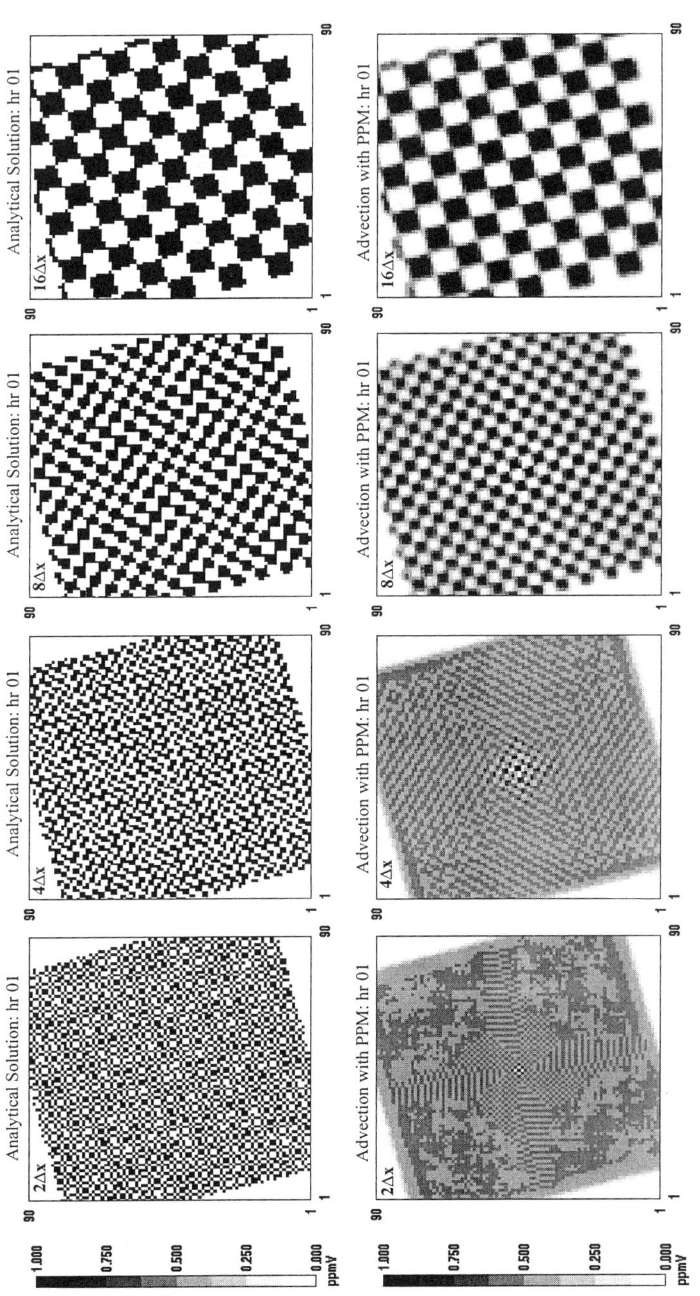

FIG. 5a. *Advection of checker patterns with various wavelengths. It compares the numerical results with the piecewise parabolic method (PPM) and the analytical solutions for the rotational flow at hour 1 (after 60 time steps). The short wave signals are smoothed out in the numerical solutions while long wave signals are relatively well maintained. When the relative position of the signal to the grid is retained (at every 6 hours for the given rotational flow), the analytical solution recovers the initial pattern perfectly.*

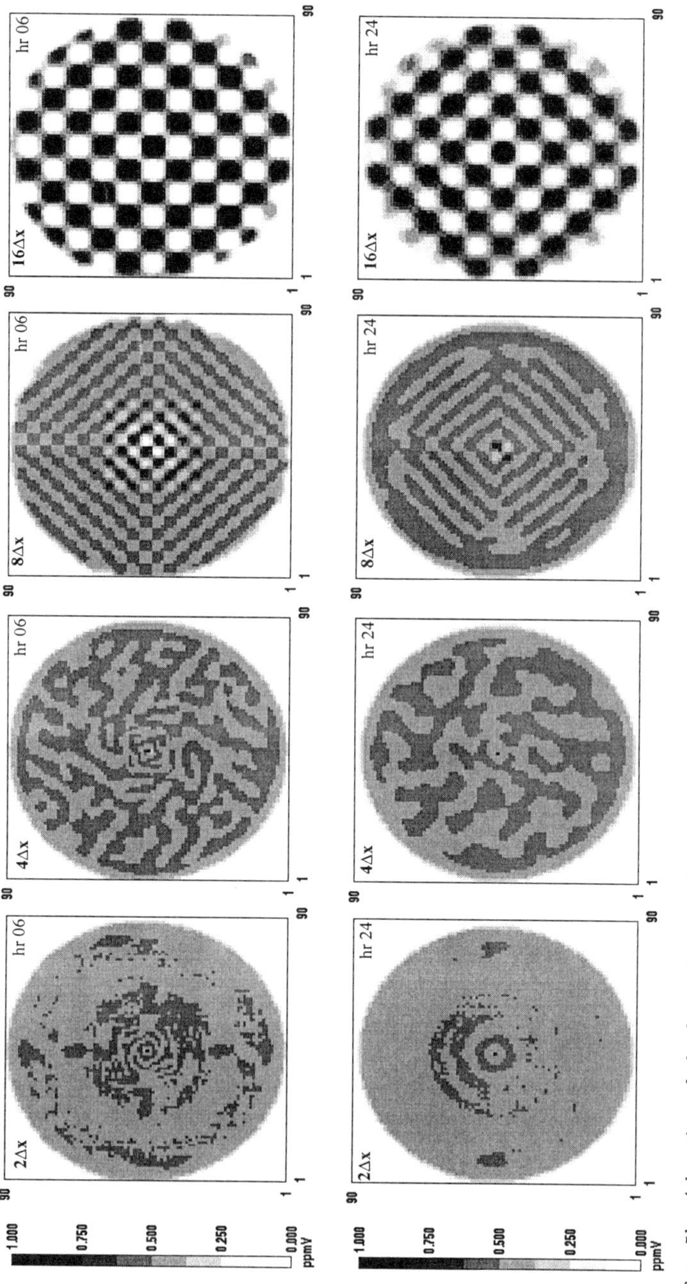

FIG. 5b. Advection of checker patterns with various wavelengths. It compares results of numerical advection with PPM at hour 6 and hour 24 (one full rotation). The signals are losing peaks as the integration continues due to the numerical diffusion. Because the Courant numbers are larger near the boundaries, the signals are mored diffused.

TABLE 2

Summary of the performance measures showing effects of numerical advection. q_i^e is the exact mixing ratio for the tracer species i., and the three signals satisfy $q_i^A = q_i^B + q_i^C$ under ideal conditions. ρ^* is the grid-value adjusted air density.

Performance Measure	Formula	Description
peak ratio (PEAK)	$\dfrac{\max\{\rho^*\|q_i\}}{\max\{\rho^* q_i^e\}}$	measure of peak preservation (best when 1.0)
mass ratio (MASS)	$\dfrac{\sum(\rho^* q_i)}{\sum(\rho^* q_i^e)}$	measure of mass conservation characteristic (best when 1.0)
distribution ratio (DIST)	$\dfrac{\sum(\rho^* q_i)^2}{\sum(\rho^* q_i^e)^2}$	measure of shape retention (best when 1.0)
average absolute error (AVAE)	$\dfrac{1}{N}\sum \rho^*\|q_i - q_i^e\|$	measure of absolute difference (best when 0.0)
root-mean square error (RMSE)	$\sqrt{\dfrac{1}{N}\sum\left(\dfrac{q_i - q_i^e}{q_i^e}\right)^2}$	measure of distribution error (best when 0.0)
linearity measure (LINR)	$\dfrac{\max\{\rho^*\|q_i^A - q_i^B - q_i^C\|\}}{\max\{\rho^* q_i^e\}}$	measure of linearity of advection process (best when 0.0)

the discretization is relatively small and predictable, we used the analytical solutions as surrogate exact solutions to assess the performance measures (see Table 2).

5. Effects of mass adjustment process for steady-state non-divergent flows. It has been known that numerical advection for a realistic flow with non-uniform density distributions has difficulties in maintaining mass conservation and/or constant mixing ratios. For a mass consistent flow, numerical schemes should conserve mass of the trace species. Because we are using 1-D mass conserving numerical schemes here, the problem with mass conservation, if any, will be caused by the application of time splitting. Occasionally, flow fields used for the trace species transport are not mass consistent. In such cases, conservation of mixing ratio, at least, can be achieved by the application of a mass adjustment scheme. In this section, we study the effectiveness of different mass adjustment schemes under various flow conditions. We quantify mass conservation and linearity of the numerical advection process for the two-dimensional non-divergent flows listed in Table 3.

5.1. Idealized flows with uniform density (Type-0 flow). For a steady-state non-divergent flow with uniform density, all of the five correction schemes become identical for a Cartesian grid system. The non-

TABLE 3

Summary of test runs for steady-state non-divergent flow patterns. (Refer to Table 1 for the adjustment scheme symbols.)

Density Distribution Flow Pattern	**Type-0;** uniform ρ, uniform J_s	**Type-1;** uniform ρJ_s but ρ and J_s individually not uniform (hydrostatic)	**Type-2;** mass consistent, but non-uniform ρ, and uniform J_s (non-hydrostatic)	**Type-3;** *Type-2* flow with density slope added to make mass inconsistent (non-hydrostatic)
stretching deformation (st)	*A0, A1, A2, A5*	*A0, A1, A2, A5*	*A0, A1, A5*	*A0, A1, A2, A3, A5*
shearing deformation (sh)	*A0, A5*	*A0, A1, A2, A5*	*A0, A1, A5*	*A0, A1, A2, A3, A5*
vorticity (vo)	*A0, A5*	*A0, A1, A2, A5*	*A0, A1, A5*	*A0, A1, A2, A3, A5*
stretching and shearing deformation (stsh)	*A0, A5*		*A0, A1, A5*	*A0, A1, A5*
stretching deformation and vorticity (stvo)	*A0, A5*		*A0, A1, A5*	
shearing deformation and vorticity (shvo)	*A0, A5*		*A0, A1, A5*	
stretching, shearing, and vorticity, where st=sh=vo (stshvo)	*A0, A5*		*A0, A1, A5*	
stretching, shearing, and vorticity, where st=sh, vo = 4 × sh (stshvoa)		*A0, A2, A5*	*A0, A1, A5*	

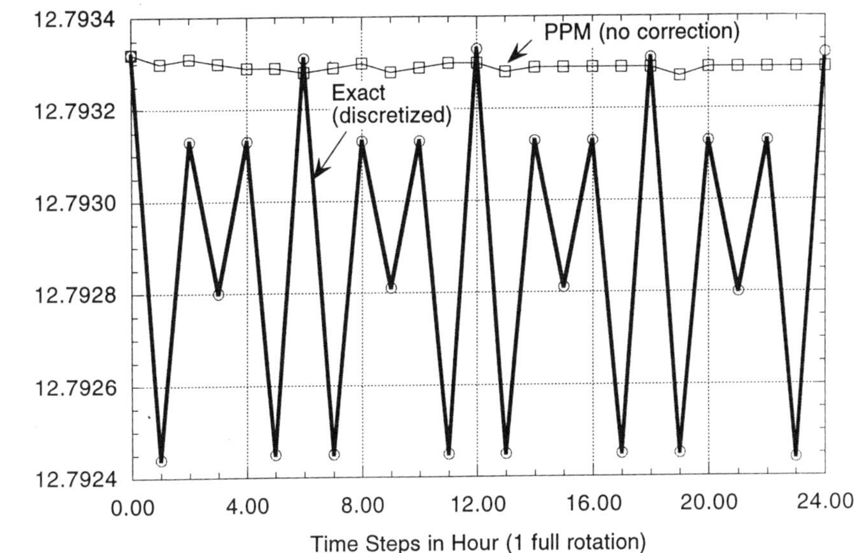

FIG. 6. *Cyclic fluctuation of domain average mixing ratios when signal with three cones are advected by a rotational flow with the angular speed of one revolution per day. The original values are recovered at every 6 hours when the relative position of the signal to the grid is maintained. PPM scheme (without adjustment) shows a high degree of mass conservation.*

divergent wind fields are mass consistent and therefore the only error we need to fix is the inaccurate numerical solution associated with the flow patterns. For the pure rotation (**vo**) and shearing deformation (**sh**) flows, the numerical integration preserves the uniform field very well. Essentially no correctional step is needed. Also, for flow pattern **shvo**, which is a bi-directional flow as shown in Figure 1, no cross-term effect is present. The mass conserving PPM scheme has no problem maintaining the uniform field. However, for the stretching (**st**) flow, several performance measures such as peak ratio, mass ratio, distribution, and linearity show that the numerical advection failed to maintain the initial signal. Appendix B explains why the cross-term effect is large for the stretching flow for the time-splitting application.

As expected, uniform field tests (not presented here) show that the correction schemes *A1*, *A2*, *A3*, *A4*, and *A5* behave the same way in the Cartesian grid system maintaining the uniform field. It seems that most of the errors originated from the inaccuracy in the time splitting procedure, and the correction step removed this problem. The adjustment also improved the linearity for the stretching flow, in particular those accentuated by the cross-term errors. As expected, the hill signals lose peak values due to the numerical diffusion associated with the PPM scheme.

The adjustment provides a marginal improvement of the peak values. For the composite flows without the **st** component, the simulations conserve mass and maintain linearity with very small relative error $O(10^{-5})$ for the single-precision computation. Other composite flows with the **st** component experience similar difficulties showing noticeable linearity errors $O(10^{-3})$. Because the density is uniform for the **Type-0** flows, the errors are purely of kinetics origin. The superposition test shows that the PPM scheme has the accuracy suitable to study the effects of mass inconsistency on the numerical integration of trace species.

5.2. Idealized flows with uniform Jacobian-weighted density (Type-1 flow). Flows studied here have uniform ρ^* distributions. For a Cartesian grid, this is identical to the uniform density case, making the scheme *A1* the same as *A5* and *A3* (for the steady state case). Therefore, we compared *A0* (no adjustment) with *A1* (or *A5*) and *A2* adjustment schemes (Figure 7). Like in the **Type-0** case, advection tests without correction preserve the uniform mixing ratios for rotational and shearing flows. The linearity is well maintained for both uniform field and hill signals. However, when a wrong adjustment scheme (in this case scheme *A2*) is used, the results get worse. Errors in the total mass and the distribution become larger as the integration continues. The adjustment scheme *A1* does not corrupt the good simulation result for shear and rotational flow. For the stretching (**st**) flow, advection with no adjustment (*A0*) encounters some problems in preserving uniform fields.

Figure 8 provides advection test results for the **stshvoa** flow (with large vorticity component) showing that correct adjustment (*A1* or *A5*) is necessary to maintain the constant mixing ratio. Without correction, signals become badly corrupted and in turn lead to severe non-linearity in the solutions. Errors at the boundaries propagate into the inner domain as well. The adjustment scheme *A2* could not correct this problem. *A1* (thus *A5*), improves the linearity of the numerical solution, essentially by removing the cross-term errors. The small non-linearity originates from the nonlinear character of the PPM itself. The simulated peak value (with the correction step) is less than that of the exact solution, which is not unexpected because PPM is a scheme with some numerical diffusion.

5.3. Idealized flows with nonuniform, but mass consistent density distribution (Type-2 distribution). For **Type-2** flows, the Jacobian-weighted density distribution is not uniform. For the convenience of setting up meteorological fields, we made the Jacobian (J_s) uniform in the horizontal direction. In an operational meteorological model with a normalized terrain-influenced height coordinate, J_s varies with the topography. Depending on the vertical grid structure of the model, the distribution of J_s will change with height as well. The latter condition is not relevant in the present two-dimensional flow tests. Under these conditions, the adjustment scheme *A2* becomes identical to *A5*. For shearing (**sh**)

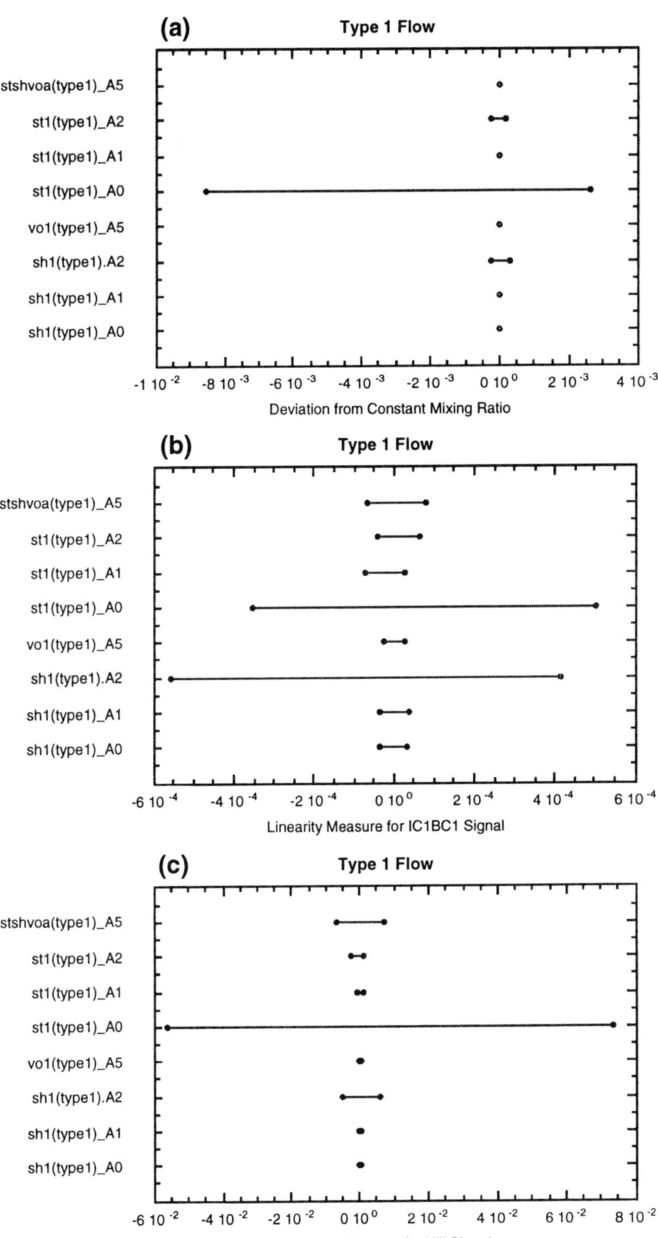

FIG. 7. *Summary graph comparing numerical performance of mass adjustment schemes for* **Type-1** *density distributions: (a) deviation from constant mixing ratio 1.0; (b) linearity measure for the* $IC = 1.0$, $BC = 1.0$ *signal; and (c) linearity measure for hill superposition test.*

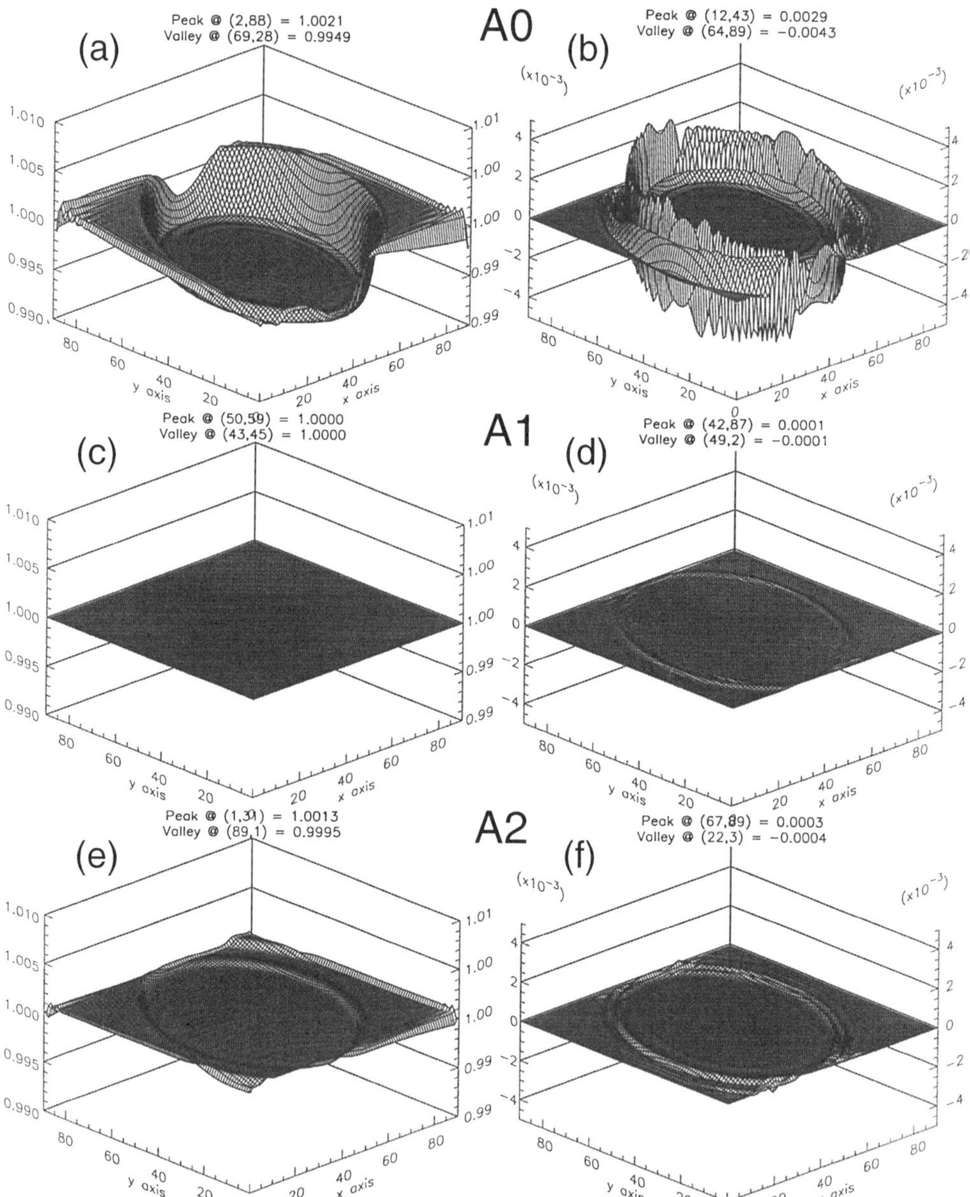

FIG. 8. *Advection test results for the* stshvoa *flow (with large vorticity component for a* **Type-1** *density distribution: (a) advection of constant mixing ratio 1.0 without correction (A0); (b) linearity measure for the IC = 1.0, BC = 1.0 signal for A0; (c) the same as (a), but with A1 scheme; (d) the same as (b), but with A1 scheme; (e) the same as (a), but with A2 scheme; and (f) the same as (b), but with A2 scheme. (Frames on left side show results from advection of IC = 1.0, and BC = 1.0 while those on right side represent linearity errors.)*

and rotational (**vo**) flows, the PPM algorithm without correction could not handle the advection of the uniform field as precisely as the **Type-0** and **Type-1** density distribution cases. This demonstrates that the cross-term error is not purely kinematic, thus it does not vanish even for the **sh** and/or **vo** flows when the air density distribution is non-uniform. The mass-ratio (MASS) and distribution-ratio (DIST) errors were small but the root-mean square error (RMSE) was large. The *A5* adjustment recovers all other performance measures except for the linearity measure (LINR). For flows with **st** component, test results exhibit problems in maintaining peak values and mass mixing ratio, etc. It appears that the density field interacts with the velocity field making the cross-term effect more pronounced. The linearity error is caused by the combined effects of the non-linearity in the 1-D algorithm and the cross-term error associated with the time splitting. Figure 9 summarizes these characteristics for the **Type-2** flows with the stretching wind component. Figure 10 demonstrates advection of IC/BC signals for the **Type-2 stsh** flow. Although started with the mass-consistent density distribution, the advection without correction cannot maintain the constant mixing ratio and suffers significant loss along the axis of effluent flows. The adjustment scheme *A1* improves the performance measures but not perfectly while *A5* scheme fixes most of the problem, except for those caused by inherent non-linearity and numerical diffusion of the algorithm.

5.4. Idealized flows with nonuniform and mass inconsistent density distribution (Type-3 Flow). When flow fields are provided at a regular time interval, say hourly, from a wind field model using observations or from an operational weather forecasting model, one should expect some kind of mass inconsistency in the data. The **Type-3** flow represents a situation where mass consistent error is present in the data. In the real atmosphere, it is impossible to have a steady state flow with this type of baroclinic thermal flow. Here, we intentionally added density slope to **Type-2** distributions to investigate the possible effects of mass inconsistency on numerical advection. Addition of an incremental density slope causes $\hat{\mathbf{V}}_s \bullet \nabla_s \rho^* \neq 0$ and the error term Q_ρ becomes nonzero for the non-divergent steady flow. Because air density and wind fields are not consistent for the **Type-3** flows, we cannot expect conservation of species mixing ratio if not corrected.

We have tested performance of the numerical advection with the adjustment process under these conditions (Figure 11). Without any adjustment the numerical advection cannot preserve constancy of mixing ratio for the **Type-3** flows. An incorrect scheme, for example the *A1* scheme, cannot maintain uniform mixing ratio even for **sh** and **vo** flows. Compared with a **Type-2** distribution, a **Type-3** distribution causes a much larger error for shearing flow without adjustment. The recommended scheme *A5* (and *A2* because the original density field follows the **Type-2** distribution) fixes the problem. Figure 12 suggests that *A5* can detect and correct the

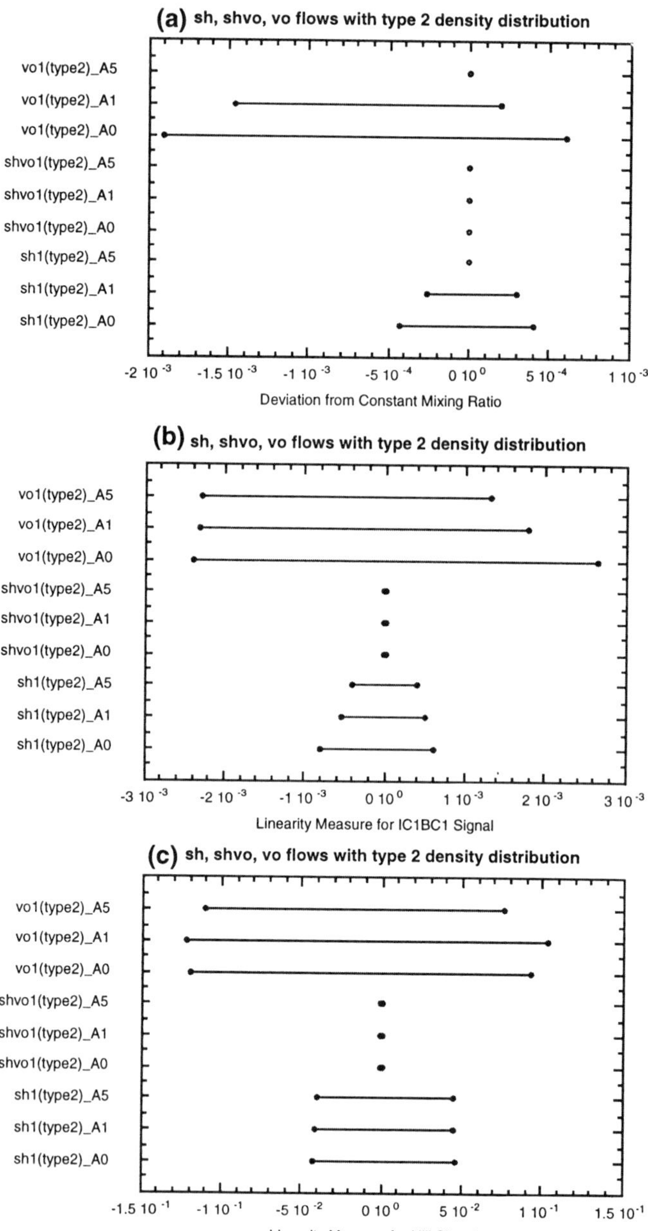

FIG. 9. *Summary graph comparing numerical performance of mass adjustment schemes for* sh, shvo, *and* vo *flows under* **Type-2** *density distributions: (a) deviation from constant mixing ratio 1.0; (b) linearity measure for the* $IC = 1.0$, $BC = 1.0$ *signal; and (c) linearity measure for hill superposition test.*

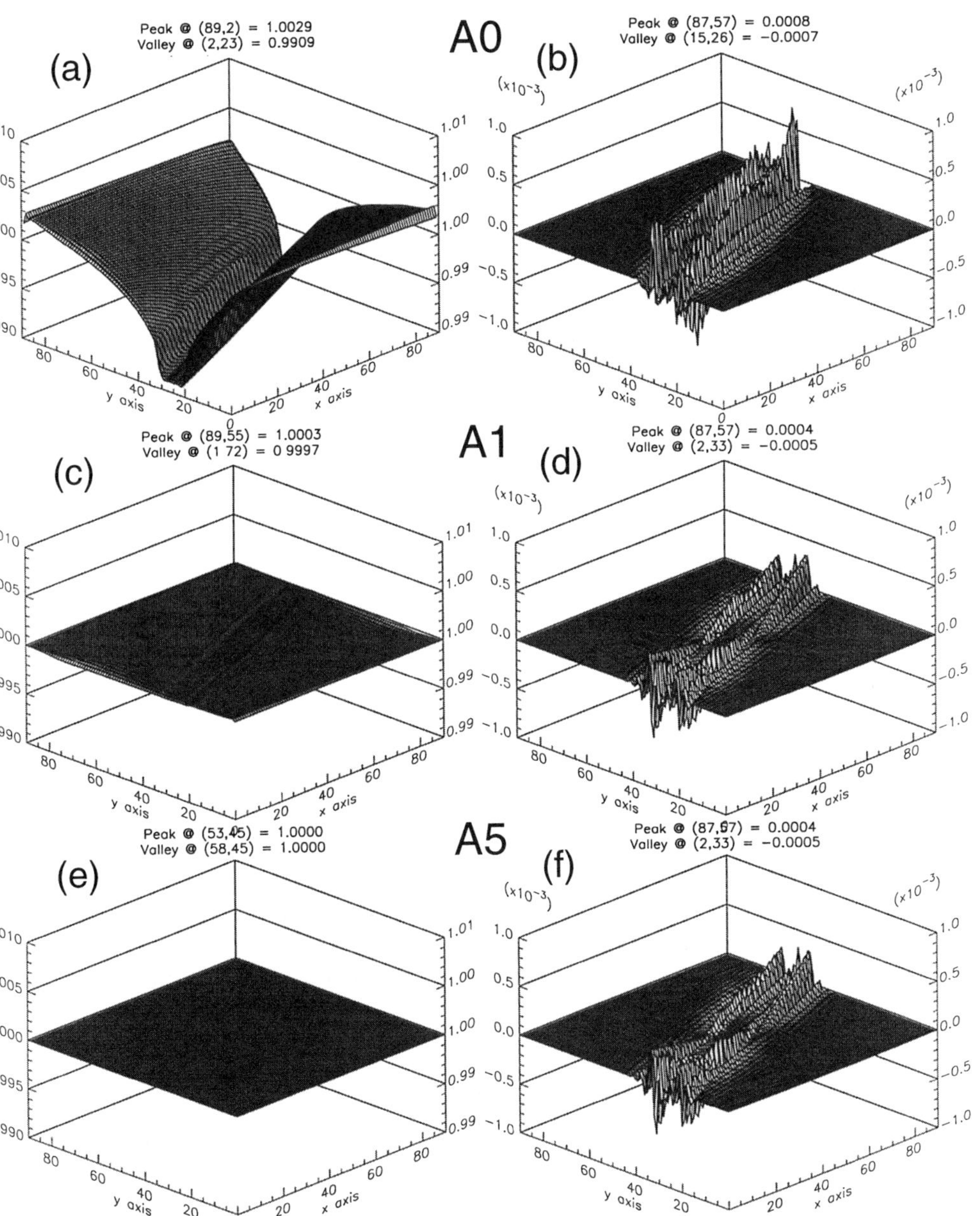

FIG. 10. *Advection test results for the* stsh *flow with a* **Type-2** *density distribution: (a) advection of constant mixing ratio 1.0 without correction (**A0**); (b) linearity measure for the IC = 1.0, BC = 1.0 signal for **A0**; (c) the same as (a), but with **A1** scheme; (d) the same as (b), but with **A1** scheme; (e) the same as (a), but with **A5** scheme; and (f) the same as (b), but with **A5** scheme. (Frames on left side show results from advection of IC = 1.0, and BC = 1.0 while those on right side represent linearity errors.)*

combined errors both in the meteorology data and the numerical scheme. Test results for **st** flow and other complex flows, **stvo**, **stsh**, **shvo**, reveal results similar to the other simple flow cases. Without adjustment, the system is continuously losing mass: the $A1$ scheme improves it somewhat, and the $A5$ scheme conserves mass very well. The linearity is not much improved with the adjustment. However, the non-linearity problem is not so significant for PPM. The $A1$ scheme for **stshvo** flow produces worse results than without correction for the advection of the uniform mixing ratio. Figure 12 compares the results after 24-hour advection of three cones under st flow. $A0$ suffers the effect of mass inconsistency in the flow and application of $A1$ cannot improve the result. $A5$ removes the effect of mass inconsistency. Even after correction, numerical solutions reveal typical behavior of PPM that produces flatter peaks compared with the analytical solution.

6. Experiment with divergent flows. There have been relatively few systematic tests of numerical algorithms with divergent flows in the literature. However, two-dimensional divergent flows are common in the atmosphere and therefore, their impact on numerical advection of trace species must be studied. Unless there is a continuous supply of air, divergent density flows cannot be in a steady state. If steady state of air density is assumed for a 2-D horizontal case, the mixing ratio should decrease with time because the trace species mass is lost with the effluent flow at the rate of the divergence while influx of fresh air (with no trace species) continues.

In this section, we study the numerical advection and adjustment algorithms with several kinematic divergent flows where the air density decreases (increases for convergent flows) over time consistently with strength S_Δ. Without the mass correction scheme, the system loses mass at a much higher rate than that expected from the divergent flow. Figure 13 demonstrates this for a non-steady divergent flow. The Jacobian was set to be a constant with time (i.e., a **Type-2** density distribution), but air density decreases corresponding to the amount of the divergent flow, resulting in about 1% change for the 24-hour period. It is not unusual that the mass inconsistency originating from the meteorology data and from the time-splitting error have magnitudes similar to the weak divergence used in the test. Although the flow is weakly divergent, the feedback with the mass inconsistency causes exponential loss of trace species mass. Without a correction, about 95% of the mixing ratio is lost within the 24-hour period. Only when the appropriate mass adjustment process is applied, the numerical advection results become consistent with the expected mass flux divergence.

Figure 14 summarizes numerical characteristics of the adjustment schemes for the various divergent flows and density distribution types. It demonstrates that the numerical advection under the divergent flow is essentially linear for the IC/BC test and nominally nonlinear for the hill

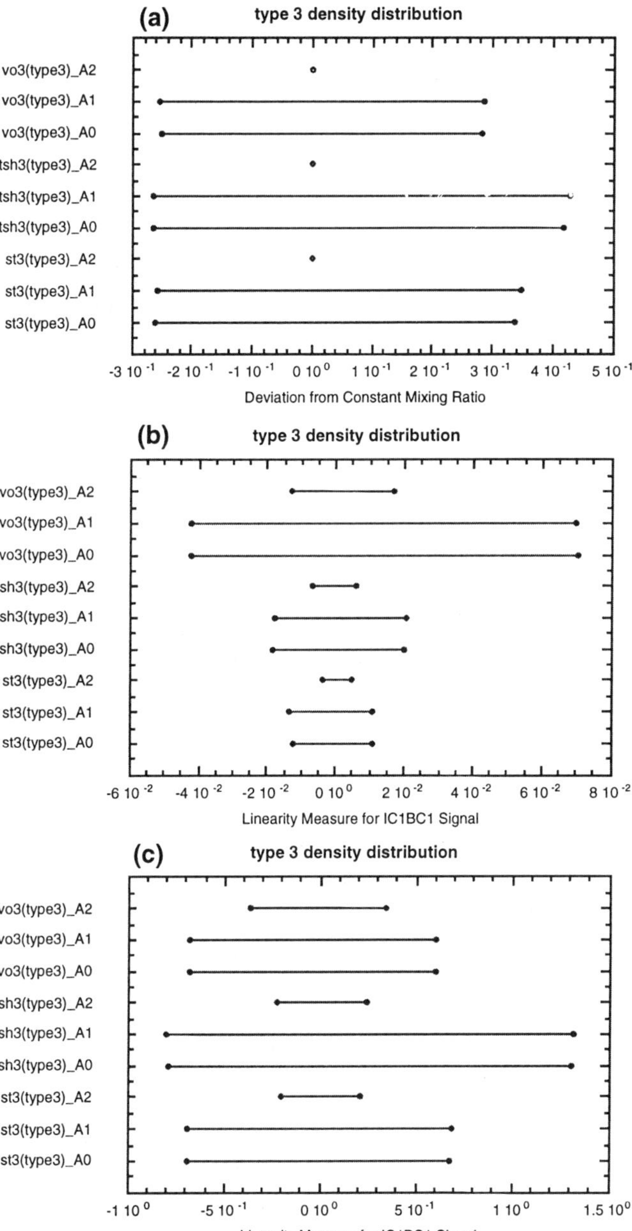

FIG. 11. *Summary graph comparing numerical performance of mass adjustment schemes for* **Type-3** *density distributions: (a) deviation from constant mixing ratio 1.0; (b) linearity measure for the $IC = 1.0$, $BC = 1.0$ signal; and (c) linearity measure for hill superposition test.*

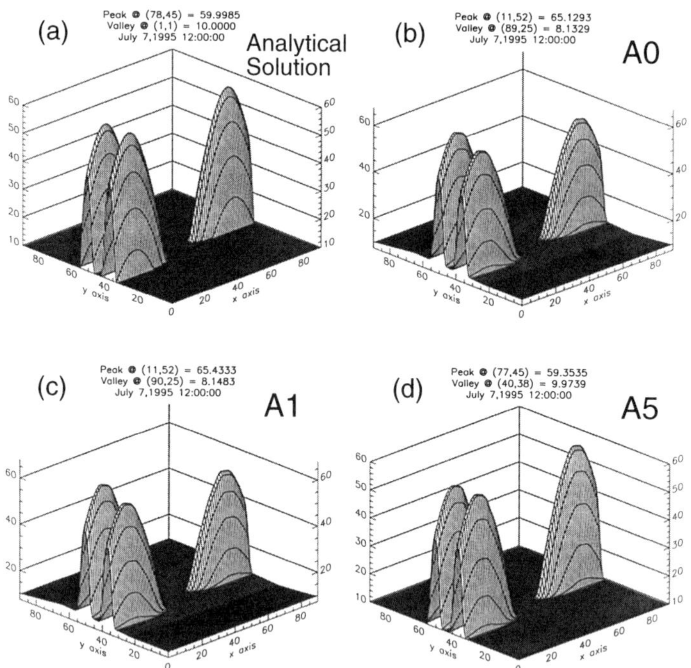

FIG. 12. *Advection test results for the* st *flow with* **Type-3** *density distributions: (a) discretized exact solution; (b) without correction (**A0**); (c) with **A1** scheme; and (d) with **A5** scheme.*

TABLE 4
Summary of test runs for divergent flow patterns.

Density Distribution Flow Pattern	**Type-0**; uniform ρ, uniform J_s	**Type-1**; uniform ρJ_s but ρ and J_s individually not uniform (hydrostatic)	**Type-2**; mass consistent, but non-uniform ρ, and uniform J_s (non-hydrostatic)
dv-steady	*A0, A1, A2, A5*	*A0, A1, A2, A5*	*A0, A1, A5*
dv-nonsteady	*A0, A1, A2, A5*	*A0, A1, A2, A5*	*A0, A1, A5*
shvodv-steady			*A0, A1, A5*
shvodv-nonsteady			*A0, A1, A5*
stshdv-steady			*A0, A1, A5*
stshdv-nonsteady			*A0, A1, A5*
stvodv-steady			*A0, A1, A5*
stvodv-nonsteady			*A0, A1, A5*
stshvodv-steady			*A0, A1, A2, A5*
stshvodv-nonsteady		*A0, A1, A2, A5*	*A0, A1, A2, A5*

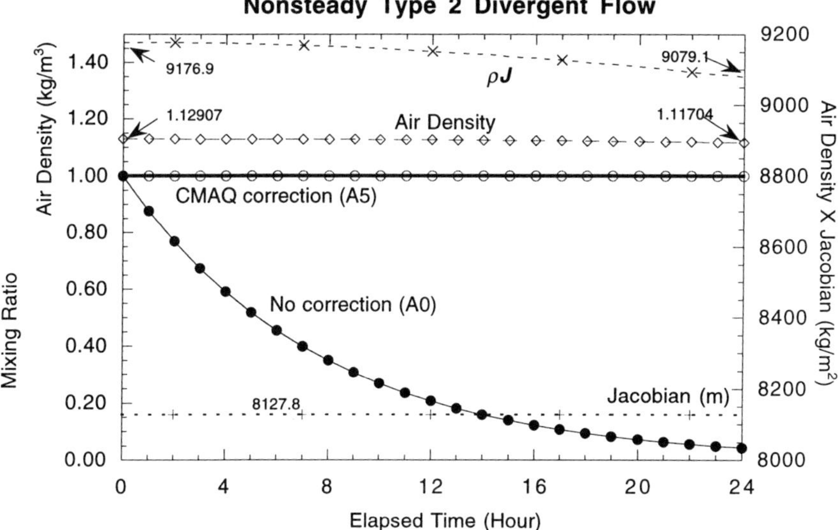

FIG. 13. *Change in mixing ratio for the* **Type-2** *divergent flow. The Jacobian was set to be a constant with time (i.e., a* **Type-2** *density distribution), but air density decreases corresponding to the amount of the divergent flow, resulting in about 1% change for the 24-hour period. Without the correction, the system loses mass at much higher rate than the amount expected from the divergent flow. With the correction* **A5**, *the mixing ratio is conserved.*

superposition test. The seemingly larger non-linearity shown with the divergent flow for the adjustment schemes **A1**, **A2** than that of **A0** is due to the difference in magnitude of the resulting signal left in the domain. The numerical advection under divergent flow without correction essentially loses most of its signal and the absolute differences among the signals become very small. For the constant (and uniform) density case, **A1**, **A2** and **A5** become identical. Similar results are obtained for the **stshvodv** flow. As expected, the **A2** scheme is inferior to **A1** for the **Type-1** density distribution, and numerical advection without correction shows a serious problem in maintaining mixing ratio as well as linearity.

Figure 15 contrasts numerical advection results between non-divergent and divergent flows. Although the numerical advection without correction performs relatively well with only minor problems for the **stvo Type-2** flow, the results become unsatisfactory for a flow with even a weak divergence component. For the **stvodv** flow, the difference between the corrected and uncorrected applications is striking. Figure 16 summarizes numerical test results for the divergent flow with the **Type-2** density distributions. As before, the **A1** correction improves the performance of numerical advection significantly although it is inferior to **A5**.

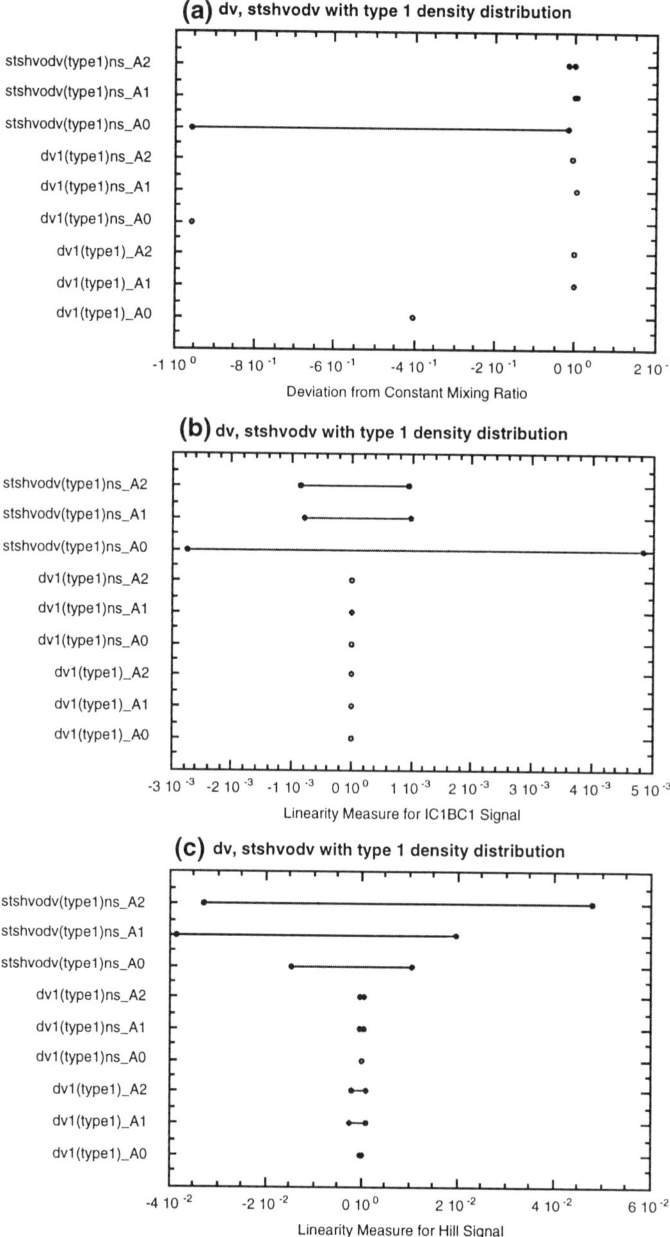

FIG. 14. *Summary graph comparing numerical performance of mass adjustment schemes for* **Type-3** *density distributions: (a) deviation from constant mixing ratio 1.0; (b) linearity measure for the $IC = 1.0$, $BC = 1.0$ signal; and (c) linearity measure for hill superposition test. The suffix 'ns' stands for the case with non-steady mass consistent density distribution and a density type without the suffix is with constant but mass-inconsistent density.*

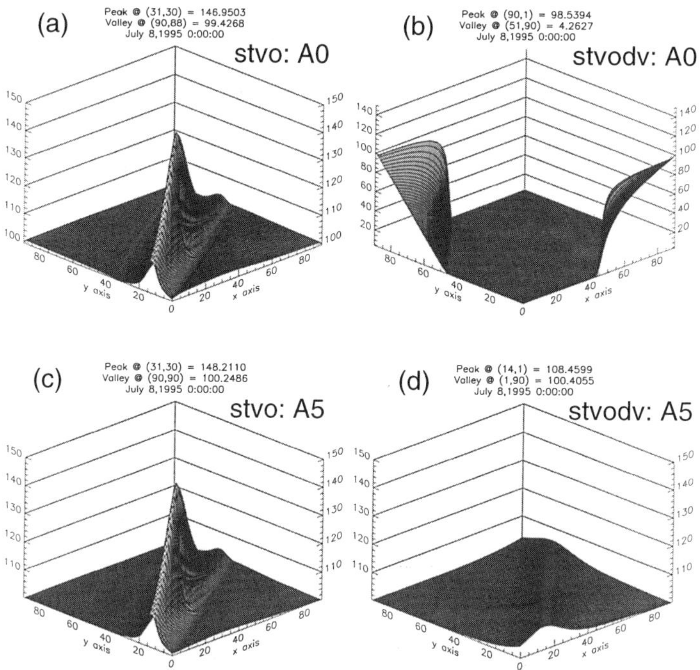

FIG. 15. *Advection test results for the* stvo *and* stvodv *flows with* **Type-2** *density distributions: (a) without correction (A0) for* stvo*; (b) without correction (A0) for* stvodv*; (c) with* **A5** *scheme for* stvo*; and (d) with* **A5** *scheme for* stvodv*.*

7. Conclusions. We have studied numerical integration of the trace species advection equation under various linear wind flows (non-divergent and divergent) and density distributions (mass consistent and inconsistent). Several mass (air density) adjustment schemes are tested for their effectiveness of removing some of the problems associated with numerical advection. The results can be summarized as:

(1) Advection under stretching (**st**) flow is subject to a large cross-term error while tests for shearing (**sh**) and vortex (**vo**) flows show little errors under uniform density (or Jacobian weighted density) distributions. The cross-term error does not vanish even for the **sh** and/or **vo** flows when the air density distribution is non-uniform. A correction mass adjustment scheme can effectively eliminate these errors.

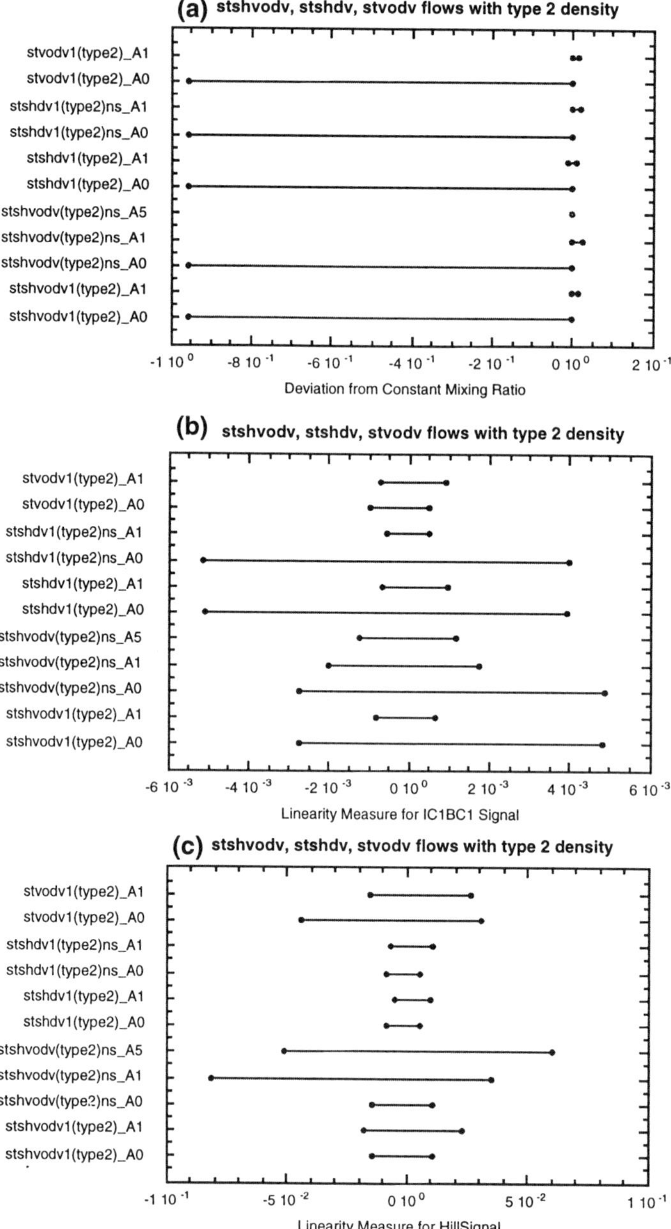

FIG. 16. *Summary graph comparing numerical performance of mass adjustment schemes for various divergent flows with* **Type-2** *density distributions: (a) deviation from constant mixing ratio 1.0; (b) linearity measure for the $IC = 1.0$, $BC = 1.0$ signal; and (c) linearity measure for hill superposition test.*

(2) $A5$ is a general correct adjustment method that conserves mixing ratio. Depending on the flow types and density distribution patterns, simpler adjustment methods become the same as $A5$. An incorrect adjustment algorithm can make the numerical result worse than no adjustment case.

(3) When the linearity error originates from the cross-term error, the adjustment scheme can improve the linearity of the advection process. However, the adjustment scheme cannot improve the inherent non-linearity with a numerical algorithm.

(4) Most flux-form advection algorithms suffer numerical diffusion. The diffusive characteristics are dependent on the flow (i.e., Courant number and the relative compositions of the fundamental flows), density distributions, and the wavelengths of signals. The Mass adjustment process does not change this. Application of a correct adjustment improves the numerical diffusion characteristics by minimizing cross-term errors and by eliminating errors caused by the mass inconsistent flow.

(5) If air density does not change corresponding to the divergence in the flow, it becomes a source of error (steady divergent flow is mass inconsistent by definition). Application of a correct adjustment process is necessary because the cumulative effects of divergence in the flow and errors in the numerical advection deteriorate the solution rapidly.

(6) Even a weak divergence in the flow with a non-steady but mass consistent density distribution results in serious errors in the numerical solution. The divergence flow itself is source of the cross-term error and exponential loss of trace species mass. Application of the correct adjustment scheme is necessary to obtain the solution consistent with expected mass flux divergence.

In summary, application of a correct mass adjustment process, $A5$ in general, improves the performance of existing numerical advection. There are several areas that need to be studied further. For example, analytical forms of phase, amplitude, and Courant number relations can usually be obtained by the utilization of von Neuman spectral analysis to the discretized form of an advection algorithm. However, the method is not applicable for many nonlinear schemes, especially those that rely on flux-correction methods. For those algorithms, these relationships must be tested numerically with a set of linear flows and signals of different wavelengths. Future research should also quantify mass inconsistencies in the data from operational meteorological models and the effectiveness of the adjustment scheme. A hydrostatic meteorological model with a pressure (or normalized terrain-influenced pressure) vertical coordinate provides meteorological data representing a **Type-1** density distribution with temporal variations. A non-hydrostatic model with constant height (or normalized terrain-influenced height) vertical coordinate provides meteorological data for a **Type-2** density distribution with temporal variations. Due to complications such as underlying topography, effects of moisture variables on the air density, and numerical problems, there is a strong need for a mass adjustment step in these meteorological models. There is even a bigger need for mass adjustment in an air quality model that relies on archived meteorological data obtained from either a prognostic meteorological model or a diagnostic objective analysis model based on observations.

Acknowledgments. The authors express sincere appreciation to Mr. Kenneth Schere, Drs. Jeff Young, Frank Binkowski of the U.S. Environmental Protection Agency (EPA), and Dr. David Chock of Ford Research for contributing their insights on the issue.

Disclaimer. The information in this document has been funded wholly or in part by the United States Environmental Protection Agency. It has been subjected to Agency review and approved for publication. Mention of trade names or commercial products does not constitute endorsement or recommendation for use.

APPENDIX

A. Streamline equations for linear velocity fields.

A.1. Derivation of streamline equations for linear velocity fields.
Starting from Eqs. (10a) and (10b), one can express the streamline function of a linear flow as

$$\text{(A.1)} \quad \frac{dy}{dx} = \frac{dy/dt}{dx/dt} = \frac{v}{u} = \frac{v_c + (S_\Delta + S_\Omega)(x - x_c) + (S_\Delta - S_\Gamma)(y - y_c)}{u_c + (S_\Delta + S_\Gamma)(x - x_c) + (S_\Lambda - S_\Omega)(y - y_c)}.$$

For the nontrivial case where at least one of the elementary streamline patterns other than the translation (i.e., $u = u_c$, and $v = v_c$) exists, Eq. (A.1) is apparently a non-homogeneous differential equation. However, for $S_\Delta^2 + S_\Omega^2 \neq S_\Lambda^2 + S_\Gamma^2$, it can be transformed into a homogeneous one by substituting (x, y) with (p, q), which are defined in terms of the intersecting point of the two linear equations for u and v.

$$\text{(A.2)} \quad (p, q) = (x - x_c - x_p, y - y_c - y_q)$$

where coordinates of the intersection point are given as

$$\text{(A.3)} \quad (x_p, y_q) = \left(\frac{(S_\Delta - S_\Gamma)u_c - (S_\Lambda - S_\Omega)v_c}{S_\Lambda^2 + S_\Gamma^2 - (S_\Delta^2 + S_\Omega^2)}, \frac{(S_\Delta + S_\Gamma)v_c - (S_\Lambda + S_\Omega)u_c}{S_\Lambda^2 + S_\Gamma^2 - (S_\Delta^2 + S_\Omega^2)} \right).$$

Then, the differential equation is simplified as

$$\text{(A.4)} \quad \frac{dq}{dp} = \frac{(S_\Lambda + S_\Omega)p + (S_\Delta - S_\Gamma)q}{(S_\Delta + S_\Gamma)p + (S_\Lambda - S_\Omega)q}.$$

The homogeneous differential equation can be solved easily by substituting $r = q/p$. Using $\frac{dq}{dp} = r + p\frac{dr}{dp}$, Eq. (A.4) can be rewritten in the form where the variables are separated

$$(A.5) \qquad \frac{dp}{p} = \frac{(S_\Delta + S_\Gamma) + (S_\Lambda - S_\Omega)r}{(S_\Lambda + S_\Omega) - 2S_\Gamma r - (S_\Lambda - S_\Omega)r^2} \ .$$

By integrating Eq. (A.5), one can obtain a streamline equation for the general linear wind field. For a flow where vorticity dominates, i.e., $S_\Omega^2 > S_\Lambda^2 + S_\Gamma^2$, the streamline equation is given as

$$(A.6) \qquad \begin{aligned} &\frac{S_\Delta}{\sqrt{S_\Omega^2 - (S_\Lambda^2 + S_\Gamma^2)}} \tan^{-1}\left[\frac{-(S_\Lambda - S_\Omega)r - S_\Gamma}{\sqrt{S_\Omega^2 - (S_\Lambda^2 + S_\Gamma^2)}}\right] \\ &- \frac{1}{2}\ln\left|(S_\Lambda + S_\Omega) - 2S_\Gamma r - (S_\Lambda - S_\Omega)r^2\right| = \ln p + \text{const.} \end{aligned}$$

For a deformation dominant flow, i.e., $S_\Omega^2 < S_\Lambda^2 + S_\Gamma^2$, we have

$$(A.7) \qquad \begin{aligned} &\frac{S_\Delta}{2\sqrt{S_\Lambda^2 + S_\Gamma^2 - S_\Omega^2}} \ln\left|\frac{-(S_\Lambda - S_\Omega)r - S_\Gamma - \sqrt{S_\Lambda^2 + S_\Gamma^2 - S_\Omega^2}}{-(S_\Lambda - S_\Omega)r - S_\Gamma + \sqrt{S_\Lambda^2 + S_\Gamma^2 - S_\Omega^2}}\right| \\ &- \frac{1}{2}\ln\left|(S_\Lambda + S_\Omega) - 2S_\Gamma r - (S_\Lambda - S_\Omega)r^2\right| = \ln p + \text{const.} \end{aligned}$$

For the case when the vorticity and deformation are balanced, i.e., $S_\Omega^2 = S_\Lambda^2 + S_\Gamma^2$, we get

$$(A.8) \qquad \frac{S_\Delta}{S_\Gamma + (S_\Lambda - S_\Omega)r} = \ln p + \text{const.}$$

When there is no divergence in the flow, Eqs. (A.6) and (A.7) become identical and the streamlines follow Eq. (13). Finally for the special case $S_\Delta^2 + S_\Omega^2 = S_\Lambda^2 + S_\Gamma^2$, and $u_c = v_c = 0$ the streamlines are given as

$$(A.9) \qquad (S_\Lambda - S_\Omega)(y - y_c) - (S_\Delta - S_\Gamma)(x - x_c) = \text{const.}$$

A.2. Computation of the exact concentration distribution with the back-trajectory equations. Under steady flow conditions, the streamline equations derived above also represent a set of trajectories of particles released at different starting points at the initial time. However, for finding the back trajectory of each particle advected, it is more convenient to describe the trajectories in terms of position and time rather than the phase relation (i.e., relation between x- and y-coordinates) as in the streamline. Eq. (A.4) can be separately written in terms of a set of linear differential equations with respect to time as,

$$(A.10a) \qquad \frac{dp}{dt} = (S_\Delta + S_\Gamma)p + (S_\Lambda - S_\Omega)q = u,$$

$$(A.10b) \qquad \frac{dq}{dt} = (S_\Lambda + S_\Omega)p + (S_\Delta - S_\Gamma)q = v \ .$$

These equations can be separated as second order linear differential equations for p or q:

$$\text{(A.11a)} \qquad \frac{d^2 p}{dt^2} - 2S_\Delta \frac{dp}{dt} - \left[(S_\Lambda^2 + S_\Gamma^2) - (S_\Omega^2 + S_\Delta^2) \right] p = 0,$$

or

$$\text{(A.11b)} \qquad \frac{d^2 q}{dt^2} - 2S_\Delta \frac{dq}{dt} - \left[(S_\Lambda^2 + S_\Gamma^2) - (S_\Omega^2 + S_\Delta^2) \right] q = 0$$

For $S_\Omega^2 < S_\Lambda^2 + S_\Gamma^2$, the general solutions for above equations are

$$\text{(A.12a)} \qquad p = c_1 e^{m_1 t} + c_2 e^{m_2 t},$$

$$\text{(A.12b)} \qquad q = c_3 e^{m_1 t} + c_4 e^{m_2 t},$$

with real parameters $m_1 = S_\Delta + \sqrt{(S_\Lambda^2 + S_\Gamma^2) - S_\Omega^2}$ and $m_2 = S_\Delta - \sqrt{(S_\Lambda^2 + S_\Gamma^2) - S_\Omega^2}$.

For $S_\Omega^2 > S_\Lambda^2 + S_\Gamma^2$, the general solutions are given as

$$\text{(A.13a)} \qquad p = c_1' e^{mt} \cos nt + c_2' e^{mt} \sin nt,$$

$$\text{(A.13b)} \qquad q = c_3' e^{mt} \cos nt + c_4' e^{mt} \sin nt,$$

where $m = S_\Delta$ and the imaginary part n is defined as $n = \sqrt{|S_\Omega^2 - S_\Lambda^2 - S_\Gamma^2|}$. For $S_\Omega^2 = S_\Lambda^2 + S_\Gamma^2$, we have a double root for the characteristic equation. Then, the general solutions are written as

$$\text{(A.14a)} \qquad p = c_1' e^{mt} + c_2' t e^{mt},$$

$$\text{(A.14b)} \qquad q = c_3' e^{mt} + c_4' t e^{mt}.$$

Because our objective is to find the back trajectory of a particle positioned at $(x_\alpha, y_\alpha) = (p_\alpha + x_c + x_p, q_\alpha + y_c + y_p)$ for a given time t_α, the coefficients of the general solutions should be found using some kind of initial conditions. One of the initial conditions is the particle position at the current time. Also, we can use the fact that we have steady flow (i.e., the wind speed at a position stays the same during the particle advection) as another condition. Then, one can establish a set of linear equations for the coefficients. For $S_\Omega^2 < S_\Lambda^2 + S_\Gamma^2$, we have

$$\text{(A.15a)} \qquad h_1 c_1 + h_2 c_2 = p_\alpha$$

$$\text{(A.15b)} \qquad h_1 c_3 + h_2 c_4 = q_\alpha$$

$$\text{(A.15c)} \qquad m_1 h_1 c_1 + m_2 h_2 c_2 = u_\alpha$$

$$\text{(A.15d)} \qquad m_1 h_1 c_3 + m_2 h_2 c_4 = v_\alpha$$

where $h_1 = e^{m_1 t_\alpha}$ and $h_2 = e^{m_2 t_\alpha}$. Parameters with the subscript α represent values at the position of interest at time t_α. The coefficients can be solved in terms of the parameters defined at the current position (x_α, y_α) and time t_α:

$$(\text{A.16}) \qquad \begin{pmatrix} c_1 \\ c_2 \\ c_3 \\ c_4 \end{pmatrix} = \frac{1}{h_1 h_2 (m_2 - m_1)} \begin{pmatrix} -h_2 u_\alpha + h_2 m_2 p_\alpha \\ h_1 u_\alpha - h_1 m_1 p_\alpha \\ -h_2 v_\alpha + h_2 m_2 q_\alpha \\ -h_1 v_\alpha + h_1 m_1 q_\alpha \end{pmatrix}$$

For $S_\Omega^2 > S_\Lambda^2 + S_\Gamma^2$, the linear set of equations is given as

$$(\text{A.17a}) \qquad a_1 c_1' + a_2 c_2' = p_\alpha / h$$

$$(\text{A.17b}) \qquad a_1 c_3' + a_2 c_4' = q_\alpha / h$$

$$(\text{A.17c}) \qquad a_3 c_1' + a_4 c_2' = u_\alpha / h$$

$$(\text{A.17d}) \qquad a_3 c_3' + a_4 c_4' = v_\alpha / h,$$

where $a_1 = \cos nt_\alpha$, $a_2 = \sin nt_\alpha$, $a_3 = m \cos nt_\alpha - n \sin nt_\alpha$, $a_4 = m \sin nt_\alpha + n \cos nt_\alpha$, and $h = e^{mt_\alpha}$. The coefficients are solved to give

$$(\text{A.18}) \qquad \begin{pmatrix} c_1' \\ c_2' \\ c_3' \\ c_4' \end{pmatrix} = -\frac{1}{nh} \begin{pmatrix} a_2 u_\alpha - a_4 p_\alpha \\ -a_1 u_\alpha + a_3 p_\alpha \\ a_2 v_\alpha - a_4 q_\alpha \\ -a_1 v_\alpha + a_3 q_\alpha \end{pmatrix},$$

where $a_2 a_3 - a_1 a_4 = -n$ is used. Similarly, for $S_\Omega^2 = S_\Lambda^2 + S_\Gamma^2$, a set of linear equations for the coefficients is given as

$$(\text{A.19a}) \qquad c_1' + t_\alpha c_2' = p_\alpha / h,$$

$$(\text{A.19b}) \qquad c_3' + t_\alpha c_4' = q_\alpha / h,$$

$$(\text{A.19c}) \qquad m c_1' + (1 + m t_\alpha) c_2' = u_\alpha / h,$$

$$(\text{A.19d}) \qquad m c_3' + (1 + m t_\alpha) c_4' = v_\alpha / h.$$

Then the coefficients are expressed as

$$(\text{A.20}) \qquad \begin{pmatrix} c_1' \\ c_2' \\ c_3' \\ c_4' \end{pmatrix} = -\frac{1}{h} \begin{pmatrix} t_\alpha u_\alpha - (1 + m t_\alpha) p_\alpha \\ -u_\alpha + m p_\alpha \\ t_\alpha v_\alpha - (1 + m t_\alpha) q_\alpha \\ -v_\alpha + m q_\alpha \end{pmatrix}.$$

Once the coefficients are known, we can relate the signal at the current time t_α, $q_i^e(x_\alpha, y_\alpha, t_\alpha)$, to the initial distribution of the trace species mixing ratio, q_i^o. For $S_\Omega^2 < S_\Lambda^2 + S_\Gamma^2$, we have

$$(\text{A.21}) \qquad q_i^e(x_\alpha, y_\alpha, t_\alpha) = q_i^o(c_1 + c_2 + x_c + x_p, c_3 + c_4 + y_c + y_p),$$

and, for $S_\Omega^2 \geq S_\Lambda^2 + S_\Gamma^2$, we get

$$(\text{A.22}) \qquad q_i^e(x_\alpha, y_\alpha, t_\alpha) = q_i^o(c_1' + x_c + x_p, c_3' + y_c + y_p).$$

A.3. Computation of the back-trajectory concentration values for a limited-domain simulation with a non-uniform boundary condition. In a numerical simulation of the advection process, the computational domain is usually limited and the boundary conditions must be specified. For the test cases used, the inflow boundary conditions are fixed throughout the simulation. When the back trajectory falls out of the modeling domain, q_i^e must be equal to the inflow boundary condition. We want to capture the fixed (constant with time, but spatially varying) value from the boundary for the analytical solutions. This issue is important because we want to evaluate numerical characteristics of advection schemes with the performance measures listed in Table 2. Most other numerical advection tests in the literature avoid this issue by specifying uniform boundary conditions with zero boundary flux conditions. However, for the hill superposition test we do not impose such a restrictive assumption. We are interested in the case when a back trajectory intersects one of the four boundary walls ($x=0$, $x = x_L$, $y=0$, $y = y_L$) for the rectangular modeling domain, where x_L and y_L represent the size of the computational domain. By finding out the time to reach the boundary and knowing the exact location, one can determine the current concentration value from the boundary conditions. Because steady state of the wind field is an essential requirement, we only study the non-divergent case ($m = 0$) here.

To determine the minimum possible time a back trajectory intersects the boundary, we solve the back trajectory equations for the travel time in terms of the coefficients for the general solutions. For $S_\Omega^2 < S_\Lambda^2 + S_\Gamma^2$ we have, when c_1 and c_3 are not zero,

$$(A.23a) \qquad \tau_x = \frac{1}{n} \ln \left[\frac{p \pm \sqrt{p^2 - 4c_1 c_2}}{2c_1} \right],$$

$$(A.23b) \qquad \tau_y = \frac{1}{n} \ln \left[\frac{q \pm \sqrt{q^2 - 4c_3 c_4}}{2c_3} \right],$$

where τ_x and τ_y are the travel time scales for x- and y-directions, respectively. When $c_1 = c_3 = 0$,

$$(A.23'a) \qquad \tau_x = \frac{1}{n} \ln \left[c_2/p \right],$$

$$(A.24'a) \qquad \tau_y = \frac{1}{n} \ln \left[c_4/q \right].$$

Similarly, for $S_\Omega^2 > S_\Lambda^2 + S_\Gamma^2$, we get

$$(A.24a) \qquad \tau_x = \frac{1}{n} \left[\arcsin \left(\frac{p}{\lambda_x} \right) - \arcsin \left(\frac{c_1}{\lambda_x} \right) \right]$$

$$(A.24b) \qquad \tau_y = \frac{1}{n} \left[\arcsin \left(\frac{q}{\lambda_y} \right) - \arcsin \left(\frac{c_3}{\lambda_y} \right) \right]$$

where $\lambda_x = \pm\sqrt{c_1^2 + c_2^2}$ and $\lambda_y = \pm\sqrt{c_3^2 + c_4^2}$. For the case $S_\Omega^2 = S_\Lambda^2 + S_\Gamma^2$ (when c_2' and c_4' are not zero), we obtain

$$(A.25a) \qquad \tau_x = \frac{p - c_1'}{c_2'}$$

$$(A.25b) \qquad \tau_y = \frac{q - c_3'}{c_4'}.$$

When $c_2' = c_4' = 0$, the travel time is indeterminate.

Once we obtain the coefficients for the trajectories at the given cell for the initial time $(t = 0)$, we can compute the travel time to a boundary using the above equations. Given the magnitude of the elementary flows, the minimum absolute travel time can be found for the four cases possible: (1) west side boundary $(x = 0,\ p = -x_c - x_p)$, (2) east side boundary $(x = x_L,\ p = x_L - x_c - x_p)$, (3) south side boundary $(y = 0,\ q = -y_c - x_q)$, and (4) north side boundary $(y = y_L,\ q = y_L - y_c - x_q)$. When the travel time to a boundary cannot be determined, or the computed travel time is positive, the back trajectory does not intersect the boundary and thus is discarded. When the current time t_α is larger than the minimum absolute travel time to reach a boundary, we can relate the signal at the given position $q_i^e(x_\alpha, y_\alpha, t_\alpha)$, to the boundary value.

B. Cross-term error analysis of two dimensional numerical advection with the time splitting method. To understand why the cross-term effects are most apparent with the stretching deformation component of the flow, we use a two dimensional Taylor series expansion of concentration $c(x, y, t)$ about the reference point (x_c, y_c) as suggested by Aijun Xiu (personal communication):

$$(B.1) \qquad \begin{aligned} c_{lm}^{n+1} = c_{lm}^n &- u\Delta t \left.\frac{\partial c}{\partial x}\right|_{lm}^n - v\Delta t \left.\frac{\partial c}{\partial y}\right|_{lm}^n + \frac{u^2\Delta t^2}{2}\left.\frac{\partial^2 c}{\partial x^2}\right|_{lm}^n \\ &+ \frac{v^2\Delta t^2}{2}\left.\frac{\partial^2 c}{\partial y^2}\right|_{lm}^n + \frac{uv\Delta t^2}{2}\left.\frac{\partial^2 c}{\partial x\partial y}\right|_{lm}^n + \text{HOT}, \end{aligned}$$

where index n stands for the current time step, l and m are indices for the reference position, and HOT represents higher order terms. Eq. (B.1) can be written in a symbolic operator form, neglecting HOT:

$$(B.2) \qquad c_{lm}^{n+1} = (I - A_x - B_y + C_{xy})c_{lm}^n,$$

where I is the unit matrix, and differential operators A_x, B_y and C_{xy} are defined as

$$A_x = u\Delta t \left.\frac{\partial}{\partial x}\right|_{lm}^n - \frac{u^2\Delta t^2}{2}\left.\frac{\partial^2}{\partial x^2}\right|_{lm}^n,$$

$$B_y = \nu \Delta t \left. \frac{\partial}{\partial y} \right|_{lm}^{n} - \frac{\nu^2 \Delta t^2}{2} \left. \frac{\partial^2}{\partial y^2} \right|_{lm}^{n},$$

$$C_{xy} = \frac{u\nu \Delta t^2}{2} \left. \frac{\partial^2}{\partial x \partial y} \right|_{lm}^{n}.$$

A_x and B_y are proportional to the net fluxes in $x-$ and $y-$directions, respectively. C_{xy} is the operator responsible for the cross-term. The time splitting application of a second-order one-dimensional numerical scheme can be expressed as

$$\text{(B.3a)} \qquad\qquad c_{lm}^{*} = c_{lm}^{n} - A_x c_{lm}^{n},$$

$$\text{(B.3b)} \qquad\qquad c_{lm}^{n+1} = c_{lm}^{*} - B_y c_{lm}^{*} .$$

The above equations can be rewritten in the following form

$$\text{(B.4)} \qquad\qquad c_{lm}^{n+1} = (I - A_x - B_y + A_x B_y)c_{lm}^{n} .$$

The so-called cross-term error for the time splitting method is the difference between Eqs. (B.2) and (B.4);

$$A_x B_y - C_{xy} =$$

$$\text{(B.5)} \quad -\frac{u\nu^2 \Delta t^3}{2} \left. \frac{\partial}{\partial x} \frac{\partial^2}{\partial y^2} \right|_{lm}^{n} - \frac{u^2 \nu \Delta t^3}{2} \left. \frac{\partial}{\partial y} \frac{\partial^2}{\partial x^2} \right|_{lm}^{n} + \frac{u^2 \nu^2 \Delta t^4}{4} \left. \frac{\partial^4}{\partial x^2 \partial y^2} \right|_{lm}^{n}.$$

When we neglect the higher order term $O(\Delta t^4)$ for a moment, and substitute u and ν with Eqs. (10.a) and (10.b), Eq. (B.5) can be rewritten as

$$\text{(B.6)} \quad \begin{aligned} A_x B_y - C_{xy} = \quad & -\frac{u\nu \Delta t^3}{2} \left\{ S_\Lambda \left(\delta_x \left. \frac{\partial^2}{\partial y^2} \right|_{lm}^{n} + \delta_y \left. \frac{\partial^2}{\partial x^2} \right|_{lm}^{n} \right) \right. \\ & + S_\Omega \left(\delta_x \left. \frac{\partial^2}{\partial y^2} \right|_{lm}^{n} - \delta_y \left. \frac{\partial^2}{\partial x^2} \right|_{lm}^{n} \right) \\ & \left. + S_\Gamma \left(-\delta_y \left. \frac{\partial^2}{\partial x \partial y} \right|_{lm}^{n} + \delta_x \left. \frac{\partial^2}{\partial x \partial y} \right|_{lm}^{n} \right) \right\} + O(\Delta t^4), \end{aligned}$$

where the standard difference notation $\delta_x f = f(x + \Delta x/2) - f(x - \Delta x/2)$, and $\delta_y f = f(y + \Delta y/2) - f(y - \Delta y/2)$ are used. From Eq. (B.6), one can expect that the operator related with the stretching deformation is mostly responsible for the cross-term effects while the operators involved with the shearing deformation and rotation are semi-linear in each direction when the finite differencing approximation is applied.

REFERENCES

BYUN, D.W., 1999a: Dynamically consistent formulations in meteorological and air quality models for multi-scale atmospheric applications: Part I. Governing Equations in Generalized Coordinate System. *J. Atmos. Sci.*, **56**, 3789–3807.

BYUN, D.W., 1999b: Dynamically consistent formulations in meteorological and air quality models for multi-scale atmospheric applications: Part II. Mass conservation issues. *J. Atmos. Sci.*, **56**, 3808–3820.

BYUN, D.W. AND J.K.S. CHING, 1999: *Science Algorithms of the EPA Models-3 Community Multiscale Air Quality (CMAQ) Modeling System.* Editors. National Exposure Research Laboratory, U.S. Environmental Protection Agency, Research Triangle Park, NC. (Available from National Exposure Research Laboratory, U.S. Environmental Protection Agency, Research Triangle Park, NC 27711.)

CHANG J.S., S. JIN, Y. LI, M. BEAUHARNOIS, C.-H. LU, H.-C. HUANG, S. TANRIKULU, AND J. DAMASSA, 1997: *The SARMAP air quality model. Final Report*, SJVAQS/AUSPEX Regional Modeling Adaptation Project, p. 53. (Available from California Air Resources Board, 2020 L Street, Sacramento, California 95814.)

COLELLA, P. AND P.R. WOODWARD, 1984: The piecewise parabolic method (PPM) for gas-dynamical simulations. *J. Comp. Phys.*, **54**, 174–201.

KITADA, T., 1987: Effect of non-zero divergence wind fields on atmospheric transport calculations. *Atmos. Environ.*, **21**, 785–788.

ODMAN, M.T., 1998: *Research on Numerical Transport Algorithms for Air Quality Simulation Models.* EPA Report. EPA/660/R-97/142. (Available from National Exposure Research Laboratory, U.S. Environmental Protection Agency, Research Triangle Park, NC 27711.)

SCHEFFE, R.D AND R.E. MORRIS, 1993: A review of the development and application of the Urban Airshed Model. *Atmos. Env.*, **27**B, 23–39.

STANIFORTH, A.J., J. COTE, AND J. PUDYKIEWCZ, 1987: Comments on Smolarkiewicz's deformation flow. *Mon. Wea. Rev.*, **115**, 894–900.

WILLIAMSON, D.L. AND P.J. RASCH, 1989: Two-dimensional semi-Lagrangian transport with shape-preserving interpolation. *Mon. Wea. Rev.*, **117**, 102–117.

COUPLED TRANSPORT-CHEMISTRY COMPUTATIONS IN 4D-VAR DATA ASSIMILATION FOR AIR POLLUTION MODELS

DACIAN DAESCU* AND GREGORY R. CARMICHAEL[†]

Abstract. The 4D variational data assimilation for large scale air quality models is a very intensive computational process. In this paper we analyze a coupled numerical treatment of the transport-chemistry operators in the 4D-var context which may reduce the CPU time and the memory storage requirements of the assimilation process. The integration of the forward model is based on a 2-stage Rosenbrock method with approximate Jacobian (W-method). The advantage over operator splitting is that the stiff transients in the chemical system are eliminated, which allows for large step sizes and reduces the storage requirements of the forward integration. Implementation of the adjoint code is done by combining automatic differentiation tools (TAMC) with symbolic preprocessing (KPP), which leads to exact computation of the gradients and allows flexibility for the chemical model. Numerical results are presented for a 1-D air pollution model.

Key words. Data assimilation, adjoint modeling, stiff equations, operator splitting, W-method.

AMS(MOS) subject classifications. 65K10, 65L06, 49J15.

1. Introduction. The increasing amount of data received from satellite observations and the development of powerful computing machines make feasible the data assimilation for atmospheric models. Data assimilation uses observational data and the forecast model to provide an optimal analysis state of the atmosphere. A good guide through the theory of parameter estimation and data assimilation techniques can be found in [14, 28]. The variational methods (3D- var, 4D- var) have been successfully applied in meteorological data assimilation and show promising results for atmospheric chemistry models. The 4D- var method was used by *Fisher* [7] for a photochemical box model with trajectories (Lagrangian model). *Khattatov et al.* [15] used a similar model to implement both the variational and a Kalman filter method. A box model for tropospheric chemistry was used by *Elbern et al.* [6] for variational data assimilation, then extended to an Eulerian transport-chemistry model [5] in a fully 4D- var data assimilation experiment. In the 4D- var data assimilation approach a cost function is defined as the weighted least squares distance between model predictions and observations over the assimilation window. A minimization algorithm is then used to find the set of control variables that minimizes the cost function. Most of the powerful minimization methods require the evalua-

---

*Program in Applied Mathematical and Computational Sciences, The University of Iowa, 14 MLH, Iowa City, IA 52242.

[†]Center for Global and Regional Environmental Research and The Department of Chemical and Biochemical Engineering, The University of Iowa, Iowa City, IA 52242.

tion of the gradient of the cost function which for large scale models is a very demanding computational task. Using the adjoint method, the gradient of the cost function can be computed at the expense of few function evaluations, making the optimization process very efficient. The theory of the adjoint equations associated with linear models is presented in [18], whereas for non-linear problems we will refer to [19]. A derivation of the adjoint model for the continuous and discrete case is presented in [7, 27]. When chemical transformations are considered in an atmospheric model, the complexity of the implementation and the computational cost of the adjoint model are highly increased. Two aspects must be emphasized: the non-linearity and the stiffness introduced in the model by the chemical reactions. For non-linear problems the adjoint equations depend on the forward trajectory (obtained by direct integration of the model) [19]. In order to perform the adjoint computations (also called backward or reverse integration), the forward trajectory must be available in reverse order, such that a large amount of memory must be allocated for storage during the forward run. Since the variational method relies on the linearization of the forecast model, the non-linearity introduced by the chemistry has direct impact on the interval time length and the qualitative aspects of the assimilation process. The existence of multiple local minima of the cost function may lead to ambiguous assimilation results. The stiff differential equations generated by the chemical reactions must be integrated with a highly stable numerical method. For consistency between the computed cost function and its gradient in the minimization process, the adjoint model is usually generated from the numerical method used to integrate the forward model. Explicit methods are easy to implement, but are not practical since they may take prohibitive small step size and the state of the model after all these steps must be stored before the backward integration can be started. Linear-implicit and implicit methods [12] may easily integrate the stiff ODE systems, but their implementation both in forward and reverse mode is more challenging. From the family of linear-implicit solvers, Rosenbrock methods [12] have been proved to be reliable chemistry solvers [24, 29] due to their outstanding stability properties and conservation of the linear invariants of the system. In [3] was shown that the adjoint computations for Rosenbrock methods can be efficiently implemented.

The numerical method used at the operator level has also direct implications in the efficiency of the assimilation process. In this paper we present the implementation of the adjoint of a 2-stage W-method and discuss the potential advantages over the operator splitting in the data assimilation context. In the W-method splitting is done at the linear algebra level such that processes such as advection, diffusion and chemical reactions are coupled at the operator level. Numerical results obtained with a 1-D air pollution test model show that the W-method may significantly reduce the CPU time of the assimilation process. The paper is organized as follows: in Section 2 we present the variational data assimilation problem for a transport-chemistry model; in Section 3 we present the adjoint computations and implementation for a 2-stage W-method; the test model and the numerical results are presented in Section 4; conclusions and further work are discussed in Section 5.

2. 4D variational data assimilation for a transport-chemistry model.

The time evolution of the concentration vector $\mathbf{c}(t, \mathbf{x}) \in \mathbb{R}^s$, $\mathbf{c} = (c_1, \dots, c_s)^T$ of the chemical species considered in a transport-chemistry model is determined by processes such as advection and diffusion, chemical transformations, emissions, and depositions. The mass balance can be expressed as the PDE system:

$$(2.1) \quad \frac{\partial}{\partial t} c_i = -\nabla \cdot (\mathbf{u} c_i) + \nabla \cdot (\mathbf{K} \cdot \nabla c_i) + f_i(\mathbf{c}) + S_i, \quad i = 1, \dots, s.$$

In (2.1) $\mathbf{u}(t, \mathbf{x})$ is the wind field, $\mathbf{K}(t, \mathbf{x})$ is the diffusion tensor, and the source/sink terms are represented by $S_i(t, \mathbf{x})$. The chemical production and loss processes are modeled by the nonlinear stiff functions $f_i(\mathbf{c}) = P_i(\mathbf{c}) - L_i(\mathbf{c}) c_i$. Depositions may be included in the boundary conditions at earth's surface. With appropriate boundary values and the initial condition $\mathbf{c}(t_0) = \mathbf{c}_0$, system (2.1) represents the continuum forward (forecast, direct) model. For practical implementation, a discrete version of (2.1) must be considered. After semi-discretization of the model (2.1) on a spatial grid (N_x, N_y, N_z), the resulting ODE system can be written:

$$(2.2) \quad \begin{cases} \dfrac{d\mathbf{c}}{dt} = F(\mathbf{c}) = F_A(\mathbf{c}) + F_D(\mathbf{c}) + F_R(\mathbf{c}) + S \\ \mathbf{c}(t_0) = \mathbf{c}_0 \end{cases}$$

where F_A represents the advection and horizontal diffusion terms, F_D represents the vertical diffusion, and the chemical transformations are given by F_R. The dimension of the problem (2.2) is $n = s \times N_x \times N_y \times N_z$. We assume that a previous analysis provides a "background" estimate $\mathbf{c}^b$ of $\mathbf{c}_0$ and observations (measurements) $\mathbf{c}_k^o \in \mathbb{R}^{n_k}$ are taken at time $t_k, k = 1, 2, \dots, m$. Mapping from the state space into the observations space at moment t_k is done by a state independent linear operator $\mathbf{H}_k : \mathbb{R}^n \to \mathbb{R}^{n_k}$. The 4D-var data assimilation finds the set of states $\mathbf{c}_0, \mathbf{c}_1, \dots, \mathbf{c}_m$ that minimizes the cost function $\mathcal{F}$ defined as:

$$(2.3) \quad \begin{aligned} \mathcal{F} &= \frac{1}{2}(\mathbf{c}_0 - \mathbf{c}^b)^T \mathbf{B}^{-1}(\mathbf{c}_0 - \mathbf{c}^b) \\ &+ \frac{1}{2} \sum_{k=1}^{m} (\mathbf{H}_k \mathbf{c}_k - \mathbf{c}_k^o)^T \mathbf{R}_k^{-1}(\mathbf{H}_k \mathbf{c}_k - \mathbf{c}_k^o). \end{aligned}$$

In the deterministic least-squares approach, the matrices $\mathbf{B} \in \mathbb{R}^{n \times n}$, $\mathbf{R_k} \in \mathbb{R}^{n_k \times n_k}$ are diagonal and positive definite, without any probabilistic meaning, and are used to provide appropiate weights for curve-fitting of the model trajectory to the measurements. In the stochastic approach, $\mathbf{B}$ is the covariance matrix of the errors in the background estimate and $\mathbf{R_k}$ are the covariance matrices of the errors in measurements and model representativeness [17]. Under several assumptions [2], it can be shown [14] that the minima of $\mathcal{F}$ provides the maximum likelihood estimate of the state (maximize the conditional density $p(\mathbf{c}_0, \mathbf{c}_1, \ldots, \mathbf{c}_m | \mathbf{c}_1^o, \mathbf{c}_2^o, \ldots, \mathbf{c}_m^o)$). For comprehensive atmospheric models, the dimension n of the state vector can be easily of order 10^6 such that minimization of the functional $\mathcal{F}$ is an intensive computational process. The state $\mathbf{c}_k = \mathbf{c}(t_k, \mathbf{x})$ is determined by the values of various parameters. In atmospheric data assimilation the set of control variables may include the boundary values, initial state of the model, emissions and deposition rates. For the purpose of this paper, we consider *only* the initial state $\mathbf{c}_0$ as the set of control parameters. In this context, we can view $\mathbf{c}$ as a function of the initial condition, $\mathbf{c} = \mathbf{c}(t, \mathbf{x}, \mathbf{c}_0)$ such that $\mathcal{F} : \mathbb{R}^n \to \mathbb{R}, \mathcal{F} = \mathcal{F}(\mathbf{c}_0)$. Since $\mathbf{B}$ and $\mathbf{R}_k$ are symmetric, the gradient of the cost function is:

$$(2.4) \quad \nabla_{\mathbf{c}_0} \mathcal{F}(\mathbf{c}_0) = \mathbf{B}^{-1}(\mathbf{c}_0 - \mathbf{c}^b) + \sum_{k=1}^{m} \left(\frac{\partial \mathbf{c}_k}{\partial \mathbf{c}_0} \right)^T \mathbf{H}_k^T \mathbf{R}_k^{-1} (\mathbf{H}_k \mathbf{c}_k - \mathbf{c}_k^o).$$

An efficient way to compute the gradient is provided by the adjoint algorithm (backward, reverse integration):

(2.5)   ADJOINT ALGORITHM

Step 1. Initialize $\nabla \mathcal{F} = 0$
Step 2. for $k = m, 1, -1$ do

$$\nabla \mathcal{F} = \left(\frac{\partial \mathbf{c}_k}{\partial \mathbf{c}_{k-1}} \right)^T [\mathbf{H}_k^T \mathbf{R}_k^{-1} (\mathbf{H}_k \mathbf{c}_k - \mathbf{c}_k^o) + \nabla \mathcal{F}]$$

Step 3. $\nabla \mathcal{F} = \mathbf{B}^{-1}(\mathbf{c}_0 - \mathbf{c}^b) + \nabla \mathcal{F}$.

The advantage of the adjoint algorithm as compared to forward methods is that it avoids matrix multiplication, such that the *matrix * vector* products can be computed directly at *Step 2*. However, in order to start the adjoint computations the forward trajectory must be available in reverse order such that a large amount of memory must be allocated for storage during the forward run. Intensive research is focused on developing optimal storage strategies and checkpointing schemes [11]. If the time integration of the problem (2.2) may be performed with large step size, the number of intermediate states of the forward run may be considerably reduced. As a consequence, the storage requirements are reduced. The performance of a second order Rosenbrock method (ROS2) in the context of various types of operator splitting and implemented as a W-method was analyzed in [1, 29]. For the 3D model LOTOS the second order W-method using step size of 15 to 20 min. showed good results (in forward mode) both in the qualitative and quantitative aspects. In the next section we show that the adjoint code may be efficiently implemented and discuss the potential advantage over the operator splitting techniques.

3. The adjoint of a second order W-method. Time integration of problem (2.2) in the interval $[t_{k-1}, t_k]$ generates a sequence of intermediate states $c_{k-1} \to c_k^1 \to \ldots \to c_k^q \to c_k$. In order to perform the backward integration in (2.5) from t_k to t_{k-1} this intermediate trajectory must be available in reverse order, such that it needs to be stored during the forward run. The numerical method used in the forward integration has then direct impact on the efficiency of the adjoint code. To illustrate the process, is enough to analyze the forward/backward computations related with one time step integration, from c_0 to $c_h = c(t_0 + h)$.

A popular way to integrate the forward model is the operator splitting, where different processes such as advection, diffusion, and chemical reactions are integrated separately with different numerical methods. If second order Strang splitting [26] is used, the solution c_h is expressed as:

$$(3.1) \quad c_h = \overline{F}_A \left(t_{\frac{h}{2}}, \frac{h}{2} \right) \overline{F}_D \left(t_{\frac{h}{2}}, \frac{h}{2} \right) \overline{F}_R(t_0, h) \overline{F}_D \left(t_0, \frac{h}{2} \right) \overline{F}_A \left(t_0, \frac{h}{2} \right) c_0$$

where the operators $\overline{F}$ are given by the numerical method used to solve each process. During the integration source terms S may be included in F_D or in F_R. The step size h is bounded by stability and accuracy constraints, and in practice usually ranges from 10 min. to 1 hour. In the context of the data assimilation, operator splitting has some drawbacks. By introducing intermediate states (after integration of each process) the forward trajectory is considerably increased. The stiff transients introduced in the chemical model by the operator splitting affect the performance of the chemical integration, such that even with a stiff solver a of small step size must be taken at the beginning of the chemistry integration. This fragmentation of the forward trajectory is reflected in the increased storage requirements for the backward integration such that in general at least a 2-level checkpointing scheme must be considered [5]. Splitting at the operator level may be avoided by implementing methods with approximate Jacobian (W- methods [12]).

We consider a 2-stage linear-implicit W-method applied to problem (2.2) which solution c_h can be written as :

$$(3.2) \qquad c_h = c_0 + m_1 k_1 + m_2 k_2$$

where the intermediate stages k_1 and k_2 are obtained by solving the linear systems:

$$(3.3) \qquad (A - \frac{1}{\gamma h} I) k_1 = F(c_0)$$

$$(3.4) \qquad (\mathbf{A} - \frac{1}{\gamma h}\mathbf{I})\mathbf{k}_2 = F(\mathbf{c}_0 + \alpha\mathbf{k}_1) + \frac{\beta}{h}\mathbf{k}_1$$

with $\mathbf{A}$ an approximation of the Jacobian matrix $\mathbf{J}_0 \equiv F'(\mathbf{c}_0) \equiv \mathbf{J}_A + \mathbf{J}_D + \mathbf{J}_R$ and $\mathbf{I}$ the $n \times n$ identity matrix. The method (3.2-3.4) is second order accurate and is L-stable if $\mathbf{A} \equiv \mathbf{J}_0$ and $\gamma = 1 \pm 1/\sqrt{2}$, $\alpha = -1/\gamma$, $\beta = 2/\gamma$, $m_1 = -3/(2\gamma)$, $m_2 = -1/(2\gamma)$. Stability and positivity properties with this choice of parameters are analyzed in [29]. A matrix factorization

$$(3.5) \qquad (\mathbf{A} - \frac{1}{\gamma h}\mathbf{I}) = (\mathbf{I} - \gamma h \mathbf{J}_D)(\mathbf{J}_R - \frac{1}{\gamma h}\mathbf{I})$$

as used in [1, 29] can be obtained by choosing

$$(3.6) \qquad \mathbf{A} = \mathbf{J}_D + (\mathbf{I} - \gamma h \mathbf{J}_D)\mathbf{J}_R = \mathbf{J}_D + \mathbf{J}_R - \gamma h \mathbf{J}_D \mathbf{J}_R.$$

From (3.5) it results that in (3.2)-(3.4) the vertical diffusion and chemical reactions are integrated linearly implicitly while the advection is treated explicitly. As memory usage is one of the main concerns, a direct implementation of method (3.2)-(3.4) has a significant drawback since requires the storage of the $n \times n$ matrix $\mathbf{A}$. To avoid this problem, in our implementation and numerical experiments presented in Section 4, we *recompute* $\mathbf{A}$ during the second stage (3.4). Solving the systems (3.3) and (3.4) decouples in the first step over the horizontal dimension and chemical species and in the second step over the spatial grid such that the advantages of operator splitting are preserved. The adjoint algorithm and implementation corresponding to (3.2)-(3.4) with $\mathbf{A} \equiv \mathbf{J}_0$ in a box model are described in detail in [3]. Here we will focus on the role played by (3.5) in the adjoint computations. Implementation of the adjoint algorithm (2.5) requires computation of products of the form $\left(\frac{\partial \mathbf{c}_k}{\partial \mathbf{c}_{k-1}}\right)^T \mathbf{s}$, with $\mathbf{s}$ an arbitrary seed vector. Corresponding to (3.2), we have:

$$(3.7) \qquad \left(\frac{\partial \mathbf{c}_h}{\partial \mathbf{c}_0}\right)^T \mathbf{s} = \mathbf{s} + \left(\frac{\partial}{\partial \mathbf{c}_0}(m_1\mathbf{k}_1 + m_2\mathbf{k}_2)\right)^T \mathbf{s}.$$

To symplify the notation we will use the symbol $'$ to denote the operator $\frac{\partial}{\partial \mathbf{c}_0}$. To evaluate the righthand side term in (3.7) first we multiply (3.3) by m_1 and (3.4) by m_2 then add the results:

$$\left(\mathbf{A} - \frac{1}{\gamma h}\mathbf{I}\right)(m_1\mathbf{k}_1 + m_2\mathbf{k}_2) = m_1 F(\mathbf{c}_0) + m_2 F(\mathbf{c}_0 + \alpha\mathbf{k}_1) + \frac{m_2\beta}{h}\mathbf{k}_1$$

next we differentiate with respect to $\mathbf{c}_0$:

$$(3.8) \qquad \begin{aligned} \mathbf{A}'(m_1\mathbf{k}_1 + m_2\mathbf{k}_2) &+ \left(\mathbf{A} - \frac{1}{\gamma h}\mathbf{I}\right)(m_1\mathbf{k}_1 + m_2\mathbf{k}_2)' \\ &= m_1\mathbf{J}_0 + m_2\mathbf{J}_1(\mathbf{I} + \alpha\mathbf{k}_1') + \frac{m_2\beta}{h}\mathbf{k}_1' \end{aligned}$$

where $\mathbf{J}_1 \equiv F'(\mathbf{c}_0 + \alpha\mathbf{k}_1)$. It is important to notice that in the first term on the lefthand side of relation (3.8) only the *numerical value* of $m_1\mathbf{k}_1 + m_2\mathbf{k}_2$ is required. Assuming that $\mathbf{c}_h$ was stored during the forward run, we simply set $m_1\mathbf{k}_1 + m_2\mathbf{k}_2 = \mathbf{c}_h - \mathbf{c}_0$. In particular, the numerical value of $\mathbf{k}_2$ is not required, such that there is no need to recompute it via (3.4) during the backward integration. This is a considerable saving in CPU time. Observe that the value of $\mathbf{k}_1$ must be recomputed since is used in the righthand side of (3.8) to evaluate $\mathbf{J}_1$ (see also (3.10) below). After arranging the terms and taking the transpose in (3.8), it results:

$$(m_1\mathbf{k}_1 + m_2\mathbf{k}_2)'^T =$$

$$(3.9) \quad \left(m_1\mathbf{J}_0 + m_2\mathbf{J}_1 + (\alpha m_2\mathbf{J}_1 + \frac{m_2\beta}{h}\,\mathbf{I})\,\mathbf{k}_1' - \mathbf{A}'(\mathbf{c}_h - \mathbf{c}_0)\right)^T \left(\mathbf{A} - \frac{1}{\gamma h}\mathbf{I}\right)^{-1\,T}.$$

The explicit formulae can be completed by differentiating (3.3) to obtain

$$(3.10) \quad \mathbf{k}_1' = \left(\mathbf{A} - \frac{1}{\gamma h}\mathbf{I}\right)^{-1}(\mathbf{J}_0 - \mathbf{A}'\mathbf{k}_1).$$

The implementation of the adjoint equations (3.7), (3.9), (3.10) requires the solution $\mathbf{v}$ of linear systems of the form:

$$(3.11) \quad (\mathbf{J}_R - \frac{1}{\gamma h}\mathbf{I})^T(\mathbf{I} - \gamma h\mathbf{J}_D)^T\mathbf{v} = \mathbf{w}$$

with given $\mathbf{w}$, computations of products of the form

$$(3.12) \quad \mathbf{J}^T\mathbf{s} = \mathbf{J}_A^T\mathbf{s} + \mathbf{J}_D^T\mathbf{s} + \mathbf{J}_R^T\mathbf{s}$$

with $\mathbf{s}$ a seed vector, and evaluation of the terms

$$(3.13) \quad (\mathbf{A}'\mathbf{v})^T\mathbf{w}$$

with $\mathbf{v}$ and $\mathbf{w}$ given vectors. If a central difference formula is used to discretize the vertical diffusion, $\mathbf{J}_D$ is a block tridiagonal state independent matrix such that the adjoint diffusion computations are relatively easy to implement. The adjoint advection operator has a sparse regular structure given by the space discretization scheme, but may be state dependent (for example if a flux-limiting is used). The computational load of (3.11)-(3.13) is given by the chemistry computations. We used an enhanced version of KPP [4] to perform sparse adjoint chemistry computations using the sparse structure of $\mathbf{J}_R$, and we generated the adjoint of the advection terms with the adjoint compiler TAMC [9, 10]. Implementation of (3.13) is the most CPU time consuming process since requires second order derivatives. Replacing (3.6) in (3.13) and using the fact that $\mathbf{J}_D$ does not depend on $\mathbf{c}_0$ we have:

$$(3.14) \quad (\mathbf{A}'\mathbf{v})^T\mathbf{w} = ((\mathbf{I} - \gamma h\mathbf{J}_D)\mathbf{J}_R'\mathbf{v})^T\mathbf{w} = (\mathbf{J}_R'\mathbf{v})^T(\mathbf{I} - \gamma h\mathbf{J}_D)^T\mathbf{w}.$$

In (3.14) first we compute $\mathbf{z} = (\mathbf{I} - \gamma h \mathbf{J}_D)^T \mathbf{w}$ then we use the property

$$(3.15) \qquad\qquad (\mathbf{J}'_R \mathbf{v})^T \mathbf{z} = (\mathbf{J}^T_R \mathbf{z})' \mathbf{v}.$$

Property (3.15) is a consequence of the symmetry of $(\mathbf{J}^T_R \mathbf{z})'$ for any vector $\mathbf{z}$ and is proved in [3]. In (3.15) we have then to evaluate a Jacobian vector product. We use KPP for sparse generation of $\mathbf{J}^T_R \mathbf{z}$ then apply *forward automatic differentiation* to compute $(\mathbf{J}^T_R \mathbf{z})' \mathbf{v}$. This process (forward over reverse) leads to an efficient implementation since full advantage is taken of the sparsity of $\mathbf{J}_R$.

4. Numerical experiments. The adjoint computations described in the previous section were implemented using symbolic preprocessing software (KPP) and automatic differentiation tools (TAMC). This allows flexibility for the chemical model and leads to exact computations of the gradients. The numerical experiments we present in this section were performed with a test 1-D horizontal model, such that F_D should be interpreted as the horizontal diffusion. We consider a spatial domain $[0, 500]$ Km with a uniform grid $\Delta x = 5$ Km. The chemical mechanism is CBM-IV [8] with 32 variable species and one fixed (H_2O). A highly polluted region is considered between 100–200 Km. The advection is discretized using a limited upwind flux interpolation scheme as presented in [13] and the diffusion operator using the central differences formula. The wind field is taken constant $\mathbf{u} = 10$ Km/h and the diffusion coefficient is $\mathbf{K} = 0.001$ Km^2/sec. Time dependent Dirichlet boundary values are specified at x=0, and at the right boundary we impose $\frac{\partial \mathbf{c}}{\partial x} = 0$. In order to check the stability of the W-method we artificially increased the diffusion by a factor 10, 50 and 100. The initial state and the state after a 6 hours simulation are shown for O_3 in Figure 1. With a 15 min step size no instabilities were noticed.

The data assimilation is performed in the twin experiments framework using both operator splitting and the W-method. The time splitting step is 15 min, whereas for the W-method we restricted the maximum step size to 15 min. For 10 species in the model first guess initial conditions $\mathbf{c}_0^p$ were taken of the form $\mathbf{c}_0^p = \mathbf{c}_0(1 + r(x)\alpha)$, with $r(x) \in (0, 1)$ random numbers and α as shown in Table 1.

The optimization routine used is the limited memory L-BFGS [16], and for the optimization process we imposed a reduction in the cost function $\mathcal{F}_{opt}/\mathcal{F}_{init} < 10^{-4}$ or a maximum of 100 iterations. A 6 hours assimilation interval $[4:30, 10:30]$LT is considered with "measurements" provided only for O_3 and NO_2 as follows: in Run 1 every 15 min, at all grid points; in Run 2 every 15 min, 10 grid points apart (50 Km); in Run 3 every hour, 10 grid points apart. The adjoint code for the operator splitting method was implemented with a 2-level checkpointing scheme. A first forward run was used to store the trajectory after each time split step. The number of chemistry steps taken (ROS2, exact Jacobian) was in the range 5–12. A second forward run was used to store the states inside the splitting step

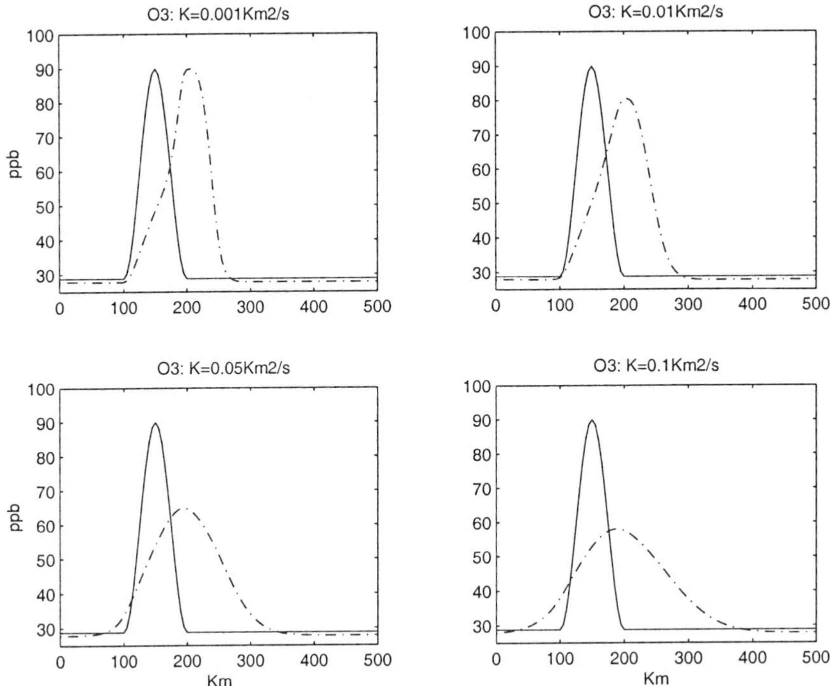

FIG. 1. *In order to check the stability of the W-method, the diffusion coefficient* **K**=*0.001Km²/s is increased by a factor of 10, 50, 100;* **u**=*10 Km/h, constant; with solid line the initial O_3 state (LT=4:30), with dash-dot line the O_3 state at LT=10:30.*

TABLE 1

Qualitative analysis of the data assimilation using the W-method. "Measurements" were provided for O_3 and NO_2 only.

Chemical Species	α	$\mathcal{F}^i_{opt}/\mathcal{F}^i_{init}$		
		Run 1	Run 2	Run 3
O_3	0.3	1.6×10^{-4}	2.2×10^{-3}	3.1×10^{-3}
OH	2.0	1.0×10^{-2}	4.7×10^{-2}	5.2×10^{-2}
NO	2.0	1.7×10^{-3}	1.0×10^{-1}	1.2×10^{-1}
NO_2	1.0	9.3×10^{-5}	7.6×10^{-2}	8.6×10^{-2}
NO_3	1.0	7.0×10^{-3}	3.3×10^{-2}	3.6×10^{-2}
HO_2	1.0	4.9×10^{-1}	6.4×10^{-1}	6.4×10^{-1}
N_2O_5	2.0	3.1×10^{-3}	5.1×10^{-2}	6.0×10^{-2}
HNO_3	0.3	4.7×10^{-1}	5.0×10^{-1}	5.0×10^{-1}
$HONO$	1.0	2.6×10^{-1}	3.0×10^{-1}	3.1×10^{-1}
$HCHO$	1.0	6.4×10^{-1}	8.8×10^{-1}	9.0×10^{-1}

(after advection-diffusion and all chemistry steps). The adjoint of the W-method was implemented with only one forward run used to store the state after each step (all taken steps were 15 min), with recomputation of the matrix $\mathbf{A}$ as pointed out in Section 3. Both methods require additional forward recomputations during the backward integration due to the non-linearities. The average CPU time of the forward run using the operator splitting during the optimization was $cpu(\mathcal{F}) \approx 1.9$ sec, whereas for the W-method we obtained an average $cpu(\mathcal{F}) \approx 0.8$ sec. The technical report of the optimization process and the performance of the adjoint code are outlined in Table 2.

TABLE 2

Performance of the adjoint code and optimization process using operator splitting and the W-method. Computations were done on a HP-UX B.10.20 A 9000/778 machine. The I/O operations were not included in the CPU time.

Method	Operator splitting			W-method		
Run	1.0	2.0	3.0	1.0	2.0	3.0
No. iter.	43.0	23.0	16.0	44.0	27.0	18.0
$10^5 \times \mathcal{F}_{opt}/\mathcal{F}_{init}$	9.9	9.8	9.1	9.7	9.7	8.6
$\approx cpu/\text{iter(sec)}$	7.5	7.9	7.8	2.4	2.4	2.6
$\approx cpu(\mathcal{F} + \nabla\mathcal{F})/cpu(\mathcal{F})$	3.8	3.8	3.8	2.8	2.8	2.8

Both methods perform similarly in terms of the ratio between the CPU time of the *pure adjoint* computations to the CPU time of the forward run (splitting uses an additional forward run), but the optimization using the W-method is much faster since the forward W-method is more efficient. We performed a qualitative analysis of the assimilation results by defining a monitor cost function $\mathcal{F}^i$ for each species i in a similar way as $\mathcal{F}$, and updating it every 15 min *at all grid points*. Both operator splitting and the W-method produced a similar analysis and in Table 1 we present the results obtained with the W-method. It can be seen that a good analysis for was obtained for the species chemical related with O_3 and NO_2.

5. Conclusions and further work. Data assimilation for comprehensive atmospheric models is a relatively young discipline and intensive research is focused on improving the qualitative and quantitative aspects of the assimilation process. The dimension of the state variables of order $10^6 - 10^7$ is (still) a barrier in the full implementation of many of the statistical data analysis methods. Variational methods implemented with adjoint techniques may be used to provide a viable analysis state of the atmosphere. In practice the adjoint code is often implemented directly from the numerical method used to integrate the forward model, such that the efficiency of the adjoint code is highly dependent on the method used to to perform the forward integration. In this paper we analyzed the adjoint of a 2 stage W-method which may be used as an alternative to traditional splitting methods and emphasized the particular properties of the method which can improve the performance of the adjoint code and data assimilation process. Our test experiments indicate that coupling the transport and chemistry processes at the operator level may lead to an efficient adjoint model and considerable savings in CPU time. Further work must be done for testing and validation on real-life models and observational data.

REFERENCES

[1] BLOM, J.G. AND VERWER, J.G. *A comparison of integration methods for atmospheric transport-chemistry problems*. CWI Report, MAS-R9717, 1997.

[2] COHN, S. E. *An introduction to estimation theory*. J. of the Meteorol. Soc. of Japan, **75**, No. 1B, 1997.

[3] DAESCU, D., CARMICHAEL, G. R., AND SANDU, A. *Adjoint implementation of Rosenbrock methods applied to variational data assimilation problems*. J. of Computational Physics, **165**, 496–510, 2000.

[4] DAMIAN-IORDACHE AND V., SANDU, A.: *KPP-A symbolic preprocessor for chemistry kinetics-User's guide*. Tech. Rep., Univ. of Iowa, Department of Mathematics, 1995.

[5] ELBERN, H. AND SCHMIDT, H. *A four-dimensional variational chemistry data assimilation scheme for Eulerian chemistry transport modeling*. J. of Geophysical Research, **104-D15**, 18583–18589, 1999.

[6] ELBERN, H., SCHMIDT, H., AND EBEL, A. *Variational data assimilation for tropospheric chemistry modeling*. J. of Geophysical Research, **102-D13**, 15967–15985, 1997.

[7] FISHER, M. AND LARY, D. J. *Lagrangian four-dimensional variational data assimilation of chemical species*. O.J.R. Meteorol. soc., **121**, 1681–1704, 1995.

[8] GERY, M.W., WHITTEN, G.Z., KILLUS, J.P., AND DODGE, M.C. *A photochemical kinetics mechanism for urban and regional scale computer modeling*. J. of Geophysical Research, **94**, 12925–12956, 1989.

[9] GIERING, R. *Tangent linear and adjoint model compiler, Users manual 1.2* . "*http* : //*puddle.mit.edu*/ ∼ *ralf/tamc*", 1997.

[10] GIERING, R. AND KAMINSKI, T. *Recipes for adjoint code construction*. ACM Trans. Math. Software, 1998.

[11] GRIEWANK, A. *Evaluating Derivatives: Principles and Techniques of Algorithmic Differentiation*. Frontiers in Applied Mathematics 19, 2000.

[12] HAIRER, E. AND WANNER, G.: *Solving Ordinary Differential Equations II. Stiff and Differential-Algebraic Problems*. Springer-Verlag, Berlin, 1991.

[13] HUNDSDORFER, D., KOREN, B., LOON, M., AND VERWER, J.G. *A positive finite-difference advection scheme* J. of Comput. Physics, **117**, 35–46, 1995.

[14] JAZWINSKI, A. H. *Stochastic Processes and Filtering Theory*. Academic Press, 1970.

[15] KHATTATOV, B. V.*et al. Assimilation of photochemically active species and a case analysis of UARS data*. J. of Geophysical Research, **104-D15**, 18715–18737, 1999.

[16] LIU, D.C. AND NOCEDAL, J. *On the limited memory BFGS method for large scale minimization*. Math. Prog., **45**, 503–528, 1989.

[17] LORENC, A. C. *Analysis methods for numerical weather prediction*. Q.J.R. Meteorol. Soc., **112**, 1177–1194, 1986.

[18] MARCHUK, G.I. *Adjoint Equations and Analysis of Complex Systems*. Kluwer Academic Publishers, 1995.

[19] MARCHUK, G.I., AGOSHKOV, I.V., AND SHUTYAEV, P.V.: *Adjoint Equations and Perturbation Algorithms in Nonlinear Problems*. CRC Press, 1996.

[20] ROSTAING, N., DALMAS, S., AND GALLIGO, A. *Automatic differentiation in Odyssée.* Tellus, 45, 558–568, 1993.

[21] SANDU, A., CARMICHAEL, G.R., AND POTRA, F.A. *Coupled chemistry and transport computations in air quality modeling.* International conference on air pollution models APMS'98, Paris 1998.

[22] SANDU, A., POTRA, F.A., CARMICHAEL G.R., AND DAMIAN, V. *Efficient implementation of fully implicit methods for atmospheric chemical kinetics.* J. of Comput. Physics, **129**, 101–110, 1996.

[23] SANDU, A., VERWER, J.G., LOON, M., CARMICHAEL, G.R., POTRA, A.F., DABDUB, D., AND SEINFELD, J.H. *Benchmarking stiff ODE solvers for atmospheric chemistry problems I: Implicit versus explicit.* Atmos. Environ., **31**, 3151–3166, 1997.

[24] SANDU, A., VERWER, J.G., BLOM, J.G., SPEE, E.J., AND CARMICHAEL, G.R. *Benchmarking stiff ODE solvers for atmospheric chemistry problems II: Rosenbrock solvers.* Atmos. Environ., **31**, 3459–3472, 1997.

[25] SPEE, E.J. *Numerical methods in global transport-chemistry models.* Ph.D. Thesis, Center for Mathematics and Computer Science (CWI), Amsterdam, 1998.

[26] STRANG, G. *On the construction and comparison of difference schemes.* SIAM Journal on Numerical Analysis, **5**, 506–517, 1968.

[27] TALAGRAND, O. AND COURTIER, P. *Variational assimilation of meteorological observations with the adjoint of the vorticity equations. Part I. Theory.* Q.J.R. Meteorol. Soc., **113**, 1311–1328, 1987.

[28] TARANTOLA, A. *Inverse Problem Theory.* Elsevier Science Publishers, 1987.

[29] VERWER, J.G., SPEE, E.J., BLOM, J.G., AND HUNDSDORFER, W.H. *A second order Rosenbrock method applied to photochemical dispersion problems.* CWI Report, MAS-R9717, 1997.

[30] VREUGDENHIL, C., KOREN, AND B., EDITORS *Numerical Methods for Advection-Diffusion Problems.* Vieweg, 1994.

4D–VAR DATA ASSIMILATION AND ITS NUMERICAL IMPLICATIONS FOR CASE STUDY ANALYSES

H. ELBERN* AND H. SCHMIDT*[†]

Abstract. The four–dimensional variational data assimilation method (4D–var) is applied for the assimilation of surface ozone observations. The underlying chemistry transport model (CTM) is the complex Eulerian mesoscale–α EURopean Air Dispersion pollution model (EURAD). The paper describes the parallel implementation of the assimilation system. The optimization parameters in this study are the initial values as system state variables and the emission rates. The observational basis is given by a central European ozone episode, which took place in August 1997. The improvements in analysis skill and forecast improvements are discussed.

1. Introduction. In recent years increasing research efforts and compute resources are devoted to the development of data assimilation techniques in geophysical fluid dynamics research. "The ambitious and elusive goal of data assimilation is to provide a dynamically consistent motion picture of the atmosphere and oceans, in three space dimensions, with known error bars." In its practical terms this definition from Ghil and Malanotte–Rizzoli (1989) still comprises the practical objective of any assimilation efforts in geophysics. Advanced data assimilation in the meteorological context refers principally to combining data with a numerical model to produce first a better estimate of the fluid state, and then a prediction of the later behaviour.

Data assimilation can be regarded as one objective of general inverse modelling among a variety of other aspects. Adapting from a categorization given by Bennet (1992), the following applications accrue from inverse modelling:

- estimates of states of geophysical fields,
- estimates of parameters in geophysical relations,
- design of observing systems,
- resolution of mathematically ill-posed modeling problems, and
- test of scientific hypotheses.

While applications like these are familiar in solid–earth geophysics and geophysical fluid dynamics, atmospheric chemistry has benefited only rather lately from inverse modelling.

The prominent source of information is introduced by observations. In geophysical practice, available data are incomplete, but sometimes given in terms of different observations at some special locations. Hence, the

---

*Institute for Geophysics and Meteorology, EURAD, University of Cologne, Aachener Str. 201–208, D–50931 Köln, F.R.G.
Presently at Scientific Computing and Algorithms Institute, GMD, St. Augustin, Germany.
[†]Now at Laboratoire de Météorologie Dynamique, ENS, Paris, France.

inversion problem is in general underdetermined, and, at the same time, occasionally locally overdetermined. The formal approach to overcome this situation is to introduce additional information, which may be at one's disposal prior to the analysis time. Hence with varying degree of scope and complexity, algorithms of inverse modelling seek to combine available sources of information subject to objective rules. The final result or analysis is expected to be optimal in some pertinent sense.

Apart from limited attempts, inverse modelling techniques entered reactive atmospheric chemistry simulation rather lately. This is due to several reasons:

1. With about 40 or 50 chemical species, a real world chemically reactive system, the number of constituents is about one order of magnitude larger than in meteorology and oceanography. At the same time, observations of only for very few species are available. Hence, the system is grossly underdetermined.

2. Despite its large number of constituents, the underlying chemical mechanism and consequently the coded model formulation is still incomplete and must be regarded as an open system. Further to this, observed quantities often do not comply with the actual model formulation in many cases. This can be due either to the standards and method of the observation technique, where certain classes or combinations of species are measured. Or the model formulation does only include constituents or a lumping approach to constituents, which has only a limited relation to the observed quantity.

3. Although the estimation of the chemical state of the atmosphere is a problem with many practical applications and also a matter of its own right, the incentive to obtain exact initial values for subsequent model integration is not as large as in, say, meteorological weather prediction, where correct initial values are of utmost interest. Other forcing parameters, like emission rates may be of similar or even higher importance in atmospheric chemistry.

There have been several attempts in tropospheric chemistry to use the kriging or Optimum Interpolation method (OI) to produce concentration field estimates. A comprehensive description of several algorithmic variants is provided by Fedorov (1998). Given the correct statistics, the beneficial property of pertinently designed kriging or OI algorithms is their skill to provide a Best Linear Unbiased Estimate (BLUE) of the analyzed field.

The other concept in data assimilation is based on the framework of control theory, the theoretical basis of which is given for example in Lions (1971). The by far most prominent application is adjoint modelling as a key ingredient of the variational calculus. A first introduction of this method to meteorology is provided by Penenko and Obraztsov (1971). The purely spatial variant, which pairs OI, is the three–dimensional variational data assimilation (3D–var). The principal advantage however is the ease with which 3D–var assimilates nonstandard, mostly remote sensing observations, having a complex, highly nonlinear relationship with the model variables. This feature is becoming increasingly attractive with the advent of routine remote sensing data.

Upon further elaboration on that method, the space–time variant is referred to as 4D–var, which is the target assimilation scheme for several weather services.

In atmospheric chemistry, Fisher and Lary (1995) were the first demonstrating the feasibility of 4D–var with a reduced stratospheric chemistry mechanism. For a comprehensive tropospheric gas phase mechanism Elbern et al. (1997) demonstrated the potential and limitations with the same method. Khattatov et al. (1999) gave a direct comparison between 4D–var and the Kalman filter, again in the realm of stratospheric chemistry.

While all studies on chemical 4D–var were based on box models or in a Lagrangian context, Robertson and Persson (1992) provided an investigation in assimilation of air pollution data for the variational calculus in a passive tracer model. A first full three–dimensional Eulerian model with active chemistry is used by Elbern and Schmidt (1999), where the adjoint of the CTM is developed to explore the feasibility of the variational calculus to analyze the chemical state of the troposphere with model generated data by identical twin experiments. The application of this assimilation system to an ozone case study is given in Elbern and Schmidt (2001).

The objective of the present paper is to review the features of a first implementation of an adjoint CTM within a variational data assimilation system both in terms of computational aspects and simulation improvements. The optimization parameters are the initial values and the emission rates. The data assimilation system is implemented on a massively parallel system. Special emphasis is placed on computational aspects and implementation issues.

In the next section, the theoretical foundations of the 4D–var method as an advanced data assimilation procedure will be presented. Section 3 presents the model features and discusses the parallel implementation strategy. An introductory example will be presented in Section 4. The real case initial value optimization is given in Section 5, while emission rate optimization is presented in section 6. A summary is given in Section 7.

2. Variational data assimilation. The following exposition of the theoretical aspect of variational data assimilation follows Talagrand (1998). In a general formulation the data assimilation problem evolves in the solution of an overdetermined linear system:

$$(1) \qquad \mathbf{z} = \mathbf{\Gamma}\mathbf{x}^t + \zeta,$$

where $\mathbf{z} \in \mathbf{R}^m$ denotes any information, that is a priori knowledge and observations, m is the number of information elements and n is the dimension of $\mathbf{x}$, matrix $\boldsymbol{\Gamma} \in \mathbf{R}^{m \times n}$, $m > n$, is one–to–one defining the observability condition, $\mathbf{x}^t \in \mathbf{R}^n$ is the "true" system state, $\zeta \in \mathbf{R}^n$ is the unbiased and unknown error vector of $\mathbf{z}$.

With $\mathbf{S} := \mathcal{E}(\zeta\zeta^T) \in \mathbf{R}^{m \times m}$ being the known "information" error covariance matrix, and given $\mathbf{z} \in \mathbf{R}^m$, we solve for $\mathbf{x}^a \in \mathbf{R}^n$, minimizing

$$(2) \qquad \min_{\mathbf{x}} \| \mathbf{z} - \boldsymbol{\Gamma}\mathbf{x} \|_{\mathbf{S}^{-1}}$$

with Mahalanobis scalar product $\langle \cdot, \mathbf{S}^{-1} \cdot \rangle$ and the extremal condition

$$(3) \qquad \frac{1}{2}\frac{\partial}{\partial \mathbf{x}}\langle \mathbf{z} - \boldsymbol{\Gamma}\mathbf{x}, \mathbf{S}^{-1}(\mathbf{z} - \boldsymbol{\Gamma}\mathbf{x})\rangle = -\boldsymbol{\Gamma}^T \mathbf{S}^{-1}(\mathbf{z} - \boldsymbol{\Gamma}\mathbf{x}) = 0.$$

The solution to this problem is

$$(4) \qquad \mathbf{x}^a = (\boldsymbol{\Gamma}^T \mathbf{S}^{-1} \boldsymbol{\Gamma})^{-1}\boldsymbol{\Gamma}^T \mathbf{S}^{-1}\mathbf{z}$$

where $\mathbf{x}^a$ has the property to be a Best Linear Unbiased Estimate (BLUE), that is $\mathcal{E}(\mathbf{x}^a - \mathbf{x}) = 0$. The associated analysis error covariance matrix reads

$$(5) \qquad \mathbf{P}^a := \mathcal{E}\left((\mathbf{x}^a - \mathbf{x})(\mathbf{x}^a - \mathbf{x})^T\right) = (\boldsymbol{\Gamma}^T \mathbf{S}^{-1} \boldsymbol{\Gamma})^{-1}.$$

A temporal data assimilation procedure involves a model $\mathcal{M}$, which can formally written as a stochastic differential equation

$$(6) \qquad \frac{d\mathbf{x}}{dt} = \mathcal{M}(\mathbf{x}) + \eta$$

with model error η. Upon differentiation with respect to $\mathbf{x}$ we obtain

$$(7) \qquad \frac{d\delta\mathbf{x}}{dt} = \mathcal{M}'(\delta\mathbf{x}) = \mathbf{M}\delta\mathbf{x},$$

where $\mathcal{M}' = \mathbf{M}$ is the tangent–linear model to $\mathcal{M}$. Introducing the integration operator or resolvent $\mathbf{M}(t_j, t_i)$, which propagates a perturbation $\delta\mathbf{x}(t)$ of the state variable $\mathbf{x}(t)$ from time t_i to time t_j, a stepwise tangent–linear model integration gives

$$(8) \qquad \delta\mathbf{x}(t_n) = \mathbf{M}(t_n, t_{n-1})\mathbf{M}(t_{n-1}, t_{n-2})\ldots\mathbf{M}(t_1, t_0)\delta\mathbf{x}(t_0).$$

For a direct inference of the observation–minus–model discrepancy $\mathbf{y} - \mathbf{H}_i\mathbf{x}(t_i)$ at time t_i we calculate $\mathbf{x}(t_i)$ such that

$$(9) \qquad \mathbf{H}_i\delta\mathbf{x}(t_i) = \mathbf{H}_i\mathbf{M}(t_i, t_0)\delta\mathbf{x}(t_0),$$

where $\mathbf{y} \in \mathbf{R}^p$ is the vector of observations with $p = m - n$ being the number of available observations and $\mathbf{H}_n \in \mathbf{R}^{p \times n}$ is the forward interpolator or observation operator.

The information vector $\mathbf{z}$ is now separated in terms of two categories of information sources, that is $\mathbf{x}^b$, as the a priori state, which can be obtained from climatologies or forecasts, and the vector of observations $\mathbf{y}$. Likewise, the operator $\boldsymbol{\Gamma}$ can be composed from an identity operator $\mathbf{I}$ and the model resolvents (9) for each time step, $\mathbf{G} := \mathrm{diag}(\mathbf{H}_0, \mathbf{H}_1\mathbf{M}(t_1, t_0), \ldots, \mathbf{H}_N\mathbf{M}(t_N, t_0))$, with concatenation of forward interpolator and tangent linear model resolvent. Hence it is seen that

$$(10) \qquad \boldsymbol{\Gamma} := \begin{pmatrix} \mathbf{I} \\ \mathbf{G} \end{pmatrix}, \qquad \mathbf{z} := \begin{pmatrix} \mathbf{x}^b & = & \mathbf{x} & + & \zeta^b \\ \mathbf{y} & = & \mathbf{Gx} & + & \epsilon \end{pmatrix}.$$

The error of the background estimate and the observations are ζ^b and ϵ, respectively. Assuming the errors of the background estimate and the observations being uncorrelated, we set $\mathcal{E}(\zeta^b\epsilon^T) = 0$. The background error covariance matrix is then $\mathbf{P}^b := \mathcal{E}(\zeta^b\zeta^{bT}) \in \mathbf{R}^{n\times n}$ and the observation error covariance matrix is $\mathbf{R} := \mathcal{E}(\epsilon\epsilon^T) \in \mathbf{R}^{p\times p}$. with $p = m - n$ the number of available observations. The information error covariance matrix then reads

$$\mathbf{S} := \begin{pmatrix} \mathbf{P}^b & 0 \\ 0 & \mathbf{R} \end{pmatrix}.$$

After some manipulation, the analog for (4) in the case of split and uncorrelated information sources then reads

$$(11) \qquad\qquad\qquad \mathbf{x}^a = \mathbf{x}^b + \mathbf{Kd}$$

with $\mathbf{d} := \mathbf{y} - \mathbf{Gx}^b$ being the innovation vector, and

$$\mathbf{K} := \mathbf{P}^b\mathbf{G}^T(\mathbf{GP}^b\mathbf{G}^T + \mathbf{R})^{-1} \quad \in \mathbf{R}^{n\times p}$$

the Kalman gain matrix. The analog to the analysis error covariance matrix (5) can be found to be

$$\mathbf{P}^a := \mathbf{P}^b - \mathbf{P}^b\mathbf{G}^T(\mathbf{GP}^b\mathbf{G}^T + \mathbf{R})^{-1}\mathbf{GP}^b = (\mathbf{I} - \mathbf{KG})\mathbf{P}^b \quad \in \mathbf{R}^{n\times n}.$$

In the framework of the variational calculus, the problem (2) is reformulated as the minimization of the cost function

$$\mathcal{J}(\xi) := 1/2(\boldsymbol{\Gamma}\xi - \mathbf{z})^T\mathbf{S}^{-1}(\boldsymbol{\Gamma}\xi - \mathbf{z}).$$

After transfer into the form of two information sources $\mathbf{x}^b$ and $\mathbf{y}$, the cost function then is

$$(12) \qquad \begin{aligned} J(\xi(t_0)) &= \frac{1}{2}(\mathbf{x}^b(t_0) - \xi(t_0))^T\mathbf{P}^{b^{-1}}(\mathbf{x}^b(t_0) - \xi(t_0)) \\ &\quad + \frac{1}{2}\sum_0^N (\mathbf{y}^0(t_i) - \mathbf{H}\xi(t_i))^T\mathbf{R}^{-1}(\mathbf{y}^0(t_i) - \mathbf{H}\xi(t_i)). \end{aligned}$$

The minimization of this quadratic expression is performed by classical minimization methods, which require the gradient of J with respect to the optimization parameter $\xi(t_0)$. This is easily found to be

$$(13) \quad \nabla_{\xi(t_0)} J = -{\mathbf{P}^b_0}^{-1}(\mathbf{x}^b(t_0) - \xi(t_0))$$

$$- \sum_{m=0}^{N} \mathbf{M}^T(t_m, t_0)\mathbf{H}^T \mathbf{R}^{-1}(\mathbf{y}^0(t_m) - \mathbf{H}\xi(t_m)).$$

3. Implementation issues.

3.1. Model description. The model on which the variational data assimilation algorithm is based is the CTM2 (Hass, 1991), which is an offspring of the Regional Acid Deposition Model RADM2 (Chang et al., 1987). The model's horizontal grid structure is defined by the "Arakawa C" grid stencil, where the locations of the concentrations c_i and the horizontal wind components u and v are at different positions each. The grid resolution is the same as for the MM5.

The chemistry transport model calculates the transport, diffusion, and gas phase transformation of 63 chemical species by 158 reactions. These processes are calculated sequentially by a symmetric operator splitting technique, when stepping from t to $t + \Delta t$. The method is devised by Yanenko (1971) and further popularized in air quality modeling by McRae et al. (1982). This approach is shown to minimize systematic biases introduced by a fixed sequence of operators. The following operator sequence is implemented in the configuration (Hass, 1991): $x_i(t + \Delta t) = T_h T_z D_z A D_z T_z T_h x_i(t)$, where T, D denote transport and diffusion operators in horizontal (h) or vertical (z) direction, respectively. The parameterizations of the emission sources and deposition processes are included in the gas phase chemistry module A. In the model version presented here, the dynamic time step Δt of the advection operators is 10 minutes. Bott's (1989) upstream algorithm is chosen to calculate the horizontal and vertical advection. The vertical diffusion is semi–implicitly discretized following the Crank–Nicholson scheme. The Thomas algorithm is used as solver (Lapidus and Pinder, 1982). A semi–implicit and quasi steady state approximation method (QSSA) is applied for the gas phase chemistry to solve the stiff ordinary differential equation system as derived in Hesstvedt et al. (1978). In contrast to the solvers for advection and diffusion, the chemistry time step Δt_c of the stiff ordinary differential equation solvers is highly variable in time. The geographical position of the grid follows the standard scheme for the EURAD–CTM2 forward model (see for example Hass et al., 1995), which includes the following: The integration domain applies the Lambert conformal projection centered at $50°N$ latitude. A horizontal resolution of 54 km with 77 grid points in x–direction and 67 grid points in y–direction is employed. To ease computational burden of the assimilation process, most experiments are performed in a $33 \times 27 \times 15$ sub-domain with the same resolution. In order to capture a maximal number of available observation sites the reduced grid encompasses central Europe and England. In the vertical 15 levels with terrain–following σ coordinates of Lorenz type are used for both grid configurations, with refinements at the lowest levels. The lowest model half layer, where concentrations, temperature and winds are given, is set to represent 38 m height. The isobaric level of 100 hPa which defines the top of the model, is taken as a material surface. The "first guess" model state, that is, the initial values prior to the assimilation, and boundary values are treated as described in Chang et al. (1987) for the first forward model run, covering August 1–2. This includes a seasonal mean concentration of longer lived species, dependent on latitude and height. All later model runs start with the simulated model state of the preceding run to avoid chemical spin–off problems.

3.2. The adjoint of CTM2. The adjoint operators are derived from the horizontal and vertical advection, the vertical diffusion, and the gas phase chemistry mechanism. The adjoint chemistry was coded by hand, while for the advection and diffusion routines the AMC adjoint model compiler (Giering and Kaminski, 1998) and O∂yssée differentiation system (Rostaing et al., 1993; and Faure and Papegay, 1998) were used for adjoint compilation. The correctness of the adjoint code was examined by the method proposed by *Chao and Chang* (1992). The gas phase chemistry solver and the implicit vertical diffusion operator apply adaptive time step techniques. During backward integration the same time steps are taken as determined by the forward integration.

3.3. Implementation of the adjoint algorithm. The highly demanding requirements of computer memory due to the adjoint calculus pose a principal problem for complex models. In atmospheric chemistry modeling, this is the more the case as the number of constituent state variables is one order of magnitude higher than in the case of meteorological implementations. Since the saving of all intermediate computation results during an assimilation interval is utterly impractical, storage must be traded with recalculations.

The problem size under consideration in this study entails a phase state dimension of length (grid points $\times$ prognostic species $=$) $77 \times 67 \times 15 \times 41 =$ 3172785 of which the backward integration mandates the use of a well furnished massively parallel platform. In the present implementation a Cray T3E with 128 MByte local storage per processor and a horizontal equal area partitioning strategy is applied (Elbern, 1997). The fourth order horizontal advection scheme due to Bott (1989) requires an overlap region of three grid points width in the forward mode, and consequently six grid points width during backward integration. In the present implementation a number of 42 to 120 processors must be enlisted to meet these requirements.

On the basis of these computational and storage resources a single iteration step is designed to include the following phases:

1. forward model run with storage of the model states at all dynamic time steps $t_0, \ldots, t_T$ on disks,
2. backward adjoint integration from t_T to t_0 with the following operations for each step from t_l to t_{l-1}, $\quad l = T - 1, \ldots, 0$

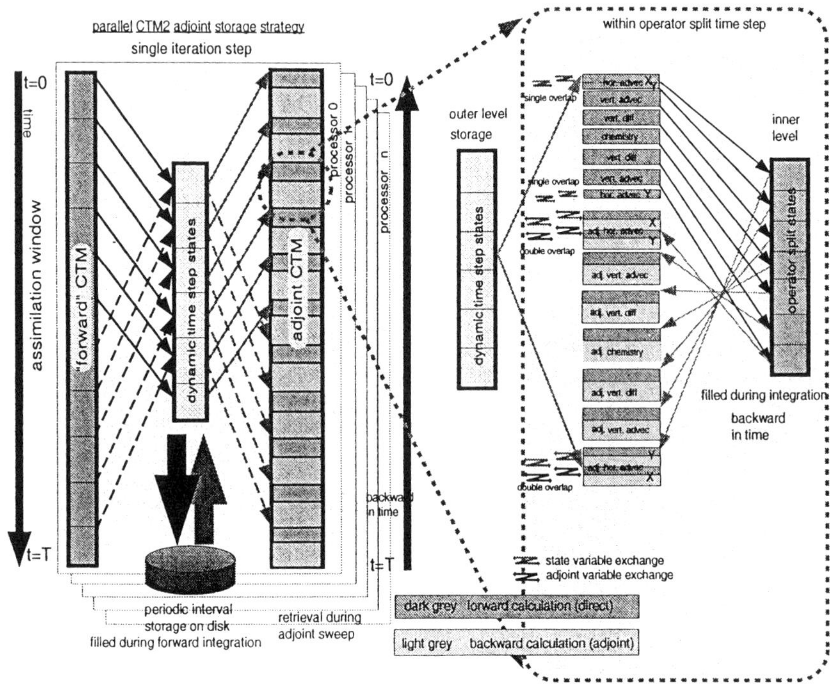

FIG. 1. *Implementation scheme for the parallel 4D-var algorithm. Left panel: Storage and retrieval for the full dynamic time step states of the entire assimilation window. Right panel: computation sequence, storage and retrieval during a single dynamic time step for the intermediate operator split states. See text for further details.*

   (a) a forward integration from t_l to t_{l+1} with storage of all operator split intermediate states in main memory,

   (b) backward integration from t_l to t_{l-1} with the adjoint operators in reverse order. The computation of the individual adjoint operator mostly includes

      i. an internal forward section to compute and store variables in the adjoint operator,

      ii. backward calculation of the adjoint section,

3. call of the L–BFGS minimization routine within the parallel implementation.

Through this procedure, the nonlinear advection and vertical diffusion operators and the chemistry solver are integrated three times forward in time per backward integration.

4. An introductory example. In the sequel an example of a box model is presented, which demonstrates the potential and limits of the 4D-var algorithm for a test case, but with artificial observations, reflecting an expectable data basis for real cases. To this end, a reference run of the model is made prior to the assimilation procedure, from which selected concentrations are taken as "observations". While these are passed to the assimilation, all further model state information of the reference run is withheld for quality control of the assimilation algorithm. Test procedures following this scheme are commonly referred to as identical twin experiments. Given are two time series of six hours length, one for ozone and another one for nitrogen dioxide. In practice, measurements of other constituents are rather rare as compared to O_3 and NO_x and consequently not introduced in this experiment. Starting at 3:00 LT and ending at 10:00 LT, the time series spans dawn and thus describes the incipient ozone production due to insolation. With given time series it is possible to select reasonably well the initial values for ozone and nitrogen dioxide, while our knowledge of all other species is speculative. In this example, our reference run is taken to have a doubled VOC concentration as compared to the first guess configuration, as the starting point of the iteration. In Figure 2 the assimilation results for O_3 and NO_2 and various VOCs are displayed for the reference run, defining the truth (bold dashed line), the first guess run (dotted line) and the analysis result (solid line). Data of O_3 and NO_2 are depicted by crosses. The first guess run exhibits an increase of ozone by 13 ppb between 3:30 LT and 9:30 LT, while observations indicate an increase of about 30 ppb. The assimilation results for both observed species O_3 and NO_2 (solid lines) compare well with the observations (crosses). Since the NO_2 values are only marginally modified, it is the VOC levels, which must have changed considerably. These conditions can be easily depicted in a schematic of the Empirical Kinetic Model Approach (EKMA) diagram, after visual inspection of Figure 2. The grey thick arrow denotes the ozone production following the first guess estimate, while the thin solid arrow spans the correctly analyzed ozone production. Upon comparison of the VOC concentration levels of the reference run and the assimilation results in Figure 2, only a slight tendency toward the true concentration levels can be observed, except for xylene, which is much too large. Taking the carbon contents as a measure, only a 50% gain of the correct values could be attained (last panel of Figure 2). In effect, xylene as a quickly reacting species, supplants the oxidizing effect of other VOC to achieve a good accord of ozone with given observations.

Concentration ratios of VOCs relative to each other may be available as a priori knowledge, probably indicating that the ratios in the estimated assimilation run analyzed above may be unlikely. In many cases ratios are better known than absolute values, as for example, car traffic exhaust shows typical ratios of hydrocarbons. In this case, a definition of the entries of the traditional background error covariance matrix $\mathbf{P}^b$ is not straightforward. However, the most likely ratios of VOCs can be directly introduced in the background term of the cost function (12). Let $\mathbf{c}(t_0)$ be the subset of components of $\mathbf{x}(t_0)$ which contains the concentrations of the VOCs,

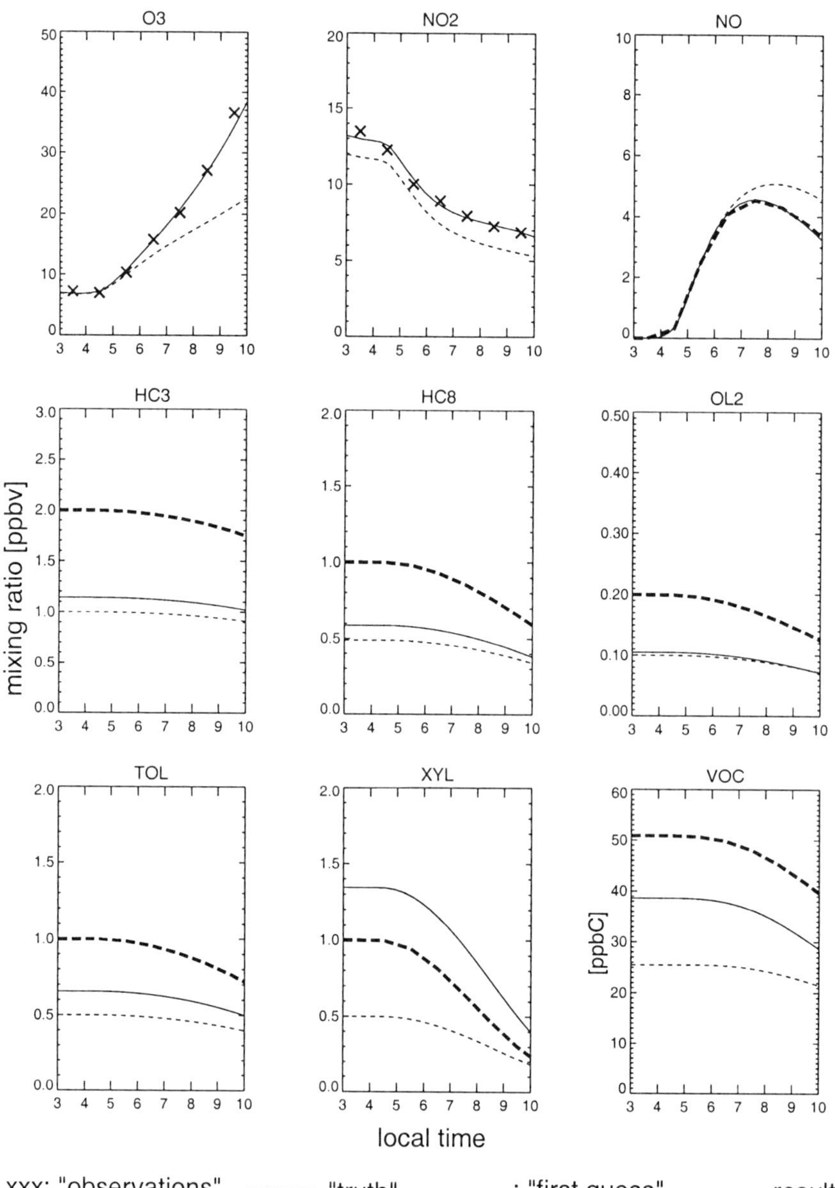

FIG. 2. *Identical twin experiments with observations for ozone (panel O3) and nitrogen dioxide as only information. Other panels show nitrogen oxide (NO), short alkanes with 3–5 carbon atoms (HC3), long alkanes with more than 8 carbon atoms (HC8), ethene (OL2), toluene (TOL), xylene (XYL), and total carbon atom concentration in terms of ppbv (VOC).*

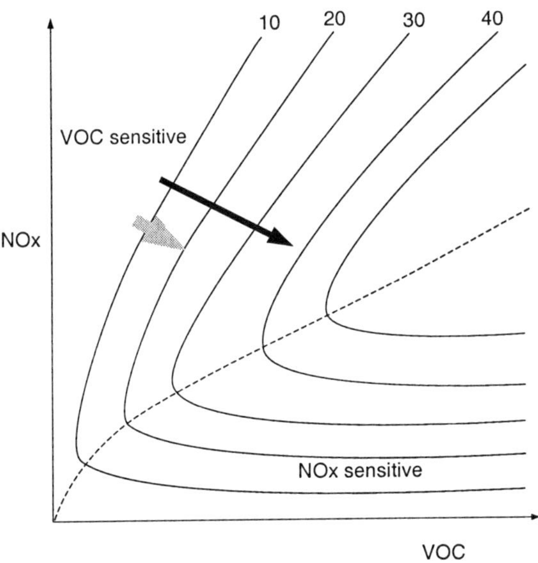

FIG. 3. *Qualitative schematic of the EKMA diagram with isopleths of ozone concentrations. Thick grey arrow symbolizes the first guess conditions, thin black arrow symbolizes the final analysis conditions.*

and $\mathbf{c}^b(t_0)$ the corresponding estimate for the background. The total VOC concentration at time t_0 is then defined $\bar{c}_0 = \sum_{VOC} c_i$ with the background estimate $\bar{c}^b$. A special background term can now be introduced

$$(14) \qquad J_c(\mathbf{c}(t_0)) = \frac{1}{2} \left(\frac{\mathbf{c}(t_0)}{\bar{c}_0} - \frac{\mathbf{c}^b}{\bar{c}^b} \right)^T \mathbf{C}^{-1} \left(\frac{\mathbf{c}(t_0)}{\bar{c}_0} - \frac{\mathbf{c}^b}{\bar{c}^b} \right)$$

with $\mathbf{C}$ now being a diagonal weighting matrix. The differentiation with respect to the ith component of the initial state vector $c_i := (\mathbf{c}(t_0))_i$, then gives

$$(15) \qquad \frac{\partial J_c}{\partial c_i} = \mathbf{C}_{(i,i)}^{-1} \frac{(c_i - \frac{c_i^b}{\bar{c}^b})(\bar{c}_0 - c_i)}{\bar{c}_0^3} - \sum_{j \neq i} \mathbf{C}_{(j,j)}^{-1} c_j \frac{(c_j - \frac{c_j^b}{\bar{c}^b})}{\bar{c}_0^3}.$$

Taking this approach, a second assimilation run has been made, where the information basis is augmented by the estimated most likely ratios of VOCs as a priori knowledge. The results are presented in Figure 4. Improvements for the observed species O_3 and NO_2 were already on a high level. However, the performance of the VOC analysis is significantly better, although no observation of any hydrocarbon is made available for the assimilation algorithm. The analysis skill of most of the VOCs was raised from about 20% increment toward the correct value up to 90%. This example demonstrates the potential benefits which can be gained from relative a priori knowledge.

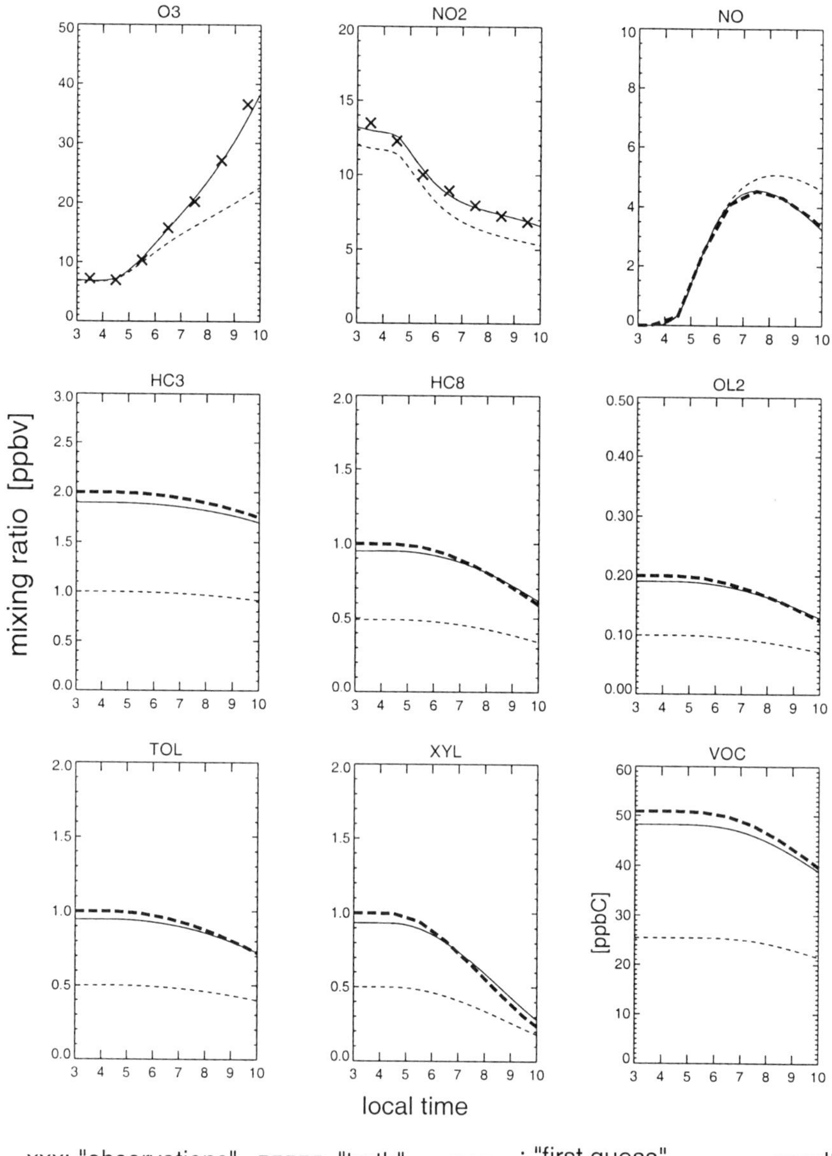

FIG. 4. *As for Figure 2, but with a priori information in terms of estimated VOC ratios.*

5. Optimization of initial values. In this section performance improvements due to variational optimization of initial values are discussed. In all cases described, a 6 hours assimilation interval is selected, starting at August 5, 2000, 0600 UT. Basically, the inclusion of observational informa-

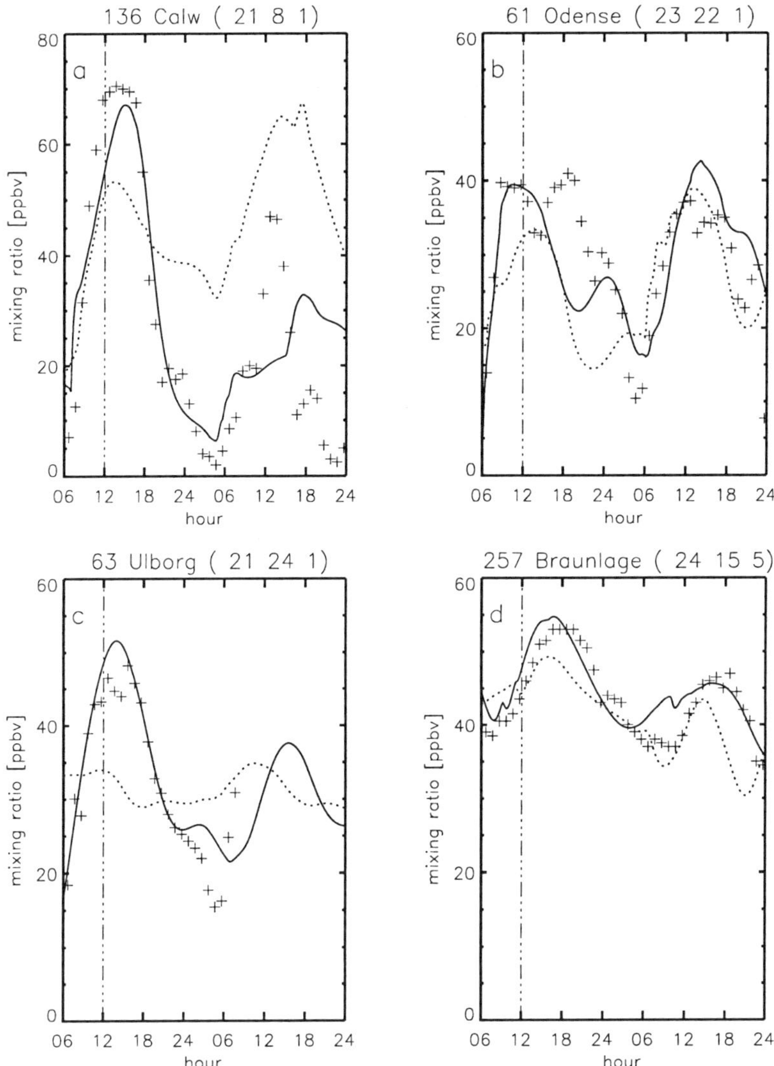

FIG. 5. *Time series of ozone evolution at 4 measurement stations (a: Calw, b: Odense, c: Ulborg, d: Braunlage) for the time span August 5, 0600 UTC to August 6, 1997, 2400 UTC. Assimilated data are confined to the 6 hours assimilation interval 0600 – 1200 UTC left of the dash–triple dot line, right of the line is forecast mode. Crosses denote observations, solid line: with assimilation of data from the first 6 hours, dotted line: "first guess" model run without assimilation.*

tion within a time interval is expected to improve the simulation skill during the well defined six hours of assimilated data. Figure 5 demonstrates the better fit with data for four stations, where comparisons are made between model runs without assimilation, based on initial values from 4 days spinup simulations (dotted lines), and assimilation based simulations, taking into account the observational data of the first 6 hours (solid line). For the central European urban station Calw, an improvement can be claimed for the assimilation interval and the subsequent rest of the day, including the substantial improvement for the afternoon ozone peak levels of the first day. A similar interpretation can be applied to the Danish stations Odense and Ulborg, which are part of a nation wide measurement net. As an example for remote mountainous conditions, observations of the central European station Braunlage in the Harz mountains is presented. The record is anticorrelated to the typical urban diurnal cycle. After application of the assimilation procedure a satisfying agreement between observations and model skill can be claimed, at least for the first simulated day.

The following examples display cases, where information gained through data assimilation at other neighbouring stations has improved the chemical analysis, such that success can be verified at the locations depicted in Figure 6. In the first case the station Maintal, which is surrounded by a dense regional observation network, benefits throughout the 42 hours simulation time from own and foreign data, that is, also the second day. A similar example can be given for the urban station Frankfurt–Griesheim. Here, the standard simulation without data assimilation is significantly too low during the daylight hours, especially in terms of the afternoon peak values, but worse than the assimilation based simulation during the night. The pronounced superiority of the assimilation based simulation for the second simulated day is entirely based on assimilated data taken from other stations. At the station Sibton the first guess run indicates simulated rural conditions, whereas the diurnal cycle in the observations clearly indicates an urban influence. The assimilation procedure was able to modify the chemical regime and now exhibits the simulation of observed urban influences. Although the model–minus–observation differences during the early morning hours of the second day are better than in the assimilation case, it is obvious that this is not based on sound simulation skill.

More problematic conditions are exhibited for the station Champs-sur–Marne east of Paris. The high variability during the first simulation day indicates the advection of heterogeneous air masses, which however, can hardly be portrayed by the assimilation based simulation. An advected secondary peak cannot be directly verified by the data, however later data indicate an afternoon peak. Comparison with other adjacent stations reveal problems of the 54 km grid to resolve local conditions, and in fact, a severe representativeness problem of this station can be stated.

6. Emission rate optimization. The optimization of the emission rates can also be made in the framework of the 4D–var calculus, after moderate modifications of the code (Elbern et al., 2000). Owing to the lack of an actual emission inventory, the emission rates in the present simulation are based on an estimate of a July 1994 working day. Diurnal variations

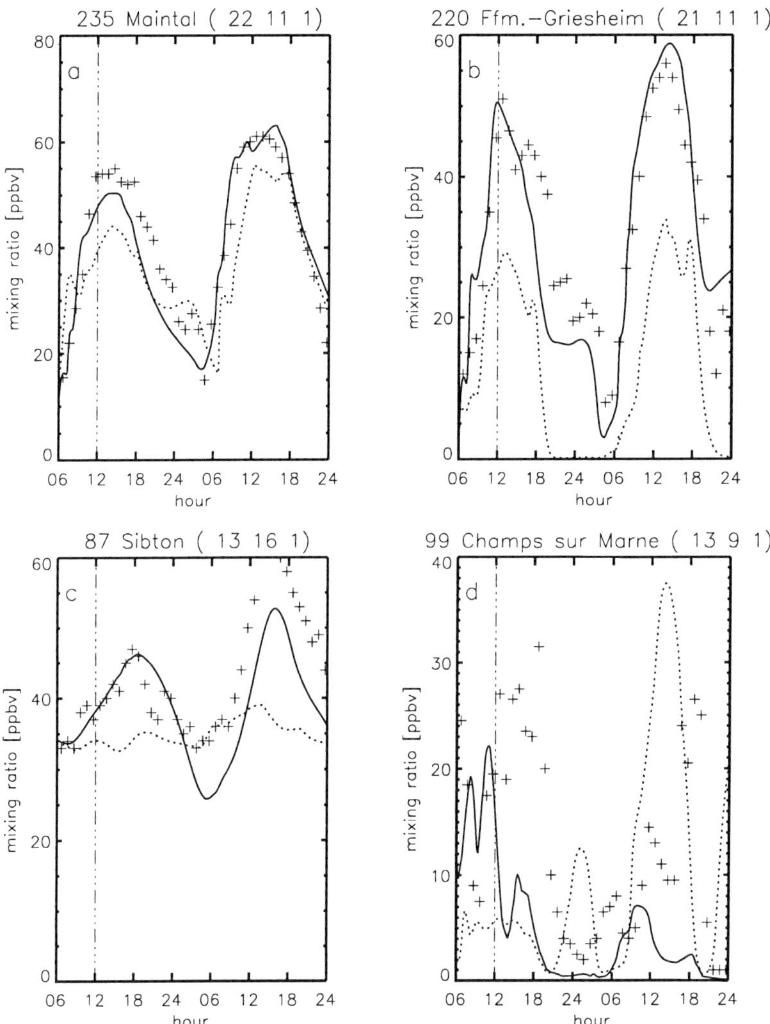

FIG. 6. *As for Figure 5 but for stations a: Maintal, b: Frankfurt–Griesheim, c: Sibton, and d: Champs sur Marne.*

of emissions, simulating car traffic, are imposed on the lowest model level. The six next layers above (that means up to 1800m) have constant emission rates, simulating industrial plants. Nineteen species are simulated to be emitted, mostly nitrogen oxides and hydrocarbons.

At some locations it was obvious that the emission inventory was erroneous. As direct optimization parameter modification factors were defined, which modify each emitted species at each emitting location, but preserve the temporal cycle of the emission source, which was taken from the emission inventory. The analysis procedure is rerun with the option of emission

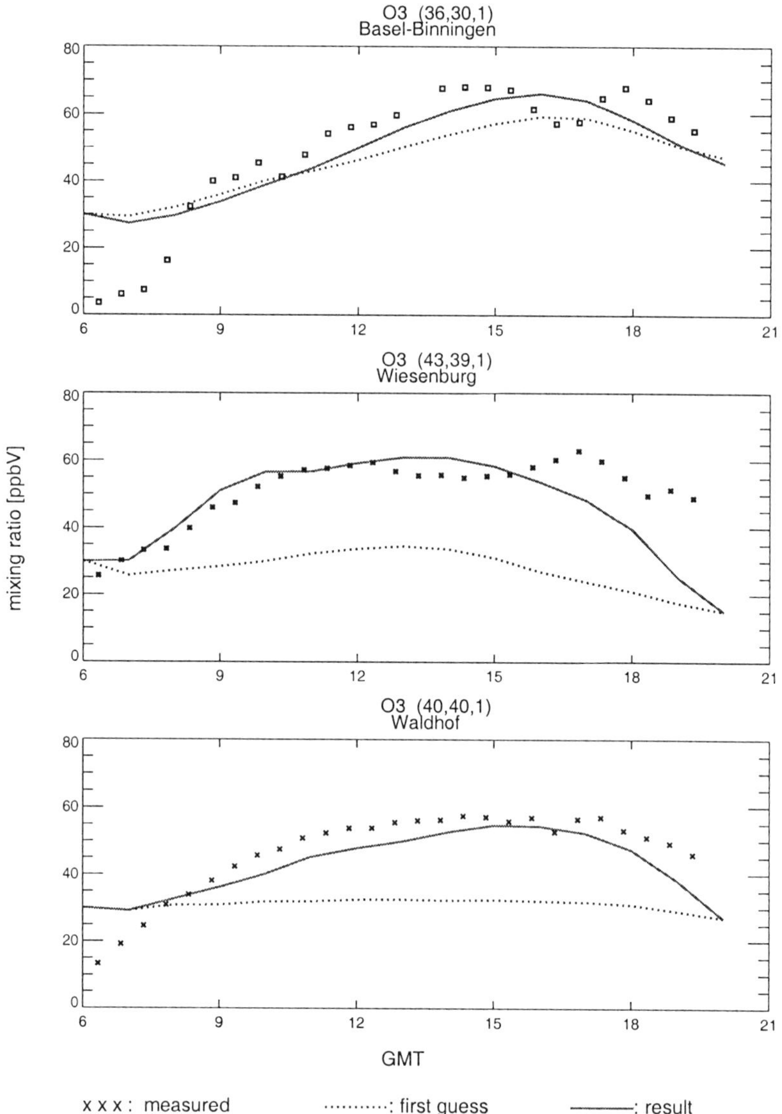

FIG. 7. *As for Figure 5 but for 3 measurement sites after mission rate optimization with 14 hours assimilation interval. Squares indicate observations.*

optimization activated. Figure 7 shows observations, first guess model performance and analysis performance again for ozone for three measurement sites. The assimilation window was defined to cover 14 hours in order to span the principal part of the diurnal emission cycle. The optimization quantity is a factor of the first guess emission rates. From the mathematical viewpoint, emission optimization based on the observed but non–

emitted product species is a highly ill–posed problem, the solution of which must take into account additional a priori knowledge. However, as demonstrated in Figure 7 once again a significant improvement can be claimed after modification of the emission rates. The performance of the first assimilated hours suffers from erroneously defined initial values. Later hours clearly exhibit a better agreement between observations and simulations after adapting the emission rates.

7. Conclusion and outlook. The 4D–var data assimilation scheme is shown to be computationally very demanding. The main reasons are the iterated model integrations for both the forward model and the adjoint model. Further to this, the storage requirements are extremely high and can only be satisfied by intermittent data swaps on disk to save the model trajectories. On the other hand, the performance improvements are significant. During assimilation intervals, the agreement between the model and the observations are close to the measurement error margins. Assimilation of morning observations improves the model skill for the simulation of afternoon ozone peak values. Owing to other model deficiencies, the benefits from optimizing the initial values vanish with time and the model performance approximates the standard run without assimilation. From other parameters, which are susceptible for variational optimization, the emission rates are taken and improved model skill could be attained by an extended data assimilation window.

The question of the pertinent optimization parameters in general goes beyond the scope of present investigations. An attempt to solve this problem is faced with the coupling of multiple subspaces of optimization parameters. The crucial point here is the proper joint metric, which balances the weights between the various parameters.

Acknowledgments. The following institutions are gratefully acknowledged for giving access to their observational data: Belgium Weather Service (IRM–KMI), German Weather Service (DWD), EMEP (co–operative programme for monitoring and evaluation of the long range transmission of air pollutants in Europe), Norway, Federal Agency of Environment (UBA), Austria, Federal Agency of Environment (UBA), Germany, Federal Agency of Environment, Forests, and Landscape (BUWaL), Switzerland, National Air Quality Information Archive, United Kingdom, National Environmental Research Institute (NERI), Denmark, Rijksinstituut voor Volksgezondheid en Milieuhygiene, (RIVM), The Netherlands, Surveillance de la qualité de l'air en Ile–de–France (AIRPARIF), France, and the Swiss Meteorological Institute. Meteorological analyses for the initial values of the MM5 were obtained from ECMWF. We also thank J. Nocedal, North–Western University, for making the L–BFGS code available. The adjoint code of the dynamic operators was developed by use of the AMC software, version 3.88, obtained from Ralf Giering, MPI of Meteorology, Hamburg, and by the Odyssee software, provided by C. Faure and Y. Papegay, INRIA,

Sophia Antipolis, France. We gratefully acknowledge the computational support by the Central Institute for Applied Mathematics (ZAM) of the Research Centre Jülich and the Regional Computing Centre of the University of Cologne (RRZK). Financial support was provided by the European Union's Climate and Environmental program project Regional dIFferences in Tropospheric Ozone (RIFTOZ) under grant ENV4–CT95–0025, the Troposphärenforschungsschwerpunkt of the German Ministry for Education, Science, Research, and Technology (BMBF) under grant 07TFS 10/LT1–C.1, and the Ministry for Science and Research (MWF) of Northrhine-Westfalia.

REFERENCES

Bennett, A., *Inverse methods in physical oceanography*, p. 346, Cambridge University Press, 1992.

Bott, A., A positive definite advection scheme obtained by nonlinear renormalization of advective fluxes. *Mon. Wea. Rev.*, **117**: 1006–1015, 1989.

Chang, J.S., R.A. Brost, I.S.A. Isaksen, S. Madronich, P. Middleton, W.R. Stockwell, and C.J. Walcek, A three-dimensional Eulerian acid deposition model: physical concepts and formulation. *J. Geophys. Res.*, **92**(14): 618–14700, 1987.

Chao, W.C. and L.-P. Chang, Development of a four-dimensional variational analysis system using the adjoint method at GLA. Part 1: Dynamics. *Mon. Wea. Rev.*, **120**: 1661–1673, 1992.

Elbern H., H. Schmidt, and A. Ebel; Variational data assimilation for tropospheric chemistry modeling. *J. Geophys. Res.*, **102**(D13): 15967–15985, 1997.

Elbern, H. and H. Schmidt, A four–dimensional variational chemistry data assimilation scheme for Eulerian chemistry transport modeling. *J. Geophys. Res.*, **104**: 18583–18598, 1999.

Elbern, H. and H. Schmidt, Ozone episode analysis by four–dimensional variational chemistry data assimilation. *J. Geophys. Res.*, **106**: 3569–3590, 2001.

Elbern, H., H. Schmidt, O. Talagrand, and A. Ebel, 4D–variational data assimilation with an adjoint air quality model for emission analysis. *Environ. Mod. Softw.*, **15**: 539–548, 2000.

Faure, C. and Y. Papegay, O∂yssée Version 1.7, the user's reference manual. *Rapport Technique N° 0224, INRIA,* Sophia Antipolis, France, 1998.

Fedorov, V., Kriging and other estimators of spatial field characteristics (with special reference to environmental studies). *Atmos. Env.*, **23**: 174–184, 1989.

Fisher, M. and D.J. Lary, Lagrangian four-dimensional variational data assimilation of chemical species. *Q.J.R. Meteorol. Soc.*, **121**: 1681–1704, 1995.

Ghil, M. and P. Malanotte–Rizzoli, Data assimilation in meteorology and oceanography. *Advances in Geophysics*, **33**: 141–266, 1989.

Giering, R. and T. Kaminski, Recipes for adjoint code construction. *ACM Trans. Math. Software*, **24**: 437–474, 1998.

Hass, H., Description of the EURAD Chemistry–Transport–Model Version 2 (CTM2), edited by A. Ebel, F.M. Neubauer, and P. Speth. *Mitt.* **83**: 100, Inst. für Geophys. und Meteorol. der Univ. zu Köln, K"oln, Germany, 1991.

Hass, H., H.J. Jakobs, and M. Memmesheimer, Analysis of a regional model (EURAD) near surface gas concentration predictions using observations from networks. *Meteorol. Atmos. Phys.*, **57**: 173–201, 1995.

Hesstvedt, E., Ø. Hov, and I.S.A. Isaksen, Quasi–steady state approximation in air pollution modeling: Comparison of two numerical schemes for oxidant prediction. *Int. J. Chem. Kinet.*, **10**: 971–994, 1978.

Khattatov, B.V., J. Gille, L. Lyjak, G. Brasseur, V. Dvortsov, A. Roche, and J. Waters, Assimilation of photochemically active species and a case analysis of UARS data. *J. Geophys. Res.*, D15: 18715–18738, 1999.

Lapidus, L. and G. Pinder, *Numerical solution of partial differential equations in science and engineering,*, p. 677, Wiley, New York, 1982.

Lions, J.-l., *Optimal control of systems governed by partial differential equations.* Springer Verlag, Berlin, 1971.

McRae, G.J., W.R. Goodin, and J.H. Seinfeld, Numerical solution of the atmospheric diffusion equation for chemically reacting flows. *J. Comput. Phys.*, 45: 1–42, 1982.

Panofsky, H, Objective weather map analysis. *J. Meteorol.*, 6: 386–392, 1949.

Penenko, V. and N.N. Obraztsov, A variational initialization method for the fields of the meteorological elements. *Meteorol. Hydrol. (Engl. Transl.)*, 11: 1–11, 1976.

Robertson, L. and Ch. Persson, On the application of four dimensional data assimilation of air pollution data using the adjoint technique. *Air Pollution Modeling and its Application, IX*, edited by H. van Dop and G. Kallos, Plenum Press, New York, 1992.

Rostaing, N., S. Dalmas, and A. Galligo, Automatic differentiation in Odyssée. *Tellus*, 45A: 558–568, 1993.

Talagrand, O., A posteriori evaluation and verification of analysis and assimilation algorithms, in *Proceeding of the ECMWF seminar "Recent developments in numerical methods for atmospheric modelling"*, Reading, UK, 7–11, 1998.

Yanenko, N.N., *The method of fractional steps.*, Springer-Verlag, New York, p. 160, 1971.

RUNNING A LARGE-SCALE AIR POLLUTION MODEL ON FAST SUPERCOMPUTERS

WOJCIECH OWCZARZ* AND Z. ZLATEV[†]

Abstract. The air pollution, and especially the reduction of the air pollution to some acceptable levels, is an important environmental problem, which will become even more important in the next 10–20 years. This problem can successfully be studied only when high-resolution comprehensive models are developed and used on a routinely basis. However, such models are very time-consuming, also when modern high-speed computers are available. Indeed, if an air pollution model is to be applied on a large space domain by using fine grids, then its discretization will always lead to huge computational problems. Assume, for example, that the space domain is discretized by using a (480x480) grid and that the number of chemical species studied by the model is 35. Then ODE systems containing 8064000 equations have to be treated at every time-step (the number of time-steps being typically several thousand). If a three-dimensional version of the air pollution model is to be used, then the above figure must be multiplied by the number of layers. It is extremely difficult to treat such large computational problems; even when the fastest computers that are available at present are used.

There is an additional great difficulty which is very often underestimated (or even neglected) when large application packages are moved from sequential computers to modern parallel machines. The high-speed computers have normally a very complicated architecture and, therefore, the task of producing an efficient code for the particular high-speed computer that is available is both extremely hard and very laborious.

Some efforts done in the process of preparation and testing of an efficient air pollution code for an IBM SMP parallel computer with two levels of parallelism will be described in this paper. The development is still in an experimental basis. Moreover, only two nodes are available on the IBM SMP computer that has actually been used. Therefore, the runs have, until now, been carried out by using a smaller problem; the two-dimensional version of the air pollution problem was discretized on a (96 × 96) grid, but the number of the chemical species was kept the same (35 species). It must be emphasized, however, that this problem, leading to the solution of ODE systems of 322560 equations during many thousands of time-steps, is still very big for many of the existing high-speed computers. This version of the model is in fact the operational version at present. It has been run by using meteorological data for 1989–1998. More details about this version and the runs performed by it can be found in the web-site of the model: http://www.dmu.dk/AtmosphericEnvironment/DEM.

Results that are obtained by using the operational version of the model on the IBM SMP computer with two nodes, each of which consists of eight processors, will be discussed in this paper. It will be shown that (i) the speed-ups achieved are very high (close to the optimal speed-ups that can be obtained on this architecture) and (ii) the cache memory is efficiently exploited.

Key words. Air pollution modelling, partial differential equations, finite element method, quasi-steady-state-approximation, ordinary differential equations, parallel computations.

AMS(MOS) subject classifications. Primary 86A10, 68C25.

---

*Danish Computing Centre for Research and Education, Technical University of Denmark - Build. 404, DK-2800 Lyngby, Denmark (`Wojciech.Owczarz@uni-c.dk`).

†National Environmental Research Institute, Department for Atmospheric Environment, Frederiksborgvej 399, P.O. Box 358, DK-4000 Roskilde, Denmark (`zz@dmu.dk`).

1. Large-scale air pollution problems. Mathematical models are an indispensable tool when different air pollution phenomena are to be studied on a large space domain (say, the whole of Europe). Such models can be used to find out when and where certain critical levels are exceeded or will soon be exceeded (since the exceedance is or will be harmful for plants, animals and human health) as well as to design optimal measures which should be taken in order to avoid the exceedance of these levels in the future (or, at least, to reduce the damaging effects caused by high pollution levels). However, all important physical and chemical processes must be taken into account in order to obtain reliable results. This leads to huge computational tasks, which have to be handled on big high-speed computers. A short description of the particular air pollution model used, the Danish Eulerian Model ([47]), will be given in this section in order to illustrate the numerical difficulties that are to be overcome when large air pollution models are treated on modern computers. We shall concentrate our attention on the mathematical backgrounds of this model, but a detailed description of the physical and chemical processes involved in the models as well as many of the validation tests and different air pollution studies that were carried out with the Danish Eulerian Model have been documented in numerous publications (see, for example, [3, 18, 21, 45, 47, 49–51] or [55]; see also the web-site of this model, [43]). The Danish Eulerian Model is described by systems of partial differential equations (PDE's):

$$
\begin{aligned}
\frac{\partial c_s}{\partial t} = &-\frac{\partial(u c_s)}{\partial x} - \frac{\partial(v c_s)}{\partial y} - \frac{\partial(w c_s)}{\partial z} \\
&+ \frac{\partial}{\partial x}\left(K_x \frac{\partial c_s}{\partial x}\right) + \frac{\partial}{\partial y}\left(K_y \frac{\partial c_s}{\partial y}\right) + \frac{\partial}{\partial z}\left(K_z \frac{\partial c_s}{\partial z}\right) \\
&+ E_s - (\kappa_{1s} + \kappa_{2s})c_s + Q_s(c_1, c_2, \ldots, c_q), \quad s = 1, 2, \ldots, q
\end{aligned}
$$

(1.1)

where (i) the concentrations of the chemical species are denoted by c_s, (ii) u, v and w are wind velocities, (iii) K_x, K_y and K_z are diffusion coefficients, (iv) the emission sources are described by E_s, (v) κ_{1s} and κ_{2s} are deposition coefficients and (vi) the chemical reactions are denoted by $Q_s(c_1, c_2, \ldots, c_q)$ (the CBM IV chemical scheme, which has been proposed in [15], is actually used in the version of the Danish Eulerian Model [47] that will be considered in this paper; it should be mentioned here that the CBM IV scheme is also used in other well-known air pollution models).

It is difficult to treat the system of PDE's (1.1) directly. This is the reason for using different kinds of splitting. A splitting procedure, based on ideas proposed in [26, 27], leads, for $s = 1, 2, \ldots, q$, to five sub-models, representing respectively the horizontal advection, the horizontal diffusion, the chemistry (together with the emission terms), the deposition and the vertical exchange:

$$(1.2) \qquad \frac{\partial c_s^{(1)}}{\partial t} = -\frac{\partial (u c_s^{(1)})}{\partial x} - \frac{\partial (v c_s^{(1)})}{\partial y}$$

$$(1.3) \qquad \frac{\partial c_s^{(2)}}{\partial t} = \frac{\partial}{\partial x}\left(K_x \frac{\partial c_s^{(2)}}{\partial x} \right) + \frac{\partial}{\partial y}\left(K_y \frac{\partial c_s^{(2)}}{\partial y} \right)$$

$$(1.4) \qquad \frac{d c_s^{(3)}}{dt} = E_s + Q_s(c_1^{(3)}, c_2^{(3)}, \ldots, c_q^{(3)})$$

$$(1.5) \qquad \frac{d c_s^{(4)}}{dt} = -(\kappa_{1s} + \kappa_{2s}) c_s^{(4)}$$

$$(1.6) \qquad \frac{\partial c_s^{(5)}}{\partial t} = -\frac{\partial (w c_s^{(5)})}{\partial z} + \frac{\partial}{\partial z}\left(K_z \frac{\partial c_s^{(5)}}{\partial z} \right).$$

If the model (1) is split into the five sub-models (1.2)–(1.6), then the discretization of the spatial derivatives in the right-hand-sides of the sub-models will lead to the solution (successively at each time-step) of five systems ($i = 1, 2, 3, 4, 5$) of ordinary differential equations (ODE's):

$$(1.7) \qquad \frac{d g^{(i)}}{dt} = f^{(i)}(t, g^{(i)}), \quad g^{(i)} \in R^N, \ f^{(i)} \in R^N,$$

where $N = N_x \times N_y \times N_z \times N_s$ and N_x, N_y while N_z are the numbers of grid-points along the coordinate axes and $N_s = q$ is the number of chemical species. The functions $f^{(i)}$, $i = 1, 2, 3, 4, 5$, depend on the particular discretization methods used in the numerical treatment of the different sub-models, while the functions $g^{(i)}$, $i = 1, 2, 3, 4, 5$, contain approximations of the concentrations at the grid-points of the space domain (more details can be found in [47]).

The size of any of the five ODE systems (1.7) is equal to the product of the number of the grid-points and the number of chemical species. It is clear, therefore, that it grows very quickly when the grid-points and/or the chemical species are increased.

Sometimes it is necessary to perform long simulation processes consisting of several hundreds of runs (see, for example, [3] or [55]). At present these problems are solved by the operational two-dimensional version of the Danish Eulerian Model (see [3, 47, 55]). In this version the following values of the parameters are used: $N_x = 96$, $N_y = 96$, $N_z = 1$, $N_s = 35$. This leads to the treatment of four ODE systems per time-step; each of them contains 322560 equations. It is desirable to solve these systems in a more efficient way. It is even more desirable to use the three-dimensional version of the model ([52]) in such runs and/or to implement chemical schemes containing more species (two other chemical schemes, with $N_s = 56$ and with $N_s = 168$, have been extensively tested, but these schemes will not be used in this paper).

More detailed information is very often needed. It is necessary to run the model on finer grids when such information is required. The use of the (96 × 96) grid, mentioned above, corresponds to grid-squares of size (50 km × 50 km) when the space domain contains the whole of Europe. A better resolution, (10 km × 10 km), leads to the application of a (480 × 480) grid if the model is applied on the same space domain. This means that four ODE systems, each of which contains 8 064 000 equations in the two-dimensional case, have to be solved at each time-step. For a typical run (with meteorological data covering one month) more than ten thousand time-steps are needed. This is why only a few runs with this fine resolution version of the model have been run until now. The task of using this version of the Danish Eulerian Model in long simulations is a very challenging problem which must be solved in the near future.

The short description of the numerical difficulties, which arise when large air pollution models are to be treated in different important for the society studies, explains why the search for more efficient numerical methods is continuing. The need of such methods is emphasized in many investigations; see, for example, [32, 47] and some of the papers in [48]). It is not necessary to describe in detail the numerical methods currently used in the Danish Eulerian Model. The major numerical methods used are linear finite element methods in the advection-diffusion part and the quasi steady state approximation (QSSA) algorithm in the chemical part.

The finite elements used have been applied in connection with air pollution models in [30, 31]. Some more details about this method can be found in [47, 14]. The systems of ODEs obtained after the finite element semi-discretization are handled by predictor-corrector methods with several different correctors (designed to increase their absolute stability regions); see [45]. Note that the advection sub-model (1.2) and the diffusion sub-model (1.3) are combined when the finite elements are used.

The QSSA algorithm used in the chemical part have been proposed in connection with air pollution modelling in [20] for box models (only one grid-point and only chemical reactions). This algorithm has been extensively tested for two more realistic examples: (i) the generalized Crowley-Molenkampf rotation test proposed in [21] (the original Crowley-Molenkampf test is described in [5, 28]) and (ii) the two-dimensional air pollution models studied in [47]. Some more advanced methods for the chemical sub-model have also been tested; in [1, 14, 37]. The latter methods will, however, not be used in this paper. The chemistry sub-model is the most time-consuming part of the model; the results given in Table 1 illustrate very clearly this fact. Therefore, the problem of finding efficient methods for this part is still open; the research efforts in this area have been very extensive during the last decade (see, for example, [4, 9, 10, 19, 23, 29, 33–36, 39–41]). Some methods, which are discussed in publications about general-purpose numerical ODE algorithms (see, for example, [6–8, 17, 25]) can also be appropriate in the efforts to improve the performance of the chemical module. Special techniques (as, for example, different implementations of sparse matrix algorithms) can also be useful (see [1, 11, 14, 37, 38, 44, 46]).

It is also very important (and much more important when the numerical methods have already been fixed) to exploit better the great potential power of the modern supercomputers. The preparation of an efficient version, by which a high computational speed can be achieved on the IBM SMP computer, will be discussed in the following part of this paper. Some ideas described in [2, 13, 14, 47] are also used in the algorithms presented here, but the main effort is put on the optimization of an operational version of the Danish Eulerian Model.

2. The IBM SMP architecture. It is very difficult to achieve high performance on a traditional, either shared memory or message passing, parallel architecture when a large application code is to be run. The code should be adjusted to the properties of the particular parallel machine which is available. Some parallelization tools can be used in the attempts to facilitate this task. Examples for such tools are the Message Passing Interface (MPI, [16]), OPEN MP ([42]) and the Parallel Virtual Machine (PVM, [12]). However, all these tools are more efficient when a large application code, which works efficiently on a given parallel computer, is to be ported to another parallel computer. When the first parallel version is prepared, one should carry out a lot of work in order to identify the parallel tasks and to reorganize some of the computations (in order to separate the parallel tasks, to achieve a better loading balance, to utilize better the cache memory, etc.).

On the IBM SMP computer (which has more complicated structure that allows us to achieve two levels of parallelization) the task of improving the performance of a large application code becomes even more difficult. A very short description of the architecture of the IBM SMP computer is given below in order to explain why the task of achieving a high computational speed (when large applications are to be run on this computer) is more difficult compared with the case where the computer is either only a shared memory computer or only a distributed memory computer.

The IBM SMP architecture is very similar to the architecture of the CEDAR computer, which has been developed at the Center for Supercomputing Research and Development (CSRD) at the University of Illinois at Urbana-Champaign in the 80ies; see, for example, [24]. The IBM SMP computer consists of several nodes (the particular computer used in these experiments has only two nodes, but this still allows us to perform many meaningful tests). Each node contains eight processors with a common memory. This means that each node works in a shared memory mode.

Each node has its own memory. This means that several nodes work in a distributed memory mode and, thus, message passing is needed in the communications between the nodes.

It is clear from this short description of the major features of the IBM SMP that the parallelization has to be achieved by efforts on two levels: (i) application of parallelization rules within each node by which the shared memory mode have to be utilized and (ii) application of parallelization rules across the nodes by which the features of message passing machines have to be exploited.

This information is quite sufficient to explain why the task of achieving high performance becomes more difficult. There are at least two reasons for this: (i) one should exploit the features of both the shared memory machines and the distributed memory machines and (ii) one should be very careful in the efforts to avoid conflicts between the devices used in the shared memory code and the devices used in the distributed memory code.

While the task of achieving high performance becomes more difficult when such a complicated architecture as the IBM SMP computer is used, it is also true that if the work is carefully done, then very good results can be achieved. This will be demonstrated in the next sections.

3. Running the code on one processor. It is intuitively clear that the first task that has to be solved is to optimize the code when it is run on one processor only (i.e. when it is run in a sequential mode). This is why we first tried to increase the performance of the code when it is run on one processor. The important task, which has to be solved here, is to exploit better the cache memory of the processor used. The efforts in this direction and the results obtained when different tests were carried out, will be discussed in this section. It should be mentioned here that the results derived in this section are valid not only for the IBM SMP computer, but also for any computer on which cache memory is available.

The two-dimensional version of the Danish Eulerian Model is used in the experiments. This version consists of the following modules:

- **Advection module.** This module treats the horizontal transport. The diffusion subroutines have been added to this module.
- **Chemistry module.** This module handles the chemical reactions involved in the model. The deposition subroutines have been added to this module.
- **Initialization module.** All operations needed to initialize arrays and to perform the first readings of input data are united in this module.
- **Input operations module.** Input data are read in this module when this is necessary; otherwise simple interpolation rules are used, both in space and in time, to produce the needed data.
- **Output operations module.** This module prepares output data. The total number of output files prepared in this version of the model is eight. The total amount of the output data is about 75 MBytes when the model is run over a rime-period of one month. For some of the concentrations (ozone and nitrogen dioxide) hourly

mean values are calculated. For other concentrations and some depositions, diurnal mean values are prepared. For all concentrations and depositions studied by the model, monthly mean values are calculated in one of the files.

The initialization module is called only once (in the beginning of the computations), while all other modules are called at every time-step.

The results obtained by the code without any attempt to optimize it are given in Table 1. The subroutines of the code were compiled by using the following options of the IBM SMP FORTRAN compiler:

$$\text{xlf\_r -O4 -qextchk -qhot -qipa -qstrict -qarch=auto.}$$

Thus, although we did not tried to optimize the code, we compiled it with a high level of optimization.

TABLE 1

Computing times (measured in seconds) obtained when the first version of the code was run on one processor of the IBM SMP computer. The parts of the computing time spent in the modules (compared with the total time for the run) are given in percent.

Module	Comp. time	Percent
Chemistry	16147	83.09
Advection	3013	15.51
Initialization	2	0.00
Input operations	50	0.26
Output operations	220	1.13
Total time	19432	100.00

The results presented in Table 1 show clearly that the chemistry module is the most time-consuming part of the code. Therefore, a lot of efforts were spent in the attempts to improve the performance of this part of the code. We started with some traditionally used actions (such as unrolling loops, reversing the order of performance the inner and the outer loops in double loops, etc.). The improvements due to these actions was minimal, because the modern compilers do automatically such optimizations. The main improvement was achieved by introducing some blocking when the chemical reactions are performed. This has been achieved by dividing the data into several chunks and performing the computations in the chunks sequentially. The algorithm based on using chunks in the chemical sub-model can be described in the following way.

Assume that a module that performs the chemical reactions at a given grid-point (such a module is normally called a box model) has been developed. Assume also that the number of grid-points is M. Then the chemical part of the code can be performed by using the following code:

```
DO I=1, M
   Call the box model
END DO
```

The important issue is that the body of this loop ("Call the box model") is in fact an inner loop in the above piece of code, because we have to carry out successively there the chemical reactions for all involved compounds: $1, 2, \ldots, NSPECIES$. This means that the above code can equivalently be rewritten as:

```
DO I=1, M
   DO J=1, NSPECIES
      Perform the chemical reactions involving
      species J in grid-point I
   END DO
END DO
```

It should be emphasized here that the inner loop consists normally of a very complicated code, where several subroutines are to be called. In fact, in most of the codes a lot of work has to be done in order to obtain the simple structure of the chemical module given above. The other important fact is that the above code is extremely inefficient on vector machines. Therefore it has to be replaced by a vectorized version by reversing the order of the loops (this is easy when the double loop is already written in the above form; the difficult task is, let us reiterate this, to rewrite the chemical module as shown above). Reversing the order of the inner and the outer loops leads to the code given below.

```
DO J=1, NSPECIES
   DO I=1, M
      Perform the chemical reactions involving
      species J in grid-point I
   END DO
END DO
```

It should be noted that the third code performs well not only on vector machines but on many other computers (because the stride in the inner loop is one). However, this code cannot exploit efficiently the cache memory of the computer used. The reason for this is that there are several big two-dimensional arrays which are referenced by rows in the inner loop. Consider, for example, the most important array, the array containing the concentrations of the chemical species. This array is denoted by $C(M, NSPECIES)$ in the code. In step I, $I = 1, 2, \ldots M$, of the inner loop $C(I, J)$ is updated, but the new value of the chemical species J depends on some of the other species K, $K = 1, 2, \ldots, J - 1, J + 1, \ldots, NSPECIES$. Thus, when we are performing the I'th step of the second code, we have to refer to some addresses in row I of array $C(M, NSPECIES)$. There are eight such arrays. It is intuitively clear that it is worthwhile to divide these arrays into chunks and to carry out the computations by chunks.

Assume that we want to use $NCHUNKS$ chunks. If M is a multiple of $NCHUNKS$, then the size of every chunks is $NSIZE = M/NCHUNKS$, and the second code given above can be modified in the following way.

```
DO ICHUNK=1, NCHUNKS
    Copy chunk ICHUNK from some of the eight
    large arrays into small two-dimensional
    arrays with leading dimension NSIZE
    DO J=1, NSPECIES
      DO I=1, NSIZE
        Perform the chemical reactions involving
        involving species J for grid-point I
      END DO
    END DO
    Copy some of the small two-dimensional
    arrays with leading dimension NSIZE
    into chunk ICHUNK of the corresponding
    large arrays
END DO
```

The work carried out in the first three codes is the same (only the order in which the operations are performed is different). Both the operations that are performed in the beginning and in the end of the first loop in the fourth code are extra. A straight-forward procedure will be to copy the current chunks of all eight arrays in the corresponding small arrays. However, this is not necessary, because some of the arrays are only used as helping arrays in the chemical module. In fact, copies from five arrays are needed in the beginning of the first loop. The situation in the end of the first loop is similar; it is necessary to copy back to the appropriate sections of the large arrays only the contents of three small arrays. The number of copies made at the end of the first loop has been reduced from five to three because some information (as, for example, the emissions) is needed in the chemical module (and has to be copied from the large arrays to the small ones), but it is not modified in the chemical module (and, thus, there is no need to copy it back to the large arrays in the end of the first loop).

The straightforward use of a box model (and running it over all grid-points) or the slight improvement given by the second piece of code corresponds to the use of chunks of minimal size (the size of each chunk being 1). The use of this version on a vector processor is, as mentioned above, a disaster. On the other hand, the vectorized version, represented above by the third piece of code, corresponds to the use of chunks with a maximal size (in fact, only one chunk, in our particular case of size 9216, being used). It is very easy to implement the former device. The latter device is very efficient on vector processors. However, on parallel machines with complicated architectures (utilizing caches on several levels) it is best to use chunks with medium sizes, while the vectorized version performs very poor in this case. Some results, which demonstrate the performance of the code when chunks of different sizes are used, are given in Table 2. These runs have been performed on three typical computers: a vector processor (Fujitsu), a parallel computer with shared memory (SGI ORIGIN 2000) and a parallel computer, which can be used in both shared memory mode and distributed memory mode (IBM SMP). All runs in Table 2 were carried out by using one processor only. It is seen that (i) the particular computer that is available has to be taken into account when the size of chunks is to be selected and (ii) the proper selection of the size of the chunks will normally lead to considerable savings in computer time and storage (storage is saved because the leading dimensions of the working arrays in the chemical part is reduced from M to $NSIZE$; $NSIZE$ is normally considerably smaller than M).

As expected, the maximal chunks are the best choice when the vector processor, Fujitsu, is used, while the minimal chunks should not be selected for this computer.

The maximal chunks should not be chosen when the SGI ORIGIN 2000 is to be used, while the difference between minimal chunks (chunks of size 1) and medium chunks (chunks of size 48) is not very large (the selection of chunks of medium size being, nevertheless, more preferable).

The maximal chunks should not be chosen either when the IBM SMP computer is available. The use of minimal chunks is also a bad choice in this case. It is best to select chunks with a medium size.

TABLE 2

Computing times (measured in seconds) obtained with different chunks on three computers (using one processor only).

Size of the chunks	Fujitsu	SGI ORIGIN 2000	IBM SMP
1	76964	14847	10313
48	2611	12114	5225
9216	494	18549	19432

In the next sections it will be shown how the selection of chunks influence the parallel runs on the eight processors within a node of the IBM SMP computer, as well as the solution of the task of achieving efficient parallel runs across the nodes of this computer.

4. Parallel runs of the code on one node. As mentioned above, each node contains eight processors with shared memory. The results given in Table 1 indicate that it is sufficient to achieve efficient parallel computations only in the advection module and in the chemistry module in order to improve the performance of the air pollution model when several processors on one node are to be used. This means that first the foremost one have to identify parallel tasks in these two modules. Moreover, it is desirable the parallel tasks to be large and of equal size. There is no problem to find such parallel tasks in the advection module, because each chemical species can be treated separately. Thus, the number of parallel tasks in the advection module is equal to the number of chemical species, and the code used is of the following type (a directive *"parallel do"* should be inserted in the beginning of this code).

```
DO J=1, NSPECIES
      Perform the advection-diffusion operations
      for the J'th chemical species
END DO
```

Furthermore, the parallel tasks in the body of the loop in the above piece of code are of equal size. Therefore, there will be a perfect loading balance when the number of chemical species is a multiple of the number of processors. However, this is not the case for our air pollution model; the number of chemical species is 35. Nevertheless, very good speed-ups can be achieved in the advection module; see the computing times and the speed-ups for 2, 4 and 8 processors under the column "Advection" in Table 3 and Table 4 respectively.

TABLE 3

Computing times (measured in seconds) obtained by using 1, 2, 4 and 8 processors in one node as well as by using all 16 processors in two nodes of the IBM SMP computer.

Processors	Advection	Chemistry	Total
1	933	4185	5225
2	478	1878	2427
4	244	1099	1405
8	144	521	799
16	62	272	424

TABLE 4

Speed ups obtained by using 2, 4 and 8 processors in one node as well as by using all 16 processors in two nodes of the IBM SMP computer.

Processors	Advection	Chemistry	Total
2	1.95 (98%)	2.23 (112%)	2.15 (108%)
4	3.82 (96%)	3.81 (95%)	3.72 (93%)
8	6.48 (81%)	8.01 (100%)	6.54 (82%)
16	15.01 (94%)	15.39 (96%)	12.32 (72%)

There is no problem to achieve efficient parallel computations also in the chemistry module. The performance of the chemical reactions at a given grid-point is a parallel task. Thus, the parallel tasks are equal to the number of grid-points, and concurrent computations are achieved by inserting a directive *"parallel do"* before the loop in which the chemical reactions are performed. This is the straightforward way of exploiting the parallelism. In the previous section it was shown that chunks can successfully be used in an attempt to exploit better the cache memory. The following parallel code can be used when the number of processors is eight (assuming here that the number of chunks is a multiple of eight).

```
Csmp$ parallel sections
   DO ICHUNK=1, NCHUNKS/8
      Call the chemical module for chunk ICHUNK
   END DO
Csmp$ section
   DO ICHUNK=NCHUNKS/8+1, 2*NCHUNKS/8
      Call the chemical module for chunk ICHUNK
   END DO
Csmp$ section
   DO ICHUNK=2*NCHUNKS/8+1, 3*NCHUNKS/8
      Call the chemical module for chunk ICHUNK
   END DO
Csmp$ section
   DO ICHUNK=3*NCHUNKS/8+1, 4*NCHUNKS/8
      Call the chemical module for chunk ICHUNK
   END DO
Csmp$ section
   DO ICHUNK=4*NCHUNKS/8+1, 5*NCHUNKS/8
      Call the chemical module for chunk ICHUNK
   END DO
Csmp$ section
   DO ICHUNK=5*NCHUNKS/8+1, 6*NCHUNKS/8
      Call the chemical module for chunk ICHUNK
   END DO
Csmp$ section
   DO ICHUNK=6*NCHUNKS/8+1, 7*NCHUNKS/8
      Call the chemical module for chunk ICHUNK
   END DO
Csmp$ section
   DO ICHUNK=7*NCHUNKS/8+1, 8*NCHUNKS/8
      Call the chemical module for chunk ICHUNK
   END DO
Csmp$ end parallel sections
```

Here *"chemical module"* refers to some subroutine which contains the body of the outer loop in the fourth code given in the previous section. The fourth code from the previous section can also be called directly (with a *"parallel do"* directive before the first loop). The experiments show that the code with parallel sections given above performs slightly better.

The loading balance is perfect, and the speed-ups are very good (in fact, they are as a rule better than those achieved in the advection module. Results can be seen under the column "Chemistry" in Table 3 and Table 4.

Of course, we are mainly interested in the results for the whole air pollution model. These results can be seen under the the the column "Total" in Table 3 and Table 4.

5. Parallel runs of the code across the nodes. A version of the parallel code for the IBM SP described in [14] has been modified for the runs on two IBM SMP nodes. MPI ([16]) is used in this code. The main ideas on which the particular version used in this paper is based can been sketched as follows. Assume that the concentrations are stored in array $C(96, 96, 35)$ and that two nodes are to be used. Array C is divided into two sub-arrays $C_1(1 : 49, 96, 35)$, and $C_2(48 : 96, 96, 35)$. For every chemical compound s, $s = 1, 2, \ldots, 35$, the sub-arrays C_1 and C_2 contain 49 rows. The advection process (united as in the previous sections with the diffusion process) can be carried out concurrently for each sub-array (assuming here that the wind velocities and the arrays containing the matrices induced by the application of the finite elements are appropriately divided into sub-arrays). It is necessary to update (both in the advection part and in the chemical part) the concentrations of rows 1–48 for C_1 and rows 49–96 of C_2. The concentrations of row 49 are only used as an inner boundary condition in the calculations involving C_1, while the concentrations of row 48 are only used as inner boundary conditions during the calculations involving C_2. Since explicit integration methods, described in [45, 47], are used to treat the systems of ODE's arising after the finite element discretization, the values of the concentrations obtained during the last application of the chemical procedure can be used in the inner boundary conditions. Both the advection part and the chemical part can be treated concurrently for every sub-domain (defined by the two sub-arrays) as in the sequential case provided that the inner boundaries are treated by taking the last updates of the concentrations in the chemical procedure. This means that the following communications must be done at each time-step after the chemical part: the contents of row 48 must be communicated from the first node to the second one, while the contents of row 49 must be from the second node to the first one. Thus, communications take place only once per time-step and, moreover, only one row per chemical species is to be communicated by each node. The same procedure, as that described in the previous section, is used within each node. Here it is assumed that (i) a preprocessor module is used to distribute the data to scratch files in the local memories of the two nodes in the very beginning of the computational process and (ii) a postprocessor module is used to collect the output data from the scratch files of the two nodes to the output files needed. More details about the preprocessor and the postprocessor can be found in [13, 14].

The situation is very similar if more nodes are used; the main difference being that for some sub-domains it is necessary to communicate to neighbouring processors two rows per chemical species (again this has to be done only once per time-step).

Results obtained by using the parallel algorithm sketched above are given in the last lines of Table 3 and Table 4.

6. Scalability of the code. It is important *to preserve* the efficiency of the code when the size of some of the involved arrays is increased (for example, as a result of refining the grid, increasing of the number of chemical compounds, the transition from the 2-D version to a 3-D version, etc.). This property is often refered to as a scalability of the code. While such a property is highly desirable (the requirements to the air pollution codes are permanently increasing), it is by no means clear in advance whether the code has such a property or not when the modern complicated computer architectures are in use.

Some experiments were performed in an attempt to check the scalability of the parallel devices discussed in the previous sections. A (288×288) grid was considered instead of the (96×96) grid considered in the previous sections. This corresponds to a transition from cells of size $(50 \ km \ \times \ 50 \ km)$ to cells of size $(16.67 \ km \ \times \ 16.67 \ km)$. The number of chemical species was kept 35. This means that in this version of the code the number of grid-points was increased by a factor of 9.

In the advection part, we had also to decrease the time-stepsize by a factor of 3. Thus, the number of arithmetic operations (or, in other words, the amount of computational work) is increased by a factor of 27 in the advection part.

There was no need to decrease the time-stepsize in the chemical part. This means that the number of arithmetic operations (or, in other words, the amount of computational work) is increased by a factor of 9 in the chemical part.

This short analysis indicates that if the code is scalable, then the computing times should be increased by factors approximately equal to 27 and 9 in the advection part and the chemical part respectively. The results given in Table 5 shows that this is precisely what happens (see the ratios between the computing times for the refined grid and the coarse grid in Table 5). The conclusion from this experiment is that the code is scalable. This conclusion was further supported (i) by some runs with the more precise version obtained by using a (480×480) grid and (ii) by performing some runs with the 3-D version obtained by using a $(96 \times 96 \times 10)$ grid.

7. When is it desirable to improve the performance. The computers are becoming bigger and bigger and faster and faster. Many of the problems which only several years ago had to be solved on big supercomputers can now be treated on work-stations or on big PCs. Therefore it is reasonable to ask the question: "Why and when is it necessary to im-

TABLE 5

Computing times obtained by using a refined (288×288) grid and a coarse (96×96) grid. The ratios of the times for the refined and coarse grids are given in the last column. The two codes were run on 16 processors of the IBM SMP computer, the size of the chunks was 48.

Process	(288 × 288)	(96 × 96)	Ratio
Advection	1523	63	24.6
Chemistry	2883	288	10.0
Total	6209	424	14.6

prove the performance?" This question is further justified by the fact that the new supercomputer architectures are becoming more and more complicated and, thus, it is more and more difficult to achieve efficiency when big application codes are run on the new and modern supercomputers. It is true, however, that if the work is properly done, then the code is becoming very efficient.

Let us reiterate here that the results from the first run of the code on the IBM SMP computer were very poor (see Table 1). The total computing time for this non-optimized code was 19432 seconds. We should like to compare this time with the best computing times obtained on the IBM SMP computer when the two versions of the code (defined on a coarse and refined grids respectively) are run. The results, computing times and ratios of improvements when compared with the non-optimized version, are given in Table 6.

TABLE 6

The best results obtained by using a refined (288 × 288) grid and a coarse (96 × 96) grid on 16 processors with chunks of size 48 The ratios of the times given in the second column and the total time in Table 1 are given in the third column.

Process	Comp. time	Ratio
(96 × 96)	424	45.8
(288 × 288)	6209	3.1

All runs in this paper were performed by using meteorological data covering a period of one month (July 1994). Runs with meteorological data covering a time-period of ten years (1989–1998) have been carried out with the Danish Eulrian Model (see [53, 54]). Recently, 24 scenarios with these data have been completed to study the influence of the biogenic emissions on the formation of ozone in different part of Europe. This is a very comprehensive task; 2880 times bigger than the simple experimental task used in the previous sections. More than 2 years CPU time will be necessary to run the 24 scenarios with meteorological data covering a timer-period of 10 years when the non-optimized version of the code is used.

While it is perhaps theoretically possible to do this, it is obvious that it is quite impossible to complete such a comprehensive task in practice. At the same time, this task was resolved, very successfully, with the optimized version in several weeks: some results about the influence of the biogenic emissions on the high ozone levels in Europe were reported at the last GLOREAM workshop held in September 2000 in Cottbus, Germany.

The results in Table 5 and Table 6 shows that even the refined, on a (288×288) grid, version can be used in the solution of some realistic tasks (but not yet in the solution of the task to run 24 scenarios over a period of 10 years).

Finally, even finer resolution is sometimes desirable. Some runs have recently been performed by using a version of the code discretized on a (480×480) grid. Also here highly optimized versions of large-scale air pollution models are required.

The main conclusion of this section is that, although there are more and more problems which can successfully be handled without optimization on work-stations and PCs, it is still necessary, when the problems are very large, to run highly optimized codes on big modern supercomputers. The set of problems which could be treated on work-stations and PCs will become bigger and bigger in the future, because the computer power of the work-stations and PCs will continue to be increasing. However, the requirements for more accuracy and obtaining more detailed information are also steadily increasing. This means that the models are becoming bigger and bigger. Of course, as illustrated by the example given below, this is a well-know, by the specialists, fact.

A. Jaffe wrote in 1984 ([22]): *"Although the fastest computers can execute millions of operations in one second, they are always too slow. This may seem a paradox, but the heart of the matter is: the bigger and better computers become, the larger are the problems scientists and engineers want to solve"*. A. Jaffe was right, is still right and will certainly continue to be right in the future: there will always exist very advanced and very close to the reality scientific problems which cannot be solved even if the fastest available computers are used.

8. Concluding remarks and plans for future work. The optimization of a large air pollution code for runs on an IBM SMP computer with two nodes have been described in the previous sections. The results illustrate the well-known fact that the computations should be carefully organized in order to exploit better the great potential power of the modern high-speed computers (compare the results in Table 1 with the results in the other tables). If this is not done, then the results can be very bad (see again the results obtained without using chunks on one processor in Table 1). On the other hand, a successful reorganization of the computations can improve the performance considerably.

The speed-ups obtained are quite satisfactory for all runs. In fact, they are close to the optimal speed-ups that can be achieved on this computer. However, the speed of computations is still not sufficient to treat the large air pollution models when (i) the grid is refined, (ii) three-dimensional versions of the models are to be used and (iii) long simulation processes containing many hundreds of scenarios are to be performed. Much faster computers (perhaps IBM SMP computers with much more nodes) are needed in the situations listed above.

The choice of numerical methods was not discussed in Sections 3–5. However, from the contents of these three sections it is clear that all devices used in this paper can also be applied when other numerical methods are used. The structure of the parallel sections will remain the same when other numerical methods are used. This is why high speed-ups will be achieved for many other numerical algorithms. The size of chunks for which best results can be achieved will in general depend both on the methods selected and on the particular computer used. It is easy to find the best size by performing several experiments where the size of the chunks is varied (similar to expreiments carried out in Sections 3–5).

The ideas used here are pretty general and could also be efficient on other computers too. Thus, it will be rather easy to port this code to other parallel computers on which MPI and OPEN MP are available.

Three major tasks have to be solved in the attempts to run efficiently a large application code (not necessarily an air pollution code): (i) one has to run the code on a fast computer, (ii) one has to select fast numerical methods and (iii) one has to organize carefully the computational process. The third task is often underestimated. In this paper we have shown (by fixing both the computer used and the numerical methods selected) that a success in the efforts to solve the third task in a more efficient manner leads to big savings in computing time. Indeed, the total time for the solution of the problem without any optimization was 19432 seconds (see Table 1), while the best achievements (obtained without changing the computer, the compiler options and the numerical methods) were 5225 seconds on one processor (see Table 2) and 424 seconds on 16 processors (see Table 3). Thus, only by improving the solution of the third task, i.e. by reorganizing the computations, it is possible to achieve results which can be achieved by the non-optimized code only if much more processors are available; the number of processors should (roughly speaking) be greater by more than a factor of two or even three if we want to obtain with the non-optimized version the same computing times as the computing times obtained when the optimized verrsion is run. The crucial role of the proper organization of the computations in the efforts to improve the performance of a big application code on parallel architectures is the most important conclusion from the results presented in Sections 3–6.

Acknowledgments. This research was supported by the NATO Scientific Programme under the projects ENVIR.CGR 930449 and OUTS.CGR.960312, by the EU ESPRIT Programme under projects WEPTEL and EUROAIR and by NMR (Nordic Council of Ministers) under a common project for performing sensitivity studies with large-scale air pollution models in which scientific groups from Denmark, Finland, Norway and Sweden are participating. Furthermore, a grant from the Danish Natural Sciences Research Council gave us access to all Danish supercomputers.

REFERENCES

[1] V. ALEXANDROV, A. SAMEH, Y. SIDDIQUE, AND Z. ZLATEV, *Numerical integration of chemical ODE problems arising in air pollution models*, Environmental Modelling and Assessment, **2** (1997), pp. 365–377.

[2] C. BENDTSEN AND Z. ZLATEV, *Running air pollution models on message passing machines*, in: Parallel Virtual Machine and Message Passing Interface. M. Bubak, J. Dongarra and J. Wasniewski (eds.), Springer, 1997, pp. 417–426.

[3] A. BASTRUP-BIRK, J. BRANDT, I. URIA, AND Z. ZLATEV, *Studying cumulative ozone exposures in Europe during a seven-year period*, Journal of Geophysical Research, **102** (1997), pp. 23917–23935.

[4] D.P. CHOCK, S.L. WINKLER, AND P. SUN, *Comparison of stiff chemistry solvers for air quality models*, Environ. Sci. Technol. **28** (1986), pp. 1882–1892.

[5] W.P. CROWLEY, *Numerical advection experiments*, Monthly Weather Review, **96** (1968), pp. 1–11.

[6] P. DEUFLHARD, *Order and stepsize control in extrapolation methods*, Numer. Math. **41** (1983), pp. 399–422.

[7] P. DEUFLHARD, *Recent progress in extrapolation methods for ordinary differential equations*, SIAM Rev. **27** (1985), pp. 505–535.

[8] P. DEUFLHARD, E. HAIRER, AND M. ZUGCH, *One step and extrapolation methods for differential-algebraic equations*, Numer. Math. **51** (1987), pp. 501–516.

[9] P. DEUFLHARD AND U. NOWAK, *Efficient numerical simulation and identification of large reaction systems*, Ber. Bensendes Phys. Chem. **90** (1986), pp. 940–946.

[10] P. DEUFLHARD, U. NOWAK, AND M. WULKOW, *Recent development in chemical computing*, Computers Chem. Engng. **14** (1990), pp. 1249–1258.

[11] I.S. DUFF, A.M. ERISMAN, AND J.K. REID, *Direct methods for sparse matrices*, Oxford University Press, 1986.

[12] A. Geist, A. Beguelin, J. Dongarra, W. Jiang, R. Manchek, and V. Sunderam, *PVM: Parallel Virtual Machine, A User's Guide and Tutorial for Networking Parallel Computing*. MIT Press, Cambridge, Massachusetts, 1994.

[13] K. GEORGIEV AND Z. ZLATEV, *Running an advection-chemistry code on message passing computers*, In: Recent Advances in Parallel Virtual Machine and Message Passing Interface. V. Alexandrov and J. Dongarra (eds.), Springer, 1998, pp. 354–363.

[14] K. GEORGIEV AND Z. ZLATEV, *Parallel Sparse Matrix Algorithms for Air Pollution Models*, Parallel and Distributed Computing Practices, **2**, No. 4 (1999), pp. 429–443.

[15] M.W. GERY, G.Z. WHITTEN, J.P. KILLUS, AND M.C. DODGE, *A photochemical kinetics mechanism for urban and regional computer modeling*, Journal of Geophysical Research, **94** (1989), pp. 12925–12956.

[16] W. GROPP, E. LUSK, AND A. SKJELLUM, *Using MPI: Portable programming with the message passing interface*, MIT Press, Cambridge, Massachusetts, 1994.

[17] E. HAIRER AND G. WANNER, *Solving ordinary differential equations, II: Stiff and differential-algebraic problems*, Springer, 1991.

[18] R.M. HARRISON, Z. ZLATEV, AND C.J. OTTLEY, *A comparison of the predictions of an Eulerian atmospheric transport chemistry model with measurements over the North Sea*, Atmospheric Environment, **28** (1994), 497–516.

[19] O. HERTEL, R. BERKOWICZ, J. CHRISTENSEN, AND Ø. HOV, *Test of two numerical schemes for use in atmospheric transport-chemistry models*, Atmospheric Environment, **27A** (1993), pp. 2591–2611.

[20] E. HESSTVEDT, Ø. HOV AND I.A. ISAKSEN, *Quasi-steady-state approximations in air pollution modelling: comparison of two numerical schemes for oxidant prediction*, International Journal of Chemical Kinetics, **10** (1978), pp. 971–994.

[21] Ø. HOV, Z. ZLATEV, R. BERKOWICZ, A. ELIASSEN, AND L.P. PRAHM, *Comparison of numerical techniques for use in air pollution models with non-linear chemical reactions*, Atmospheric Environment, **23** (1988), pp. 967–983.

[22] A. JAFFE, *Ordering of the universe: The role of mathematics*, SIAM Review, **26** (1984), pp. 473–500.

[23] L.O. JAY, A. SANDU, F.A. POTRA, AND G.R. CARMICHAEL, *Improved QSSA methods for atmospheric chemistry integration*, SIAM J. Sci. Comput. **18** (1997), pp. 182–202.

[24] D.J. KUCK, E.S. DAVIDSON, D.H. LAWRIE, AND A.H. SAMEH, *Parallel supercomputing today and the CEDAR approach*, Science, **231** (1986), pp. 967–974.

[25] J.D. LAMBERT, *Numerical methods for ordinary differential equations*, Wiley, 1991.

[26] G.I. MARCHUK, *Mathematical modeling for the problem of the environment*, Studies in Mathematics and Applications, No. 16, North-Holland, 1985.

[27] G.J. McRAE, W.R. GOODIN, AND J.H. SEINFELD, *Numerical solution of the atmospheric diffusion equations for chemically reacting flows*, Journal of Computational Physics, **45** (1984), pp. 1–42.

[28] C.R. MOLENKAMPF, *Accuracy of finite-difference methods applied to the advection equation*, Journal of Applied Meteorology, **7** (1968), pp. 160–167.

[29] M.T. ODMAN, N. KUMAR, AND A.G. RUSSELL, *A comparison of fast chemical kinetic solvers for air quality modeling*, Atmospheric Environment, **26A** (1992), pp. 1783–1789.

[30] D.W. PEPPER AND A.J. BAKER, *A simple one-dimensional finite element algorithm with multidimensional capabilities*, Numerical Heath Transfer, **3** (1979), pp. 81–95.

[31] D.W. PEPPER, C.D. KERN, AND P.E. LONG, JR., *Modelling the dispersion of atmospheric pollution using cubic splines and chapeau functions*, Atmospheric Environment, **13** (1979), pp. 223–237.

[32] L.K. PETERS, C.M. BERKOWITZ, G.R. CARMICHAEL, R.C. EASTER, FAIRWEATHER, G. GHAN, J. HALES, L. LEUNG, W. PENNELL, F. POTRA, R.D. SAYLOR, AND T. TSANG, *The current state and future direction of Eulerian models in simulating the tropospherical chemistry and transport of trace species: A review.* Atmospheric Environment, **29** (1995), pp. 189–221.

[33] A. SANDU, F.A. POTRA, G.R. CARMICHAEL, AND V. DAMIAN, *Efficient implementation of fully implicit methods for atmospheric chemical kinetics*, J. Comp. Phys., **129** (1996), pp. 101–110.

[34] A. SANDU, J. VERWER, J. BLOOM, E. SPEE, AND G. CARMICHAEL, *Benchmarking stiff ODE systems for atmospheric chemistry problems: II. Rosenbrock solvers*, Atmospheric Environment, **31** (1997), pp. 3459–3472.

[35] A. SANDU, J.G. VERWER, M. VAN LOON, G.R. CARMICHAEL, F.A. POTRA, D. DABDUB AND J.H. SEINFELD, *Benchmarking stiff ODE systems for atmospheric chemistry problems: I. Implicit versus Explicit*, Atmospheric Environment, **31** (1997), pp. 3151–3166.

[36] D. S. SHIEH, Y. CHANG AND G. R. CARMICHAEL, *The evaluation of numerical techniques for solution of stiff ordinary differential equations arising from chemical kinetic problems*, Environ. Software, **3** (1988), pp. 28–38.

[37] S. SKELBOE AND Z. ZLATEV, *Exploiting the natural partitioning in the numerical solution of ODE systems arising in atmospheric chemistry*, In: Numerical Analysis and Its Applications. L. Vulkov, J. Wasniewski, P. Yalamov, (eds.), Springer, 1997, pp. 458–465.

[38] J. SWART AND J. BLOM, *Experience with sparse matrix solvers in parallel ODE software*, Comp. Math. Appl. **31** (1996), pp. 43–55.

[39] J.G. VERWER, J.G. BLOM, M. VAN LOON, AND E.J. SPEE, *A comparison of stiff ODE solvers for atmospheric chemistry problems*, Atmospheric Environment, **30** (1996), pp. 49–58.

[40] J.G. VERWER AND M. VAN LOON, *An evaluation of explicit pseudo-steady state approximation for stiff ODE systems from chemical kinetics*, J. Comp. Phys. **113** (1996), pp. 347–352.

[41] J.G. VERWER AND D. SIMPSON, *Explicit methods for stiff ODE's from atmospheric chemistry*, Appl. Numer. Math. **18** (1995), pp. 413–430.

[42] WEB-SITE FOR OPEN MP TOOLS, http://www.openmp.org, 1999.

[43] WEB-SITE FOR THE DANISH EURLERIAN MODEL, (National Environmental Research Institute, Roskilde, Denmark), http://www.dmu.dk/AtmosphericEnvironment/DEM, 1999.

[44] R.A. WILLOUGHBY, *Sparse matrix algorithms and their relation to problem classes and computer architecture*, in Large Sparse Sets of Linear Equations. J.K. Reid, (ed.), Academic Press, 1970, pp. 255–277.

[45] Z. ZLATEV, *Application of predictor-corrector schemes with several correctors in solving air pollution problems*, BIT, **24** (1984), pp. 700–715.

[46] Z. ZLATEV, *Computational methods for general sparse matrices*, Kluwer Academic Publishers, 1991.

[47] Z. ZLATEV, *Computer treatment of large air pollution models*, Kluwer Academic Publishers, 1995.

[48] Z. ZLATEV, J. BRANDT, P.J. H. BUILTJES, G. CARMICHAEL, I. DIMOV, J. DONGARRA, H. VAN DOP, K. GEORGIEV, H. HASS, AND R. SAN JOSE, eds., *Large Scale Computations in Air Pollution Modelling*, Kluwer Academic Publishers, 1999.

[49] Z. ZLATEV, J. CHRISTENSEN, AND A. ELIASSEN, *Studying high ozone concentrations by using the Danish Eulerian Model*, Atmospheric Environment, **27A** (1993), 845–865.

[50] Z. ZLATEV, J. CHRISTENSEN, AND Ø. HOV, *An Eulerian air pollution model for Europe with nonlinear chemistry*, Journal of Atmospheric Chemistry, **15** (1992), 1–37.

[51] Z. ZLATEV, I. DIMOV, AND K. GEORGIEV, *Studying long-range transport of air pollutants*, Computational Science and Engineering, **1**, No. 3 (1994), 45–52.

[52] Z. ZLATEV, I. DIMOV, AND K. GEORGIEV, *Three-dimensional version of the Danish Eulerian Model*, Zeitschrift für Angewandte Mathematik und Mechanik, **76** (1996), pp. 473–476.

[53] Z. ZLATEV, I. DIMOV, TZ. OSTROMSKY, G. GEERNAERT, I. TZVETANOV, AND A. BASTRUP-BIRK, *Calculating Losses of Crops in Denmark Caused by High Ozone Levels*, Environmental Modeling & Assessment, **6** (2001), pp. 35–55.

[54] Z. ZLATEV, G. GEERNAERT, AND H. SKOV, *A study of ozone critical levels in Denmark over a ten-year period*, EUROSAP Newsletter, **36** (1999), pp. 1–9.

[55] Z. ZLATEV, J. FENGER, AND L. MORTENSEN, *Relationships between emission sources and excess ozone concentrations*, Computers and Mathematics with Applications, **32**, No. 11 (1996), pp. 101–123.

AN EFFICIENT 3-D MODEL FOR GLOBAL CIRCULATION, TRANSPORT AND CHEMISTRY

EDUARDO P. OLAGUER*

Table of Contents

*MCNC-Environmental Programs, 3021 Cornwallis Rd., Research Triangle Park, NC 27709.

Abstract. It has become increasingly necessary to account for global movement of chemical species in determining the environmental impact of pollution on all scales. Fast motions such as penetrating convection and tropospheric jets enhance the long-range transport of gaseous and particulate matter, allowing distant sources to contribute significantly to local concentrations. Moreover, environmental policy is tending towards a "one atmosphere" approach to coordinate multiple issues, including the deterioration of regional and global air quality, stratospheric ozone depletion, climate change, and persistent organic pollution. The Global Balance Environment (GLOBE) model was designed as a "one atmosphere" framework for analyzing such problems on regional through global scales.

GLOBE is a 3-D, 11-wave spectral, dynamical model of the lower (0–16 km) and middle (16–79 km) atmosphere coupled on-line to a grid, chemical transport model. It has a grid resolution of $6\,^\circ$ latitude $\times$ $11.25\,^\circ$ longitude $\times$ 1.2 km altitude. The model employs an enhanced version of the global balance system that conserves energy, while allowing advection of zonal mean vorticity and temperature by the mean meridional circulation. It includes a detailed radiative transfer code, efficient parameterizations of gravity waves, breaking planetary waves, equatorial Kelvin waves, cumulus convection and large-scale condensation, plus extensive inorganic and hydrocarbon chemistry. GLOBE's highly efficient design yields a fast 64-bit execution time of about 92 seconds per model day on a single processor of a Microway Alpha UP2000 Linux system, thus enabling the assessment of long-term environmental impacts of chemically reactive species.

1. Introduction. It is known from studies of Saharan dust storms (e.g., Prospero, 1999), as well as from analyses of aircraft observations of trace gases in areas remote from human activity (e.g., Schultz et al., 1999), that airborne particulates and volatile organics such as PAN can travel very large distances across the globe. There are several dynamical mechanisms that contribute to this phenomenon. Penetrating convection rapidly vents material from the boundary layer into the free troposphere (Chatfield and Delaney, 1990), where it can be coherently transported by the mid-latitude jet stream to remote areas (Stohl and Trickl, 1999). The material can then be re-entrained into the boundary layer by slow, large-scale subsidence around cumulus towers (Lelieveld and Crutzen, 1994), or by rapid descent in the vicinity of cold fronts associated with "cut-off lows" (Carmichael et al., 1998). A real episode involving all these transport mechanisms simultaneously has been demonstrated by Prados et al. (1999) in analyzing air parcel trajectories over the North American continent and the North Atlantic Ocean during the AEROCE campaign.

There is a growing consensus among scientists and policymakers that a "one atmosphere" approach to air pollution is called for, given the linkages that stem from the combination of chemical reactions, long-range transport, and climate feedbacks. It is now appreciated, for example, that the stratosphere and troposphere are highly coupled through radiative, chemical, and dynamical exchanges (Wuebbles et al., 1993), making it necessary for climate models to have their lids in the upper stratosphere and even the mesosphere, rather than the lower stratosphere, as has previously been customary. Application of the "downward control principle" has made it clear that stratosphere-troposphere exchange is not just a matter of individual cumulus towers or isolated tropopause folding events (Holton et al., 1995). Rather, an entire economy involving, among other things, the dissipation of large-scale planetary waves at relatively high altitudes must be invoked to understand this natural feature. Such an insight has implications for modeling pollution in the boundary layer, given the importance of stratospheric intrusions in air quality.

An important issue to consider is the potential influence of short-lived compounds on global climate. Acetone, organic nitrates, and other oxygenated hydrocarbons have been observed in significant quantities in the upper troposphere, and may be important in determining the tropospheric oxidation capacity (Singh et al., 1992, 2000). Indirect global warming by O_3 precursors such as CO, NO_x and non-methane hydrocarbons (NMHCs) may also be non-trivial, as indicated by 2-D model assessments (Fuglestvedt et al., 1996; Johnson and Derwent, 1996). In the case of the stratosphere, even short-lived halocarbons such as 1-bromo-propane can have significant ozone depletion potentials due to rapid transport to the stratosphere by tropical convection (Wuebbles et al., 1999).

Given the complexity of the environmental problems facing the global community, there is a need for 3-D assessment models that integrate radiative, chemical, and dynamical processes throughout the lower and middle atmospheres. This paper introduces one such tool, namely the Global Balance Environment (GLOBE) model.

2. Numerical architecture. GLOBE is a 3-D, spectral, dynamical model of the atmosphere coupled on-line to a grid, chemical transport model (CTM). It has a grid resolution of $6°$ latitude $\times$ $11.25°$ longitude $\times$ 1.2 km altitude and extends from the surface to about 79 km above ground. The dynamical model expands the fields of vorticity, temperature, and vertical velocity in a trapezoidally truncated series of spherical harmonics (denoted by Y_{mn}), with 11 zonal waves in addition to the zonal mean component (i.e., m ranges from 0 to 11). Each zonal wave has a maximum meridional mode number of 19 (i.e., n ranges from m to 19). The model thus fully resolves both planetary (m= 1 to 5) and synoptic (m= 6 to 11) waves, a valuable feature in the lower troposphere, where synoptic waves are prevalent.

The spectral decomposition of the meteorological fields allows for accurate evaluation of non-linear advection terms through the transform method. The non-aliasing Gaussian grid used for dynamical computations has 32 equally spaced longitudes and 29 nearly equally spaced latitudes. The vertical grid is equally spaced in log-pressure. The meteorological fields are vertically staggered, with temperature and vertical velocity evaluated at every full level, and vorticity at every half-level. The time-marching algorithm for spectral fields is a second-order Lorenz 4-cycle with a 15 minute time step.

The CTM employs an equally-spaced latitude grid that is staggered with respect to the Gaussian latitudes (i.e., $-87°$ to $87°$ at $6°$ intervals). The CTM longitudes, on the other hand, are identical to the Gaussian grid longitudes ($0°$ to $360°$ E at $11.25°$ intervals). The dynamical and CTM vertical grids likewise coincide. Tracer mixing ratios are evaluated at every full level. To ensure numerical mass conservation, grid values of stream function and velocity potential are derived from spectral fields, and then finite differenced to yield CTM wind fields that are staggered with respect to the tracer grid. A fourth-order Bott (1989) scheme is used for tracer advection, except for convective venting and subsidence, for which an upwind scheme is employed. Vertical and horizontal diffusion of tracers is treated using a semi-implicit Crank-Nicholson scheme, except where otherwise noted. An implicit scheme with production and loss coefficients is applied to photochemical generation. The time step used for all tracer tendency terms is 1 hour.

3. Global balance formulation. Before going on to describe the model theoretical formulation, the reader is referred to Box A for a list of symbols, and to Box B for the fundamental equations governing GLOBE. Unlike the typical General Circulation Model (GCM), which is based on the primitive equations of dynamic meteorology, GLOBE employs a filtered equation set known as the global balance system, which is computationally more efficient than the primitive equations. A price one pays for the efficiency gain is the elimination of gravity waves (except for the so-called mixed Rossby-gravity wave) and the equatorial Kelvin wave (Moura, 1976). Another limitation of most global balance models is that they do not predict the horizontal average temperature. Both shortcomings can be circumvented. In the first case, analytical parameterizations of gravity and equatorial Kelvin waves can be incorporated into the model formulation, which GLOBE in fact does. In the second case, a one-dimensional, radiative-convective model can be used to separately predict the slowly varying horizontal average temperature over longer time scales, as was done by Pitari et al. (1992). For the time being, however, GLOBE assumes a temporally invariant horizontal average temperature. Moreover, the surface air temperature is currently specified from the climatology of Legates and Willmott (1990a) to avoid simulating the ocean, geosphere, and biosphere explicitly. These compartments will be added later.

A significant new aspect of the GLOBE model is that it allows for advection of $[\zeta]$ and $[T]$ by the mean meridional circulation. Normally, the presence of velocity potential terms in the vorticity equation (B-4) other than the divergence of the planetary vorticity flux requires additional terms in the global balance equation (B-2) to conserve energy (Lorenz, 1960). However, the use of zonal averages in the third term on the right hand side of Equation B-4 avoids this complication entirely.

In solving equations B-1 through B-5, one first derives W (assuming a rigid lid at the top of the model and flow over orography at the bottom), then predicts ζ from Equation B-4. The temperature field is then obtained from the thermal wind relation, derived by combining equations B-1 and B-2. Box C shows the details of computing W. Note that the thermodynamic equation is not used to directly predict T, but is employed in deriving Equation C-1. Note also that observed, as opposed to predicted, wind fields are used to determine W at Z=0, to avoid biases in planetary wave forcing due to the use of a non-terrain following vertical coordinate and other model limitations at the surface. Values for $[u_{obs}]$ and $[v_{obs}]$ are obtained from Randel (1992) for 850 mb and Oort (1983) for 1000 mb respectively. To avoid spurious wave forcing near the South Pole, we ignore Antarctic topography and treat the terrestrial surface below about 60° S as sea ice.

Equation C-1 is technically an implicit equation for W because there are terms involving [W] and [X] on the right hand side. These terms, however, vary more slowly than W. We thus solve C-1 iteratively by treating it as an explicit equation for W, with the mean meridional circulation terms evaluated at the previous time step in the initial iteration.

4. Treatment of horizontal diffusion. Some careful attention must be paid to the treatment of horizontal diffusion, because of numerical problems associated with spectral models and the special physics of the middle atmosphere. The problem of spectral "blocking", in which wave energy is "reflected" by the "rigid lid" (truncation) in wavenumber space, is addressed by incorporating numerical fourth order diffusion, with a corresponding coefficient K_N set equal to 2×10^{17} m^4/s at pressures above 100 mb, 2×10^{18} m^4/s at pressures below 10 mb, and varying inversely with pressure at intermediate levels. The treatment of second-order diffusion, on the other hand, is summarized in Box D. The first term on the right hand side of Equation D-1 damps reflections by the artificial rigid lid, so that the air motions at lower levels are not contaminated by spurious energy.

The use of a latitude-longitude coordinate system for the CTM introduces the problem of Courant-Friedrichs-Levy (CFL) instability near the poles. Li and Chang (1996) solved this problem in the context of the Bott advection scheme by using expanded polar zones. GLOBE employs a combined technique, in which tracer concentrations immediately next to the pole (i.e., at ±87°) are required to be zonally uniform and an enhanced polar diffusion coefficient K_p is used to horizontally mix tracers between ±87° and ±81°. The use of a polar diffusion coefficient not only damps CFL instability, but also decreases the noise associated with high values of the spherical harmonic functions at polar latitudes when applied to vorticity and temperature. An implicit scheme is used to evaluate the corresponding diffusion term in the CTM. In the zonal direction, the implicit scheme is applied to the Fourier coefficients of the tracer field at ±81°, which are then transformed back to grid space.

The coefficient K_M provides added horizontal diffusion in the tropo-sphere due to unresolved mesoscale eddies generated either by deep con-vection or by processes near the surface. The coefficient K_B, on the other hand, simulates the breaking of planetary waves in the middle atmosphere due to baroclinic and barotropic instabilities in the local flow associated with a vanishing meridional gradient of potential vorticity. This highly nonlinear process is an important cause of irreversible mixing in the lower stratosphere (McIntyre and Palmer, 1983; Leovy et al., 1985), but requires fine resolution to simulate explicitly.

Our treatment of planetary wave breaking is based on Garcia (1991). Garcia's expression for the diffusivity (denoted by Γ_k) has been rewritten to take advantage of the dispersion relation for planetary waves and the assumption of wave saturation (see p. 1408 of Garcia, 1991). This was done to avoid computing the meridional and vertical wavenumbers from the phases of individual waves, as Garcia had done. The resulting expression is then applied to all zonal wavenumbers (with some vertical and horizontal smoothing away from the breaking level and latitude) and the components added together to yield K_B. A maximum diffusivity of 3×10^6 m^2/s is enforced to allow for the cascade of energy from low to high wavenumbers through explicitly resolved wave-wave interactions. The planetary wave breaking scheme, which is applied at pressures below 200 mb, is essential for the numerical stability of the model particularly during the northern hemisphere winter, when planetary waves attain large amplitudes.

5. Equatorial wave features. Certain meteorological oscillations in the equatorial middle atmosphere have important consequences for tracer transport throughout the globe. In particular, the so-called quasi-biennial oscillation (QBO) in the lower stratosphere and the mesospheric semi-annual oscillation (SAO) both affect the position of the equatorial zero wind line, where [u]=0. Because planetary Rossby waves tend to be re-flected by this line, the QBO in particular can have significant effects at high latitudes, such as influencing the intensities of the Antarctic polar vortex and spring ozone hole (Garcia and Solomon, 1987).

The QBO and SAO are the result of subtle wave-mean flow interac-tions involving equatorial Kelvin, gravity, and mixed Rossby-gravity waves (Holton and Lindzen, 1972; Dunkerton, 1997). Analytical parameteriza-tions of Kelvin and mixed-Rossby gravity waves based on the equatorial β-plane approximation have been used by Gray and Pyle (1986, 1987) to simulate tracer transport effects due to the QBO and SAO in a 2-D CTM. We adopt a similar scheme to parameterize the zonal acceleration due to the Kelvin wave, as summarized in Table E. Kelvin wave parameters are similar to those employed by Dunkerton (1997) for the two slower modes and to those by Gray and Pyle (1987) for the fastest mode. The radiative dissipation rate is also taken from Gray and Pyle (1987).

Table E also shows the model treatment of equatorial gravity waves with a continuous spectrum of phase speeds. This parameterization is adopted from Dunkerton (1997), except that the strength of the effective downward advection is set proportional to the zonal mean convective subsidence near the equatorial tropopause, as it is assumed that the waves are generated by tropical convection. Such waves provide a mechanism for equatorial shear zone descent, an essential aspect of both the QBO and SAO.

6. Discrete gravity wave turbulence. The seminal paper by Lindzen (1981) showed how turbulence results from vertically propagating gravity waves that amplify to the point of convective instability. Gravity wave saturation then causes a deceleration of the zonal flow, represented in GLOBE by a Rayleigh friction coefficient α_R. This coefficient is computed as the sum of two terms, α_O and α_I, where the subscript O denotes orographically generated waves, and the subscript I denotes waves generated by shear instability in the troposphere.

The gravity wave scheme is summarized in Box F. The parameterization of orographic waves is based on McFarlane (1987), with orographic forcing heights determined from standard deviations of high resolution topography. Because orographic waves are assumed to have zero phase speed, α_O is set to zero above a critical layer where [u] is zero or changes sign. This accounts for the absorption of the waves by the critical layer.

Gravity waves with non-zero phase speed are often generated by shear instability in the tropospheric jet stream. The parameterization of these waves is similar to the scheme of Holton (1982, 1983), except for a different determination of the phase speed c. This phase speed is selected to yield maximum deceleration of the mesospheric jet, and to avoid absorption by critical layers between the tropospheric and mesospheric jets.

7. Vertical diffusion and surface exchanges of momentum and heat. Box G displays the model treatment of vertical diffusion, including the exchange of momentum between the atmosphere and the surface. The treatment of gravity wave diffusion is consistent with the parameterization of gravity wave drag described earlier.

The coefficient K_{ZB} provides additional sponge layer damping near the model top and, moreover, simulates enhanced turbulence in the planetary boundary layer (PBL). The coefficient K_{ZC}, on the other hand, simulates turbulence due to deep convection. The three terms in brackets in Equation G-2 account for mountain torque, Ekman pumping, and surface drag respectively. The observed wind used to compute surface drag is at 850 mb (as is the case with orographic forcing), since there is a correlation between actual surface drag and the geostrophic wind above the boundary layer. The coefficient α_M is estimated from annual mean mountain torque reported by Peixoto and Oort (1992) and from observed winds. In addition to surface momentum exchange, the forcing of planetary waves by the sea-air temperature difference ΔT_{SA} is treated as sensible heat exchange with coefficient $C_H = C_D = 0.003$. Values of ΔT_{SA} are obtained from Oort (1983).

8. Radiative transfer. GLOBE employs an updated version of the detailed radiative transfer code described in Olaguer et al. (1992). Updates include a non-LTE infrared scheme for the 15 μ band of CO_2 above 56 km (Fomichev et al., 1993), improved infrared heating in the water vapor continuum band based on the parameterization of Zhong and Haigh (1995), and improved near infrared heating due to H_2O and CO_2 based on Briegleb (1992).

Radiative heating rates are computed once daily. Detailed radiative heating due to H_2O, CO_2, CH_4, and N_2O is computed only for the zonal mean. Eddy radiative heating due to H_2O and CO_2 is treated as Newtonian cooling. In the case of H_2O, the Newtonian cooling coefficient is obtained from the broadband emissivity, whereas the parameterization of Fels (1982) is employed for CO_2. Infrared heating by the 9.6 micron band of ozone and solar heating by O_2, O_3, and NO_2 are computed explicitly at each latitude and longitude.

Above the troposphere, a simple Beer's law is employed to simulate absorption and single scattering of solar radiation. Reflection by the underlying troposphere is accounted for by computing an effective lower atmospheric albedo from a multiple scattering algorithm, which also computes solar heating in the troposphere. For this purpose, the surface albedo is set to 0.07 for oceans, 0.14 for land, and 0.70 for sea ice. Multiple scattering is computed using a delta-Eddington/adding method. This algorithm is invoked for near-infrared heating by water vapor, but not for visible heating by ozone. In the latter case, the tropospheric albedo is set at 0.3. Although visible heating by ozone is negligible in the troposphere, GLOBE accounts for multiple scattering in calculating photolysis rates. Daylight average actinic fluxes are derived from solar irradiances (see Filyushkin et al., 1994) at two zenith angles after Cunnold et al. (1975).

Cloud properties required by the radiative transfer code are displayed in Box H. The expressions for cloud emissivity and optical depth in equations H-2 and H-3 are based on Kiehl et al. (1994), but ignore wavelength dependence. The scattering asymmetry parameter for cloud droplets is set at 0.85. Single scattering albedos, on the other hand, are treated as wavelength-dependent, and are computed as in Slingo (1989).

9. Hydrology and latent heating. GLOBE employs an annual and zonal mean cloud climatology from the International Satellite Cloud Climatology Project (ISCCP) D2 11-year mean dataset. Three types of cloud are specified: high, middle, and low cloud, with pressure thicknesses of 30 mb, 50 mb, and 100 mb respectively. For middle clouds, cloud top pressures correspond roughly to ISCCP 11-year mean values estimated from infrared irradiances. Low cloud tops are assumed to be a distance $2\Delta Z$ below the corresponding middle cloud top. High cloud top pressures, on the other hand, are heuristically based on cirrus cloud data presented by Newell et al. (1974).

Water vapor concentrations in the troposphere are not explicitly simulated by GLOBE, but are derived from the horizontally invariant relative humidity profile of Manabe and Wetherald (1967). This profile and other details of the model hydrology are shown in Boxes I and J. The use of a constant relative humidity is convenient in parameterizing the latent heating due to phase changes of water vapor. Latent heating is computed in two ways, corresponding to non-cumulus condensation and evaporation on the one hand, and to penetrating convection on the other.

Non-cumulus condensation and evaporation are treated through a static stability adjustment parameter σ_{lh}. This simple method takes advantage of the coupling between the specific humidity and the temperature via the Clausius-Clapeyron equation and the assumption of a constant relative humidity. Because of the importance of this particular parameterization scheme to the model, we include the derivation of the relevant heating term Q_{cond} from first principles.

We begin by considering the total derivative of the specific humidity and temperature following the motion of an air parcel in the environment outside cumulus towers, but neglecting diffusion other than that of latent heat in the vertical direction:

$$(9.1) \quad \frac{dq}{dt} = -\frac{C_p Q_{cond}}{L_v} + (q_s(T_c) - q)\max\left(\frac{\partial(PW_c)}{\partial P}, 0\right) - \frac{\partial}{\partial P}\left\{P\frac{K_Z}{H^2}\frac{\partial q}{\partial Z}\right\}$$

$$(9.2) \quad \frac{dT}{dt} = -\kappa(W - W_c)T + Q_{rad} + Q_{cond}.$$

The second term on the right hand side of Equation (9.1) represents the detrainment of water vapor from cumulus clouds to the surrounding environment.

Note that in equations (9.1) and (9.2), the total derivative includes advection by the convective subsidence W_c, which is assumed to be equal to the cumulus area fraction ($\ll 1$) times the ascent velocity within cumulus towers. The large-scale vertical velocity W can then be regarded as the grid average vertical velocity, such that $W - W_c$ is the net vertical velocity in the environment surrounding the cumulus towers. Moreover, although T in Equation (9.2) refers to the temperature of the cumulus environment, and T in B-5 refers to the grid average temperature, we shall assume that these temperatures are nearly the same, because of the small area occupied by cumulus convection (typically 2%).

The assumption of a horizontally and temporally independent relative humidity r can now be exploited as follows:

$$(9.3) \quad \frac{dq}{dt} = r\frac{dq_s}{dt} + q_s\frac{dr}{dt} = r\frac{\partial q_s}{\partial T}\frac{dT}{dt} + r\frac{\partial q_s}{\partial P}\frac{dP}{dt} + q_s(W - W_c)\frac{\partial r}{\partial Z}.$$

Recalling the definition of vertical velocity and relative humidity, and substituting the right hand side of Equation (9.2) for the total derivative of temperature, we obtain:

$$(9.4) \quad \frac{dq}{dt} = rq_s \frac{\partial(\ln q_s)}{\partial T}\left(-\kappa\left(W - W_c\right)T + Q_{rad} + Q_{cond}\right)$$
$$+ rq_s\left(W - W_c\right) + rq_s\left(W - W_c\right)\frac{\partial(\ln r)}{\partial Z}.$$

We now invoke the Clausius-Clapeyron equation for water vapor:

$$(9.5) \quad \frac{\partial(\ln q_s)}{\partial T} = \frac{0.622 L_v}{RT^2}.$$

Combining equations (9.1), (9.4), and (9.5), and noting that the derivative with respect to Z of ln r is very close to -1, we arrive at the following equation:

$$(9.6) \quad Q_{cond} = \sigma_{lh}\left(T\right)\left(W - W_c\right) - \frac{\sigma_{lh}\left(T\right)}{\kappa T}Q_{rad} + D_{lh}$$

where σ_{lh} and D_{lh} are defined in I-2. When $r=1$, σ_{lh} is approximately the static stability associated with the moist adiabatic lapse rate.

To avoid numerical problems caused by large values of σ_{lh} at high temperatures, by the fluctuating sign of W, and by distortions in eddy heating, we employ the expression in I-2 for Q_{cond} instead of Equation (9.6). The use of a reference temperature T_{ref} is numerically convenient because it allows us to stably solve the vertical velocity equation in C-1. The derivative of σ_{lh} at the reference temperature can be shown to be:

$$(9.7) \quad \frac{\partial\sigma_{lh}\left(T_{ref}\right)}{\partial T} = \sigma_{lh}\left\{\frac{0.622 L_v}{RT_{ref}^2} - \frac{1}{T_{ref}} - \frac{\sigma_{lh}}{\kappa T_{ref}}\left(\frac{0.622 L_v}{RT_{ref}^2} - \frac{2}{T_{ref}}\right)\right\}.$$

The method for computing the latent heating due to cumulus convection is a variant of the Kuo (1965) method, and is based on the external specification of total column precipitation (We assume that when convective precipitation occurs, it dominates the total precipitation compared to large-scale supersaturation). Monthly averaged precipitation rates required for the cumulus parameterization are specified as a function of latitude and longitude based on the climatology of Legates and Willmott (1990b).

Penetrating convection is allowed to occur only below the level of neutral buoyancy Z_T, where a cloud's dry static energy is equal to the moist enthalpy at the ground. The surface temperature beneath the cumulus cloud base is extrapolated from the environmental temperature at $Z=\Delta Z$ using the dry adiabatic lapse rate. The cloud lapse rate above the surface, on the other hand, is assumed to be moist adiabatic.

When the convective latent heating rates derived from J-1 are contiguously non-zero above the boundary layer, penetrating convection is assumed to occur, and the environmental subsidence associated with the cumulus mass flux is computed from J-2. Otherwise, J-2 is overridden and the convective heating and subsidence velocity are set to zero. Only the zonal mean convective latent heating is used in the thermodynamic equation (B-5), to avoid spurious forcing of stationary waves due to the use of monthly averaged precipitation and surface temperature. The eddy heating associated with sub-grid scale convection is taken to be $[W_c]\sigma^* + \langle\sigma\rangle\, W_c^*$, that is the eddy heating by adiabatic compression due to the zonal mean and eddy convective subsidence. This form of eddy heating does not generate a spuriously large flux of planetary waves, probably because of negative feedback from the first of the two eddy heating terms. The second term, on the other hand, allows the convective parameterization to generate equatorial mixed Rossby-gravity waves, as well as planetary Rossby waves.

The adjustable parameter η in J-2 roughly accounts for the fact that some of the latent heating at low temperatures is the result of fusion of cloud liquid water (i.e., water either condensed from ambient vapor or lifted from below by the convective updraft) into ice particles. Moreover, to account for sublimation of water vapor both in the cumulus cloud and in the surrounding environment, we substitute L_s for L_v whenever the appropriate temperature falls below $-25\ ^\circ$ C.

The first-order differential equation for the convective subsidence velocity is solved by invoking the boundary condition, $W_c=0$ at $Z_T=0$. The zonal mean value of this velocity is then used to compute cumulus friction in the vorticity equation (B-4), with the cloud value of zonal wind set equal to the observed zonal wind at 850 mb.

The convective venting of tracers from the boundary layer to the free troposphere, followed by slow subsidence back towards the surface is simulated by the third term on the right hand side of the tracer conservation equation (B-6). It is instructive to expand this term as follows (where we neglect the vertical variation of χ_c):

$$(9.8) \qquad \frac{\partial\{PW_c\,(\chi_c - \chi)\}}{\partial P} = (\chi_c - \chi)\,\frac{\partial\,(PW_c)}{\partial P} + W_c\frac{\partial\chi}{\partial Z}.$$

The first term after the equal sign in Equation (9.8) represents the detrainment of cloud air to the environment, provided $\partial(PW_c)/\,\partial P$ is positive, whereas the second term represents downward advection of environmental air by the convective subsidence. The cloud species mixing ratio is first set equal to that at the surface and then is replaced progressively upwards by the environmental mixing ratio wherever $\partial(PW_c)/\,\partial P$ is less than zero, so that the detrainment term vanishes whenever entrainment of environmental air into the cumulus cloud occurs. In practice, however, we use an upwind scheme to evaluate the left hand side of Equation (9.8), so that the mixing ratio at and above the current level are required to evaluate the vertical flux derivative.

10. GLOBE chemical mechanism. GLOBE employs a detailed set of gas-phase inorganic, methane, and non-methane hydrocarbon (NMHC) reactions, as well as heterogeneous chemistry pertinent to the lower stratosphere. Table 1 displays the model chemical mechanism, with abbreviations for the more complex hydrocarbons listed in Box K. Not all the listed reactions are currently employed, in particular those representing short and intermediate-lived halogenated compounds that deplete ozone in the lower stratosphere.

The NMHC scheme of Table 1 combines explicit or semi-explicit reactions for the simpler hydrocarbons (higher aldehydes, higher peroxides, and organic nitrates other than PAN are represented by surrogate species) with a mechanistic scheme for > C3 alkanes and > C2 alkenes. The mechanistic scheme is based on the extended carbon bond mechanism (CBM-X) of Gery et al. (1989). Additional peroxide and nitrate formation steps have been added for more relevance to the free troposphere, where lower NO_x concentrations allow for greater competition among radical termination pathways than in urban air. The condensed isoprene scheme is taken from Houweling et al. (1998), which is a modified version of that used in the CBM-4 mechanism (Gery et al., 1989).

The heterogeneous reactions employed by GLOBE describe the denitrification of the lower stratosphere in the presence of aerosols and polar stratospheric cloud (PSC) particles. The aerosol denitrification rates are computed as in WMO (1992). The PSC denitrification rates, on the other hand, are specified as a three-tiered step function of temperature as in Granier and Brasseur (1992), with temperature thresholds for Type I (water ice) and Type II (nitric acid trihydrate) PSCs adopted from Turco et al. (1989). No heterogeneous chemistry is currently applied in the troposphere.

Box L defines the chemical tracers that are advected in the current version of GLOBE. Some of these tracers are families in which individual member species are assumed to be in chemical equilibrium. To save computational resources, NMHC tracers are advected only at altitudes below 25 km. Non-advected families include reactive hydrogen, which is assumed to be in chemical equilibrium, and the reactive bromine family ($Br_y = Br + BrO + HOBr + HBr + BrONO_2$), the zonal mean profile of which is specified.

Daylight average photolysis rates computed under cloudless conditions are applied to daytime chemical balances for model species in equilibrium. Explicit night-time average chemistry is also simulated, particularly for nitrogen reservoir species and species that react with ozone. Diurnally averaged production and loss are then deduced for each advected tracer once each model day and applied to 24 1-hour time steps. Conservation of total reactive nitrogen is numerically enforced after applying the production and loss to the chemical tendency of individual reactive nitrogen species. This is done to avoid problems caused by rapid inter-conversion between NO_x and its various reservoirs.

Although more elaborate stiff system chemical solvers exist (e.g., Gear's method), the diurnal average/family approximation employed by GLOBE has the advantage of computational speed, and is more physically based than the polynomial fitting method of Spivakovsky et al. (1990). We demonstrate the adequacy of GLOBE photochemistry in the context of the IPCC PhotoComp scenarios (Olson et al., 1997). These scenarios assume a U.S. standard atmosphere in the computation of photolysis rates, which are then applied to a single box at a specific pressure and temperature. Ozone within the box is allowed to evolve with time assuming various initial concentrations of its precursors.

We shall consider the results of the GLOBE chemical mechanism for two different chemical solvers. Method A is the diurnal average/family technique described above, in which the reactive hydrogen family is defined as $HO_y = H + OH + HO_2 + CH_3 + CH_3O + CH_3O_2 + 2CH_3OOH + ROOH + HONO + HOCl + HCl + HOBr + HBr$. Method B, on the other hand, retains the family approximation but simulates the full diurnal variation of photolysis rates. It also employs a less constrained form of the reactive hydrogen balance (see Boxes M and N), in which the reactive hydrogen family is defined as $HO_x = H + OH + HO_2$. Method B is designed to better account for fast chemistry in heavily polluted plumes, where the transient disequilibria of CH_3O_2, CH_3OOH, ROOH, and HONO become important. To facilitate the treatment of such plumes, Method B employs a variable time step consisting of 1 second for the first minute, 1 minute for the rest of the first hour, and 1 hour for the remainder of the Photocomp simulation.

Table 2 compares diurnal average photolysis rates computed with Method A versus the results of Olson et al. (1997). A similar comparison is made for noon photolysis rates computed with Method B. In the case of Method A, deviations from the mean Photocomp results are typically within 20%, a notable exception being the photolysis of NO_2 near and at the surface. The standard deviation of the diurnal average Photocomp results, on the other hand, is always less than 15% .

Given that Method A employs only two zenith angles in the computation of diurnal average photolysis rates, and that NO_2 photolysis is sensitive to multiple scattering in the troposphere, the larger deviations from the mean Photocomp results near the surface are not surprising. Table 2 also shows, however, that when the same solar zenith angle is specified as in Photocomp, the corresponding comparisons become excellent (i.e., most deviations are less than 5%). The major exception to this is the photolysis of hydrogen peroxide, for which the deviations from the noon Photocomp mean results are about 15% . We have found that these deviations are brought to within 5% if the temperature dependence of the H_2O_2 absorption cross section is suppressed.

Figures 1 and 2 present the ozone and NO_x simulations obtained from Methods A and B for the six Photocomp scenarios (free troposphere, plume with no NMHCs, plume with NMHCs, land, land plus biogenic emissions, and marine). Figures 3, 4 display the actual Photocomp results. The ozone predictions of Method A for the free troposphere (FREE) scenario are actually better than those of Method B, the latter predicting a steeper decline with increasing time than the mean Photocomp results. The NO_x simulations obtained from the two methods in this case agree well with Photocomp, with the exception of the Method A prediction for Day 2, during which there is a transient disappearance of NO_x.

The ozone and NO_x predictions of Method B for the case of a plume with no NMHCs (PLUME-X) are in excellent agreement with the mean Photocomp results. The predictions of Method A, on the other hand, are surprisingly good, although there is somewhat less ozone and correspondingly less consumption of NO_x in the latter part of the simulation than predicted by Method B. In the case of a plume with NMHCs (PLUME-HC), the results of Method B are within the 1 rms deviation envelope, although NO_x is somewhat overpredicted during the latter part of the simulation. We suspect that the lack of reactive nitrogen conservation in some models used in Photocomp allow for overly rapid consumption of NO_x, thus skewing the results. Method A, on the other hand, predicts almost twice as much ozone as Method B in the PLUME-HC case, probably because of the lack of radical termination in Method A as compared to Method B.

In both land cases (LAND and LAND-BIO), Method B is reasonably close to the upper edge of the 1 rms deviation envelope of Photocomp, but noticeably underestimates NO_x consumption after the first day. Method A, on the other hand, does reasonably well in predicting ozone in the LAND case apart from the spike on Day 2, although there are strong oscillations in NO_x as in the FREE case. Method A does not perform as well in LAND-BIO as it does in LAND, particularly for ozone, as might be anticipated due to the very short lifetime of isoprene. In the marine case (MAR), both Method A and Method B overestimate the rate of decline of ozone with time. Method B, however, performs significantly better than Method A in simulating NO_x consumption.

On the whole, Method A seems sufficient to represent the impacts of moderately polluted air in the free troposphere. Method B significantly improves the simulation of heavily polluted plumes while avoiding the greater computational burden of a more sophisticated chemical solver. It would thus be useful in simulating boundary layer chemistry and urban plumes that penetrate into the free troposphere. Presently, however, GLOBE adopts Method A exclusively due to its computational efficiency relative to Method B.

11. Deposition and emissions. GLOBE simulates both wet and dry deposition of chemical species. A summary of the wet deposition scheme is given in Box O. Note that only the convective precipitation is used to simulate wet deposition, as the non-cumulus condensation heating can vary in sign. Convective precipitation rates are not zonally averaged when used to compute the wet deposition of soluble species. Values for Henry's law constants required for the wet deposition scheme are obtained from Panis and Seinfeld (1989) for H_2O_2, CH_2O, CH_3OOH, HNO_3, and HCl. Higher peroxides (ROOH) are assumed to rain out at the same rate as methyl hydrogen peroxide.

Dry deposition is simulated for O_3, H_2O_2, NO_2, HNO_3, PAN, and NTR. Deposition velocities assigned to these species are displayed in Table 3.

Global emissions inventories for several tracers are specified from established databases available through the Internet. Annual mean emissions of anthropogenic methane, nitrous oxide, ethane, propane, higher alkanes, ethylene, and propylene are adopted from the Emission Database for Global Atmospheric Research (EDGAR; Olivier et al., 1996). Natural and biomass burning CH_4 emissions are derived from Fung et al. (1991), as made available by NASA Goddard Institute for Space Studies. The CH_4 and N_2O emissions are not used in the current study, owing to a significant imbalance between global emissions and simulated global destruction at currently reported emission levels. Instead, their mixing ratios in the troposphere are maintained at constant values (160 ppm and 300 ppm respectively) throughout the simulation.

Monthly mean soil and lightning NO_x, methyl chloroform, and isoprene emissions are obtained from the Global Emissions Inventory Activity (GEIA) database. Lightning emissions of NO_x are vertically distributed in proportion to the convective subsidence velocity. For anthropogenic NO_x and CO, emissions from fossil fuel, biofuel, industrial processes, and agricultural waste burning are specified from annual mean values in EDGAR. Monthly mean emissions of NO_x and CO from savannah burning and deforestation are derived by seasonally distributing the annual mean values of EDGAR in the same manner as emissions of black carbon from biomass burning reported in GEIA.

Total acetone emissions are set at 25 Tg/yr, which is consistent with estimates by Singh et al. (2000). Emissions of acetone are presumed to be mostly biogenic, and are distributed spatially and seasonally in the same manner as isoprene emissions.

12. Model initialization and spin-up. The GLOBE simulations to be discussed here begin on December 1, Year 0 (Day 1) with a zonally uniform chemical composition specified from 1-D model results by Brasseur and Solomon (1986) for very long-lived species other than CO_2, the atmospheric mixing ratio of which is set at a constant level of 350 ppm. Stratospheric ozone is initialized from various satellite observations (Cunnold et al., 2000), whereas reactive nitrogen species are initialized from results of previous model versions (which in turn had been initialized using LIMS observations of NO_x and HNO_3.). Initial tropospheric values of ozone mixing ratio decrease smoothly from their values at the tropopause to a surface value of 20 ppb, guided by the tropospheric climatology of Logan (1999). The CO mixing ratio at the surface is initialized to 50 ppb below 30S and to 130 ppb above 50N, varying linearly in between. Above the surface, the CO mixing ratio is set at 10 ppb prior to spin-up.

NMHCs are assumed absent at the start of Day 1, except at the surface, where the concentrations of ethane, propane, higher alkanes, ethane, and higher olefins are based loosely on observational data of Rudolph and Johnen (1990). Corresponding values for PAN and acetone are based on model results by Wang et al. (1998abc), supplemented by acetone observations from Singh et al. (1994). Initial isoprene mixing ratios, on the other hand, are based on sparse observations summarized by Houweling et al. (1998).

Sources of reactive chlorine and bromine other than methyl chloroform are assumed to be in steady state. Monthly average reactive bromine is specified from the latest results of the Caltech 2-D model (Yuk Yung, private communication). The surface value of methyl chloroform is initially set at 100 ppt below 10S and 140 ppt above 30N, varying linearly in between. Methyl chloroform is absent in the free atmosphere prior to spin-up.

GLOBE was run for 1830 days starting from rest and a horizontally uniform temperature profile. Each model year consists of 360 days, with 30 days allotted to each "month". With 25 advected tracers, GLOBE requires only about 92 seconds per simulated day on a Microway Alpha UP2000 Linux system. This computational efficiency enables GLOBE to perform multi-decadal simulations. In the following sections, we present some results from Model Year 5 (Days 1470-1830).

13. Model general circulation. Figures 5–8 present the simulated zonal mean temperature and mean zonal wind averaged over four different months of Model Year 5. These figures can be compared to the CIRA-86 climatology (Fleming et al., 1988) displayed in Figures 9–12. The observational features that are successfully reproduced by GLOBE include:

1. the warm summer stratopause,
2. the cold spot at the equatorial tropopause,
3. the equatorial surface temperature maximum,
4. the winter westerly and summer easterly mesospheric jets,
5. the southern hemisphere mesospheric jet in April,
6. the northern and southern hemisphere tropospheric jet streams,
7. the surface trade wind pattern (equatorial easterlies, mid-latitude westerlies, and polar easterlies).

There are some observational features, however, that are not yet well simulated by the model as it is currently configured. The most important of these features occur in the lower stratosphere, namely:

1. the cold, winter polar vortex,
2. the descent during the southern hemisphere spring of the winter mesospheric jet core at mid-latitudes to the polar lower stratosphere.

While the model does predict strong winds in the southern hemisphere lower stratosphere in October that are comparable to those in the observed polar vortex, the associated cold polar temperatures occur at higher levels than observed. The cold, polar vortex serves as a containment vessel for the critical denitrification mechanism that produces the spring Antarctic ozone hole, hence the model will tend to underpredict the intensity of the ozone hole due to the lesser effectiveness of ozone depletion reactions at higher altitudes. This problem may be due either to the model's relatively coarse horizontal resolution, or to deficiencies in gravity wave and/or planetary forcing in the southern hemisphere. On the whole, however, the GLOBE simulation of the zonal mean wind and temperature is quite comparable to that of established primitive equation GCMs (e.g., Boville, 1995).

A particular success of GLOBE is presented in Figure 13, which shows an unmistakable semi-annual oscillation in the mesospheric zonal wind, although the westerly phase of the simulated SAO is somewhat weaker than observed (see Figure 14). In the lower stratosphere, there appears to be an annual oscillation with alternating easterly and westerly phases, rather than a true QBO. The lack of a QBO may be due to unrealistic temporal variations in the convective forcing of mixed Rossby-gravity waves or to deficiencies in the equatorial Kelvin wave parameterization.

At mid and high latitudes, the primary vertically propagating planetary wave mode in the middle atmosphere is the geostationary Rossby wave, which is forced by orography and by land-ocean heating contrasts. Figure 15 shows the predicted amplitude of planetary wavenumber 1 averaged over January of Model Year 5, whereas Figure 16 shows corresponding observations from Geller et al. (1983) for the northern hemisphere. As predicted by the theory of planetary wave propagation (Matsuno, 1970), the maximum predicted wave amplitude occurs in the region of intense meridional shear in the zonal wind. The simulated maximum amplitude for Model Year 5 is somewhat weaker than observed, perhaps due to sponge layer damping.

Figures 17 and 18 compare the simulated and inferred (i.e., constrained by measurements but not directly observed) Eulerian mean meridional circulation. GLOBE reproduces the Hadley, Ferrel and polar cells in each hemisphere, although model noise apparently results in extra high latitude cells. The asymmetry between the two tropical Hadley cells is also reproduced by the model, although it is not as intense as in Figure 16. The simulated Hadley cell in the winter hemisphere is also more vigorous than the inferred circulation would suggest.

Air parcel trajectories tend to follow, not the Eulerian mean circulation, but rather the Lagrangian mean circulation, which is the net transport resulting from the combination of the Eulerian zonal mean circulation and the turbulent eddies (Andrews and McIntyre, 1976). An approximation to the Lagrangian mean circulation is the so-called "residual mean" circulation (Rosenlof, 1995). Figure 19 shows the predicted residual mean circulation for January of Model Year 5, coinciding with the most active period of planetary wave propagation. Figure 20, on the other hand, presents the inferred residual mean circulation obtained by Rosenlof (1995) using various techniques. The strong Brewer-Dobson (poleward and downward) circulation in the winter hemisphere is well simulated, as is the weaker and more vertically confined cell in the summer hemisphere.

Thus far, we have presented the large-scale general circulation predicted by GLOBE. Also of interest, however, are the sub-grid scale mass fluxes produced by the cumulus parameterization. These fluxes are displayed in Figure 21, which presents results in the northern hemisphere for July of Model Year 5. Corresponding results from the Goddard Earth Observing System data assimilation system (GEOS-1 DAS) are shown in Figure 22. Below 40N, the mass fluxes generated by GLOBE in the lower troposphere are somewhat weaker than those generated by the GEOS-1 DAS. Above 40N and in the tropical upper troposphere, however, the GLOBE mass fluxes are comparable to the assimilated data.

A further test of the quality of the GLOBE model's convection scheme and general circulation is provided by the short-lived tracer ^{222}Rn. Jacob et al. (1997) presented an intercomparison of results from various CTMs in an experiment in which ^{222}Rn was advected as a passive tracer. We have reproduced this experiment with GLOBE, the results of which are illustrated in Figures 23 and 24. Figures 25 and 26, on the other hand, show corresponding results for various 3-D global models as reported in Jacob et al. (1997). The GLOBE model results are clearly within the variability displayed in Figures 25 and 26. GLOBE predicts a maximum zonal mean ^{222}Rn mixing ratio of 10^{-20} in the tropical upper troposphere as does the GISS model. The isopleths of Figure 24 and the GISS model isopleths of Figure 26 likewise show several qualitative features in common, including ridges over central Africa, northern Brazil, and eastern North America, and a trough over the eastern equatorial Pacific.

14. Stratospheric and tropospheric ozone. A demonstration of GLOBE's capability as an interactive dynamical-chemical model is provided by a comparison of model predictions of ozone in the stratosphere and troposphere with available observations. Figure 27 presents January mean stratospheric ozone mixing ratios generated by GLOBE, whereas Figure 28 shows corresponding observations from the UARS Reference Atmosphere Project (URAP) 1998 baseline standard climatology, which is a combination of data obtained from the MLS, HALOE and SAGE II instruments (Cunnold et al., 2000).

GLOBE reproduces the observed maximum in ozone mixing ratio of ~ 10 ppm at 10 mb near the equator. The shapes of the isopleths of Figures 27 and 28 also share some important features, particularly the upward bulge near the equatorial tropopause and the downward bulge at mid and high latitudes in the lower stratosphere. These bulges are caused by the upwelling and downwelling branches of the residual mean circulation (Figures 19 and 20). The model also produces an upward bulge at mid latitudes in the vicinity of 2 mb, although this feature is not as prominent as in the observations. It cannot be explained by the residual circulation, which is downward rather than upward in this region, and may instead be due to vertical gravity wave diffusion. At polar latitudes in the northern hemisphere, ozone mixing ratios are somewhat high in the vicinity of 20 mb, perhaps because of excess horizontal diffusion.

Figure 29 presents the simulated vertical profile of tropospheric ozone for various months of the year. This can be compared to the observational climatology compiled by Logan (1999), illustrated in Figure 30. Both sets of data show weak variation with latitude in the vertical ozone profile at high latitudes, and much stronger variation at low latitudes. In the tropics (22.5N), the effects of convection are obvious from the slow variation of mixing ratio with height evident in both the model results and the observations. At mid and polar latitudes, the model ozone values start increasing rapidly past 100 ppb somewhere between 300 mb and 400 mb, as is typically observed. However, the inflection points of the simulated profiles occur at higher values of ozone than the climatology of Logan (1999) would suggest.

Figures 31 and 32 present the model-predicted total ozone column averaged over January as compared to 11-year average satellite data derived from the Total Ozone Mapping Spectrometer (TOMS) on Nimbus-7 (Stanford et al., 1995). As January is a period of intense planetary wave activity in the northern hemisphere, a wavelike structure is clearly visible in the simulated ozone column at northern mid and high latitudes. The observed ridge in column ozone near Siberia is shifted southwest by the model towards northern China, whereas the observed southern hemisphere ridge near 60S is displaced by the model significantly to the east. In general, however, the simulated column values are quite comparable to the TOMS measurements, although the tropical and south polar column values are somewhat lower than observed by the TOMS instrument. Figures 33 and 34 present the simulated and observed standard deviation of column ozone from the zonal mean as a function of latitude, as originally recommended by Jiang et al. (1998). Both datasets show standard deviations below 10 Dobsons throughout the southern hemisphere and tropics, and a significant increase in standard deviation above 30N.

Finally, Figures 35 and 36 compare the simulated and observed (Stanford et al., 1995) seasonal variation in zonal mean total column ozone. Both datasets show minimum values throughout the year in the tropics and high values in the northern hemisphere spring at high latitudes. GLOBE produces an Antarctic ozone hole (and a corresponding ridge at mid latitudes), with minimum column values around 200 Dobsons. The simulated hole and accompanying mid-latitude ridge, however, appear about one month too early compared to observations. This defect may be caused by GLOBE's imperfect simulation of the dynamics of the Antarctic spring polar vortex.

15. Summary and conclusion. This paper has presented some initial results of the GLOBE model. While a more extensive evaluation involving a wider variety of tracers cannot be attempted here, it has nevertheless been demonstrated that the primary features of the atmospheric general circulation and of the distribution of ozone in the stratosphere and troposphere are reasonably well simulated by GLOBE. There are, of course, outstanding areas for improvement, such as the lack of a QBO and of a sufficiently developed spring Antarctic polar vortex in the lower stratosphere.

In spite of its current shortcomings, GLOBE occupies an important niche in the hierarchy of models that assess the impact of human activity on the global atmosphere. A valuable aspect of GLOBE that sets it apart from other models is its ability to combine the most important features of both GCMs and 3-D CTMs in a single, cost-effective platform. This is due to a number of simplifying approximations, including the use of the global balance system of hydrodynamic equations and diurnal average photochemistry, and the parameterization of the hydrological cycle based on an externally specified, horizontally uniform relative humidity profile. Because of the increasing power and affordability of high-performance workstations, GLOBE is an inexpensive tool with which to conduct decadal simulations of atmospheric pollution.

GLOBE's comprehensive architecture makes it useful for a variety of purposes, such as the evaluation of ozone depletion potentials and indirect global warming potentials of short-lived compounds, and the simulation of the long-range transport of persistent organic pollutants. It is especially appropriate as the global component of a one-atmosphere model consisting of a series of nested models spanning urban to global scales. It can also be combined with soil and water compartments to yield a powerful multimedia assessment tool. Given the increasing importance of multi-scale pollution problems and the assessment of collateral impacts of environmental policies and regulations, such a one-atmosphere model may play an important role in the development of more effective pollution control strategies in the future.

Acknowledgements. I would like to acknowledge Prof. K.K. Tung of the University of Washington for his suggestion to incorporate equatorial and breaking planetary wave parameterizations in the global balance scheme. I am also grateful to Drs. Adel Hanna, Rohit Mathur, Kiran Alapaty, and Carlie Coats of MCNC for their help in critiquing and implementing various aspects of the model. Thanks also to Prof. Peter Stone of MIT, Prof. Don Wuebbles of the University of Illinois at Urbana-Champaign, and Dr. Joseph Pinto of the U.S. EPA for their helpful suggestions and comments. Above all, A.M.D.G.

List of Boxes

A. Symbol glossary
B. Fundamental equations
C. Computation of vertical velocity
D. Horizontal diffusion
E. Equatorial wave parameterizations
F. Discrete gravity wave parameterization
G. Vertical diffusion and boundary layer turbulence
H. Cloud radiative properties
I. Non-cumulus condensation and evaporation
J. Cumulus parameterization
K. Chemical abbreviations
L. Advected tracers
M. Method A reactive hydrogen balance
N. Method B reactive hydrogen balance
O. Wet deposition scheme

Box A: Symbol Glossary

$< > =$ horizontal average
$(\ )' =$ deviation from horizontal average
$[\ ] =$ zonal average
$(\ )^* =$ deviation from zonal average
$(\ )_s =$ surface value
$(\ )_{obs} =$ observed value
$(\ )_{ET} =$ value at equatorial tropopause
$(\ )_c =$ value within cumulus cloud
$a =$ radius of earth
$H =$ atmospheric scale height (7 km)
$\mathbf{i} =$ zonal unit vector
$\mathbf{j} =$ meridional unit vector
$\mathbf{k} =$ vertical unit vector
$t =$ time
$\theta =$ latitude
$\lambda =$ longitude
$P =$ pressure
$Z = -\ln P$
$z = ZH$ (approximate geometric height)
$\Delta Z =$ vertical grid spacing $(=1/6)$
$u =$ zonal velocity
$v =$ meridional velocity
$W = dZ/dt$ (vertical velocity)
$W_c =$ convective subsidence velocity
$\Phi =$ geopotential
$\psi =$ horizontal stream function
$\partial X/\partial P =$ velocity potential
$\zeta = \nabla^2 \psi$ (vorticity)
$f =$ Coriolis parameter
$\Pi =$ potential vorticity
$T =$ temperature
$T_{ref} = \min(<T>, 255\ K)$
$R =$ ideal gas constant
$C_p =$ specific heat capacity of air
$\kappa = R/C_p$
$\sigma = \partial T/\partial Z + \kappa T$ (static stability)
$\sigma_{lh} =$ condensation adjustment factor
$D_{lh} =$ latent heat detrainment/diffusion
$Q_{rad} =$ radiative heating
$Q_{conv} =$ convective latent heating
$Q_{cond} =$ non-cumulus latent heating
$\mathbf{D} =$ zonal mean surface drag
$\mathbf{E} =$ equatorial Kelvin wave forcing
$\mathbf{F} =$ forcing by equatorial gravity waves

$z_s =$ orography
$h_{SL} =$ height of surface layer $(=50\ m)$
$c =$ wave zonal phase speed
$k =$ zonal wavenumber
$A =$ Kelvin wave forcing amplitude
$N =$ Brunt-Vaisala frequency
$\rho =$ air density
$(\ )_{brk} =$ value at wave breaking level
$Fr =$ Froude number
$h_{GW} =$ gravity wave height perturbation
$\alpha_T =$ radiative dissipation rate
$\alpha_R =$ Rayleigh friction coefficient
$\alpha_O =$ orographic gravity wave dissipation
$\alpha_I =$ shear instability-generated dissipation
$\alpha_M =$ mountain torque coefficient
$C_D =$ surface drag coefficient
$K_H =$ horizontal eddy diffusion coefficient
$K_P =$ polar diffusion coefficient
$K_M =$ mesoscale horizontal diffusion
$K_B =$ breaking planetary wave diffusion
$K_N =$ fourth-order diffusion coefficient
$K_Z =$ vertical eddy diffusion coefficient
$K_{ZA} =$ gravity wave vertical diffusion
$K_{ZB} =$ background vertical diffusion
$K_{ZC} =$ vertical diffusion due to convection
$\varepsilon =$ cloud emissivity
$\tau =$ cloud optical depth
$LWP =$ liquid water path
$h_{LW} =$ liquid water scale height
$r =$ relative humidity
$q =$ specific humidity
$q_s =$ saturation specific humidity
$L_v =$ latent heat of vaporization
$L_s =$ latent heat of sublimation
$L_f =$ latent heat of fusion
$\mathcal{P} =$ column precipitation (mm/day)
$Z_T =$ level of neutral buoyancy
$r_i =$ rainout rate of chemical species i
$H_i =$ Henry's law constant for species i
$R^* =$ molar ideal gas constant
$\mathcal{L} =$ volume fraction of liquid water
$\rho_w =$ density of liquid water
$\chi =$ chemical species volume mixing ratio
$G =$ photochemical generation

Box B: Fundamental Equations

(B-1) Hydrostatic Approximation

$$RT' = \frac{\partial \Phi'}{\partial Z}$$

(B-2) Global Quasi-Geostrophic Balance

$$\nabla^2 \Phi' = \nabla \bullet f\nabla \psi$$

(B-3) Mass Continuity

$$PW = \nabla^2 X$$

(B-4) Vorticity Equation

$$\frac{\partial \zeta}{\partial t} = -\mathbf{k} \times \nabla\psi \bullet \nabla(f + \zeta) - \nabla \bullet f\nabla \frac{\partial X}{\partial P} - \nabla \bullet \left([\zeta]\nabla \frac{\partial [X]}{\partial P} + [W]\frac{\partial \nabla[\psi]}{\partial Z} \right)$$

$$- \nabla \bullet \left(\mathbf{k} \times \mathbf{D} + \mathbf{E} + \mathbf{k} \times \mathbf{F} - \frac{\partial}{\partial P}\left\{ P[W_c]\nabla\left([\psi_c] - [\psi] \right) \right\} + \alpha_R \nabla\psi \right)$$

$$+ \nabla \bullet \frac{\partial}{\partial P}\left\{ P\frac{K_Z}{H^2} \frac{\partial(\mathbf{k} \times \nabla\psi)}{\partial Z} \right\} + \nabla \bullet K_H \nabla\zeta - \nabla^2 K_N \nabla^2 \zeta$$

(B-5) Thermodynamic Equation

$$\frac{\partial T'}{\partial t} = -\mathbf{k} \times \nabla\psi \bullet \nabla T' - \left(\langle\sigma\rangle - \sigma_{lh}(T_{ref}) \right)W - \nabla \bullet \left([T']\nabla \frac{\partial [X]}{\partial P} \right)' + \frac{\partial}{\partial P}(P[W][T'])' + Q_{rad}'$$

$$- \frac{\partial}{\partial P}\left\{ P\frac{K_Z}{H^2}\sigma \right\}' + \frac{\kappa(K_Z\sigma)'}{H^2} + \nabla \bullet K_H \nabla T' - \nabla^2 K_N \nabla^2 T' + \langle\sigma\rangle W_c^* + \left\{ [W_c]\sigma^* \right\}'$$

$$+ \left[W\frac{\partial \sigma_{lh}(T_{ref})}{\partial T}(T - T_{ref}) + D_{lh} - \sigma_{lh}(T)W_c - \left(\frac{\sigma_{lh}(T)}{\kappa T} \right)Q_{rad} + Q_{conv} \right]'$$

(B-6) Chemical Species Conservation

$$\frac{\partial \chi}{\partial t} = -\nabla \bullet \chi\left(\mathbf{k} \times \nabla\psi + \nabla \frac{\partial X}{\partial P} \right) + \frac{\partial}{\partial P}(PW\chi) + \frac{\partial}{\partial P}\left\{ PW_c(\chi_c - \chi) \right\} - \frac{\partial}{\partial P}\left\{ P\frac{K_Z}{H^2} \frac{\partial \chi}{\partial Z} \right\}$$

$$+ \nabla \bullet K_H \nabla\chi + G$$

Box C: Computation of Vertical Velocity

(C-1) Vertical Velocity Equation

$$R\left(\langle\sigma\rangle - \sigma_{lh}\left(T_{ref}\right)\right)\nabla^2 W - \left\{\left(\nabla\bullet f\nabla\right)\nabla^{-2}\right\}^2 \frac{\partial}{\partial Z}\frac{\partial\left(PW\right)}{\partial P} = R\nabla^2 G_1 - \left\{\left(\nabla\bullet f\nabla\right)\nabla^{-2}\right\}\frac{\partial G_2}{\partial Z}$$

(C-2) Thermal Forcing

$$G_1 = -\mathbf{k}\times\nabla\psi\bullet\nabla T' - \nabla\bullet\left([T']\nabla\frac{\partial[X]}{\partial P}\right)' + \frac{\partial}{\partial P}\left(P[W][T']\right)' + Q'_{rad}$$

$$+ \nabla\bullet K_H\nabla T' - \nabla^2 K_N\nabla^2 T' - \frac{\partial}{\partial P}\left\{P\frac{K_Z}{H^2}\sigma\right\}' + \frac{\kappa\left(K_Z\sigma\right)'}{H^2} + \langle\sigma\rangle W_c^* + \left\{[W_c]\sigma^*\right\}'$$

$$+ \left[W\frac{\partial\sigma_{lh}\left(T_{ref}\right)}{\partial T}\left(T - T_{ref}\right) + D_{lh} - \sigma_{lh}(T)W_c - \left(\frac{\sigma_{lh}(T)}{\kappa T}\right)Q_{rad} + Q_{conv}\right]'$$

(C-3) Mechanical Forcing

$$G_2 = -\mathbf{k}\times\nabla\psi\bullet\nabla(f+\zeta) - \nabla\bullet\left([\zeta]\nabla\frac{\partial[X]}{\partial P} + [W]\frac{\partial\nabla[\psi]}{\partial Z}\right) + \nabla\bullet K_H\nabla\zeta - \nabla^2 K_N\nabla^2\zeta$$

$$- \nabla\bullet\left(\mathbf{k}\times\mathbf{D} + \mathbf{E} + \mathbf{k}\times\mathbf{F} - \frac{\partial}{\partial P}\left\{P[W_c]\nabla\left([\psi_c] - [\psi]\right)\right\} + \alpha_R\nabla\psi - \frac{\partial}{\partial P}\left\{P\frac{K_Z}{H^2}\frac{\partial(\mathbf{k}\times\nabla\psi)}{\partial Z}\right\}\right)$$

(C-4) Bottom Boundary Condition

$$W(Z=0) = -\frac{h_{SL}}{H}\nabla\bullet[v_{obs}] + \frac{[u_{obs}]}{aH\cos\theta}\frac{\partial z_s^*}{\partial\lambda} + \frac{[v_{obs}]}{aH}\frac{\partial z_s^*}{\partial\theta}$$

Box D: Horizontal Diffusion

(D-1) Second-Order Diffusion Components

$$K_H = \left(\frac{10^2 \text{ m}^2\text{s}^{-1}}{P} \right) + K_P + K_M + K_B$$

(D-2) Polar Diffusion

$$K_P = \begin{cases} 5 \times 10^6 \text{ m}^2\text{s}^{-1} & |\theta| \geq 81° \\ 0 & |\theta| < 81° \end{cases}$$

(D-3) Mesoscale Eddy Diffusion

$$K_M = \begin{cases} 10^6 \text{ m}^2\text{s}^{-1} & [W_c] > 0 \text{ or } z \leq 1 \text{ km} \\ 0 & \text{otherwise} \end{cases}$$

(D-4) Breaking Planetary Wave Diffusion

$$K_B = \min \left\{ \sum_k \Gamma_k , 2 \times 10^6 \text{ m}^2 s^{-1} \right\}$$

$$\Gamma_k (Z_{brk}, \theta_{brk}) = \frac{\left(\dfrac{\delta_k}{k[u]} \right)\left(\dfrac{[v * v *]_k}{k[u]} \right)}{1 + \left(\dfrac{\delta_k}{k[u]} \right)^2} \quad ; \quad \frac{\partial \Pi_k^*}{\partial \theta}\bigg|_{Z=Z_{brk}, \theta=\theta_{brk}} \geq \frac{\partial [\Pi]}{\partial \theta}\bigg|_{Z=Z_{brk}, \theta=\theta_{brk}}$$

$$\Gamma_k (Z, \theta) = \begin{cases} 0 & , Z > Z_{brk} \\ \Gamma_k (Z_{brk}, \theta_{brk})(P_{brk}/P)\exp -\left| \dfrac{2(\theta - \theta_{brk})}{\Pi_k^*} \dfrac{\partial [\Pi]}{\partial \theta} \right| & , Z \leq Z_{brk} \end{cases}$$

$$\frac{\delta_k}{k[u]} = \left(\frac{fa}{NH} \right)\left(\frac{fm}{N} \right)\left\{ [u] - \frac{3}{2} \frac{\partial [u]}{\partial Z} \right\}\left(\frac{\partial [\Pi]}{\partial \theta} \right)^{-1}$$

$$\left(\frac{fm}{N} \right)^2 = \frac{1}{a[u]}\left(\frac{\partial [\Pi]}{\partial \theta} \right) - \left\{ k^2 + \frac{1}{a^2}\left(\frac{1}{\Pi_k^*} \frac{\partial [\Pi]}{\partial \theta} \right)^2 \right\} - \left(\frac{f}{2NH} \right)^2$$

$$\Pi = \zeta + f + f \frac{\partial}{\partial P}\left(\frac{PT'}{<\sigma>} \right)$$

Box E: Equatorial Wave Parameterizations

(E-1)   Kelvin Wave Forcing

$$\mathbf{E}(Z,\theta) = \mathbf{j}\frac{A}{H}\exp(Z - Z_{ET})R(Z)\exp\{-S(Z)\}\exp\left\{\frac{2\Omega a\theta^2}{([u(\theta=0)]-c)}\right\}$$

$$R(Z) = \frac{\alpha_T NH}{k([u(\theta=0)]-c)^2}$$

$$S(Z) = \int_{Z_{ET}}^{Z} R(Z')dZ'$$

	Wave 1	Wave 2	Wave 3
k	3.14×10^{-7} m^{-1}	3.147×10^{-7} m^{-1}	1.57×10^{-7} m^{-1}
A	5.0×10^{-3} m^{2}s^{-2}	2.5×10^{-3} m^{2}s^{-2}	7.5×10^{-3} m^{2}s^{-2}
c	20 m/s	30 m/s	50 m/s

(E-2)   Equatorial Gravity Waves with a Continuous Spectrum

$$\mathbf{F} = \mathbf{i}\left|\mathcal{B}_o([u])\right|\frac{\partial[u]}{\partial Z}\exp(Z - Z_{ET})$$

$$\mathcal{B}_o(c) = \frac{[W_c(Z_{ET})]}{2}\left|\frac{c}{[u(Z_{ET})]}\right|\exp\left(-\left|\frac{c}{[u(Z_{ET})]}\right|\right)$$

Box F: Discrete Gravity Wave Parameterization

(F-1) Rayleigh Friction Components

$$\alpha_R = \alpha_O + \alpha_I$$

(F-2) Orographic Gravity Wave Dissipation

$$\alpha_O = \begin{cases} \dfrac{(0.2\times10^{-10}\,\mathrm{m}^{-2})}{2N}[u]^2\,\max\left\{\dfrac{d(\ln Fr^2)}{dZ},0\right\} &, \; Z \geq Z_{brk} \\[2em] \alpha_O(Z_{brk})\exp(Z-Z_{brk}) &, \; Z \leq Z_{brk} \end{cases}$$

$$Fr(Z_{brk}) = \frac{h_{GW}\,N(Z_{brk})}{[u(Z_{brk})]}\left\{\frac{\rho(0)N(0)[u(0)]}{\rho(Z_{brk})N(Z_{brk})[u(Z_{brk})]}\right\}^{1/2} > 1$$

(F-3) Phase Speed of Gravity Waves Excited by Shear Instability

$$U_1 = \max\left(\left|\,[u(Z \geq Z_{ET})]\,\right|\right)$$

$$|\,U_1 - U_2\,| = \max|\,U_1 - [u]\,|$$

$$c = U_2 - (5\,\mathrm{m/s})\frac{(U_1 - U_2)}{|U_1 - U_2|}$$

(F-4) Dissipation of Gravity Waves Excited by Shear Instability

$$\alpha_I = \begin{cases} (0.2\times10^{-9}\,\mathrm{s/m}^2)([u]-c)^2\left\{1-\dfrac{3}{([u]-c)}\dfrac{\partial[u]}{\partial Z}\right\} &, \; Z \geq Z_{brk} \\[2em] \alpha_I(Z_{brk})\exp(Z-Z_{brk}) &, \; Z \leq Z_{brk} \end{cases}$$

$$Z_{brk} = Z_{ET} + 3\ln\left|\frac{[u]-c}{4\,\mathrm{m/s}}\right|$$

Box G: Vertical Diffusion and Boundary Layer Turbulence

(G-1)   Vertical Diffusion Components

$$K_Z = K_{ZA} + K_{ZB} + K_{ZC}$$

$$K_{ZA} = \sum_k \frac{([u] - c_k)^2}{N^2} \alpha_k$$

$$K_{ZB} = \begin{cases} (0.5\,\mathrm{m^2s^{-1}})\exp\left\{\dfrac{(z\text{ - }20\,\mathrm{km})}{16.69\,\mathrm{km}}\right\} & z \geq 20\,\mathrm{km} \\[2ex] 0.5\,\mathrm{m^2s^{-1}} & 1\,\mathrm{km} < z < 20\,\mathrm{km} \\[2ex] 10\,\mathrm{m^2s^{-1}} & z \leq 1\,\mathrm{km} \end{cases}$$

$$K_{ZC} = \begin{cases} 9.5\,\mathrm{m^2s^{-1}} & [W_c] > 0 \text{ and } z > 1\,\mathrm{km} \\[2ex] 0 & \text{otherwise} \end{cases}$$

(G-2)   Surface Momentum Exchange

$$\mathbf{D} = \left\{ \alpha_M + \left| \frac{fK_Z}{2H^2} \right|^{1/2} + \frac{2C_D V_s}{H\Delta Z} \right\} \nabla[\psi]_{Z=\Delta Z/2}$$

$$V_s = \max\left\{ \big|\,[u_{obs}]\,\big|, \big|\,[u(Z = 3\Delta Z/2)]\,\big|, \big|\,[u(Z = \Delta Z/2)]\,\big| \right\}$$

Box H: Cloud Radiative Properties

(H-1)   Cloud Liquid Water Path

$$\mathrm{LWP} = (0.18\,\mathrm{gm^{-3}})h_{\mathrm{LW}}\left\{P_{\mathrm{J+1}}^{H/h_{\mathrm{LW}}} - P_{\mathrm{J}}^{H/h_{\mathrm{LW}}}\right\}$$

$$h_{\mathrm{LW}} = 1080\,\mathrm{m} + (2000\,\mathrm{m})\cos^2\theta$$

(H-2)   Cloud Emissivity

$$\varepsilon = 1 - \exp\left\{-(0.1\,\mathrm{m^2 g^{-1}})\mathrm{LWP}\right\}$$

(H-3)   Cloud Optical Depth

$$\tau = (0.16\,\mathrm{m^2 g^{-1}})\mathrm{LWP}$$

Box I: Non- Cumulus Condensation and Evaporation

(I-1)   Relative Humidity at Pressures Greater Than 0.1 atm

$$r = 0.77\,(P - 0.02)/0.98$$

(I-2)   Non-Cumulus Latent Heating Parameterization

$$\sigma_{\mathrm{lh}}(T) = \frac{\left(r\dfrac{L_v q_s(T)}{C_p}\right)\left(\dfrac{0.622 L_v}{C_p T}\right)}{1 + r\left(\dfrac{L_v q_s(T)}{C_p}\right)\left(\dfrac{0.622 L_v}{RT^2}\right)}$$

$$D_{\mathrm{lh}} = \frac{\dfrac{L_v}{C_p}\left(q_s(T_c) - q\right)\max\left(\dfrac{\partial(PW_c)}{\partial P}, 0\right) - \dfrac{\partial}{\partial P}\left\{P\dfrac{K_z}{H^2}\dfrac{L_v}{C_p}\dfrac{\partial q}{\partial Z}\right\}}{\left\{1 + r\left(\dfrac{L_v q_s(T)}{C_p}\right)\left(\dfrac{0.622 L_v}{RT^2}\right)\right\}}$$

$$Q_{\mathrm{cond}} = \sigma_{\mathrm{lh}}(T_{\mathrm{ref}})W + \left[W\dfrac{\partial\sigma_{\mathrm{lh}}(T_{\mathrm{ref}})}{\partial T}(T - T_{\mathrm{ref}}) - \dfrac{\sigma_{\mathrm{lh}}(T)}{\kappa T}Q_{\mathrm{rad}} + D_{\mathrm{lh}} - \sigma_{\mathrm{lh}}(T)W_c\right]$$

$$T_{\mathrm{ref}} = \min(<T>, 255\,\mathrm{K})$$

Box J: Cumulus Parameterization

(J-1) Latent Heating due to Penetrating Convection

$$Q_{conv}(Z \le Z_T) = \frac{(0.25\ \text{K/mm})P\max(T_c - T, 0)}{\displaystyle\int_0^1 \max(T_c - T, 0)\,dP}$$

$$C_p T(0) + L_v q_s(0) = C_p T(Z_T) + gHZ_T$$

(J-2) Cumulus Mass Flux

$$\frac{L_v}{C_p}\frac{\partial\{PW_c q_s(T_c)\}}{\partial P} = \frac{\mathcal{F}\{\eta\max(T_c - T, 0) + (L_v / C_p)(q_s(T_c) - q)\}}{\displaystyle\int_0^{P(Z=\Delta Z)} \{\eta\max(T_c - T, 0) + (L_v / C_p)(q_s(T_c) - q)\}\,dP}$$

$$\eta = \begin{cases} 1 & T_c > 248.15\ \text{K} \\ 0.2 & T_c \le 248.15\ \text{K} \end{cases}$$

$$\mathcal{F} = P_{Z=\Delta Z}\frac{L_v}{C_p}\left\{\left(W + \frac{Q_{conv}}{\sigma_{lh}}\right)q_s - \frac{K_Z}{H^2}\frac{\partial q_s}{\partial Z}\right\}_{Z=\Delta Z}$$

Box K: Chemical Abbreviations

ALD2	Higher aldehyde
ALK4	Paraffin group
ANO2	Acetylmethylperoxy radical
AONE	Acetone
AONG	Acetone group
C2O3	Peroxyacyl radical
CRIG	Crigee biradical
DCB	Unsaturated dicarbonyls
ETE	Ethylene group
GLY	Glyoxal
ISOP	Isoprene
KET	Ketone carbonyl group
MCRG	Methyl Crigee biradical
NTR	Organic nitrate
OLE	Olefinic carbon bond
PAN	Peroxyacyl nitrate
PAR	Paraffinic carbon group
PROP	Propane group
RO2	Primary organic peroxy radical
RO2R	Secondary organic peroxy radical
ROOH	Higher peroxide
TO2	Toluene-hydroxyl radical adduct
XO2	NO to NO_2 operator
XO2N	NO to NTR operator

Box L: Advected Tracers

$O_x = O + O(^1D) + O_3$	HCl
H_2O_2	$COCl_2$
CH2O	CH_3CCl_3
CH_4	C_2H_6
$CH_xO_y = CH_3O_2 + CH_3OOH$	$PROP = C_3H_8 + n\text{-}C_3H_7O_2 + i\text{-}C_3H_7O_2$
CO	AONG = AONE + ANO2
N_2O	PAN
$NO_x = N + NO + NO_2 + NO_3 + HONO$	ALK4 = PAR + RO2 + RO2R +
N_2O_5	ROR + KET
HNO_3	$ETE = C_2H_4 + C_2H_5O_2$
HNO_4	OLE
H_2O	ISOP
$Cl_y = Cl + ClO + HOCl + ClONO_2$	NTR

Box M: Method A Reactive Hydrogen Balance

(M-1) Hydroxyl Radical Equation

$$A[OH]^2 + B[OH] - C = 0$$

$$A = 2k_{38} + 2\left\{ k_{37} + (k_{39} + k_{40})\frac{[H]}{[OH]} + k_{48}\frac{[HO_2]}{[OH]} \right\}\frac{[HO_2]}{[OH]}$$

$$B = 2k_{27}[CH_3OOH] + 2k_{65}[HONO] + 2k_{83}[HCl] + 2k_{86}[HOCl] + 2k_{102}[HBr] +$$

$$\ell_{67}[M][NO_2] + k_{69}[HNO_3] + \ell_{70}[M][NO_2]\frac{[HO_2]}{[OH]} + k_{71}[HNO_4] + k_{144}[C_2H_6] +$$

$$(k_{148} + k_{149})[C_3H_8] + (k_{158} + k_{159})[PAR] + k_{168}[ALD2] + k_{172}[AONE] +$$

$$k_{175}[KET] + k_{177}[MGLY] + k_{185}[NTR] + 1.51k_{189}[ROOH] + k_{190}[C_2H_4] +$$

$$0.15k_{202}[ISOP] + 0.9k_{210}[C_7H_8] - k_{179}[C2O3]\frac{[HO_2]}{[OH]}$$

$$C = 2k_{17}[H_2][O_2(^1\Sigma)] + 2J_{18}[H_2O] + 2(k_{19}[H_2O] + k_{20}[H_2])[O(^1D)] + 2J_{50}[H_2O_2] +$$

$$2(J_{33} + k_{31}[O] + k_{32}[Cl] + k_{108}[Br])[CH_2O] + 2(k_{21}[O(^1D)] + k_{80}[Cl])[CH_4] + 2k_{81}[Cl][H_2] +$$

$$2J_{169}[ALD2] + 2J_{170}[AONE] + J_{68}[HNO_3] + (J_{72} + j_{73})[HNO_4] + j_{166}[ROR] +$$

$$\{k_{145}[C_2H_6] + (k_{150} + k_{151})[C_3H_8]\}[Cl] + J_{171}[AONE] + J_{176}[MGLY] + J_{184}[NTR] +$$

$$J_{188}[ROOH] + J_{208}[GLY] + (0.2k_{196} + 1.6k_{197})[OLE][O_3] + 0.58k_{203}[ISOP][O_3] +$$

$$0.9k_{204}[ISOP][NO_3] + [NO]\{ k_{146}[C_2H_5O_2] + k_{152}[n\text{-}C_3H_7O_2] + k_{153}[i\text{-}C_3H_7O_2] +$$

$$k_{160}[RO2] + k_{173}[ANO2] + k_{178}[C2O3] + (k_{191} + k_{192})[C_2H_5O_3] + k_{207}[ACO2] +$$

$$1.45k_{211}[TO2] \}$$

(M-2) Equilibrium Ratios

$$\frac{[H]}{[OH]} = \frac{k_{36}[CO] + k_{41}[H_2] + k_{43}[O]}{k_{44}[O_3] + \ell_{45}[M][O_2]}$$

$$\frac{[HO_2]}{[OH]} = \frac{D}{E}$$

$$D = k_{23}[CH_4] + k_{36}[CO] + k_{41}[H_2] + k_{43}[O] + \left(k_{42} - k_{44}\frac{[H]}{[OH]}\right)[O_3] + k_{30}[CH_2O] +$$

$$k_{49}[H_2O_2] + k_{86}[HOCl] + k_{97}[BrO] + k_{144}[C_2H_6] + (k_{148} + k_{149})[C_3H_8] +$$

$$(k_{158} + k_{159})[PAR] + k_{172}[AONE] + k_{190}[C_2H_4] + 0.85k_{202}[ISOP] + k_{206}[C_2H_2] +$$

$$k_{209}[GLY] + 1.55k_{210}[C_7H_8]$$

$$E = k_{25}[CH_3O_2] + k_{46}[O] + k_{47}[O_3] + k_{57}[NO] + k_{84}[ClO] + k_{106}[BrO] + k_{147}[C_2H_5O_2] +$$

$$k_{156}[n\text{-}C_3H_7O_2] + k_{157}[i\text{-}C_3H_7O_2] + k_{164}[RO2] + k_{165}[ROR] + k_{179}[C2O3]$$

Box N: Method B Reactive Hydrogen Balance

(N-1)   Equilibrium Ratios

$$\frac{[HO_2]}{[OH]} = \frac{D}{E+F}$$

$$D = k_{23}[CH_4] + k_{36}[CO] + k_{41}[H_2] + k_{43}[O] + \left(k_{42} - k_{44}\frac{[H]}{[OH]}\right)[O_3] + k_{30}[CH_2O] +$$
$$\quad k_{49}[H_2O_2] + k_{86}[HOCl] + k_{97}[BrO] + k_{144}[C_2H_6] + (k_{148} + k_{149})[C_3H_8] +$$
$$\quad (k_{158} + k_{159})[PAR] + k_{172}[AONE] + k_{190}[C_2H_4] + 0.85k_{202}[ISOP] + k_{206}[C_2H_2] +$$
$$\quad k_{209}[GLY] + 1.55k_{210}[C_7H_8]$$

$$E = k_{25}[CH_3O_2] + k_{46}[O] + k_{47}[O_3] + k_{57}[NO] + k_{84}[ClO] + k_{106}[BrO] + k_{147}[C_2H_5O_2] +$$
$$\quad k_{156}[n\text{-}C_3H_7O_2] + k_{157}[i\text{-}C_3H_7O_2] + k_{164}[RO2] + k_{165}[ROR] + k_{179}[C2O3]$$

$$F = \ell_{70}[M][NO_2] + k_{82}[Cl] + k_{101}[Br] + k_{180}[C2O3] + k_{187}[XO2]$$

(N-2)   Hydroxyl Radical Equation

$$A[OH]^2 + B[OH] - C = 0$$

$$A = 2k_{38} + 2\left\{k_{37} + (k_{39} + k_{40})\frac{[H]}{[OH]} + k_{48}\frac{[HO_2]}{[OH]}\right\}\frac{[HO_2]}{[OH]}$$

$$B = k_{23}[CH_4] + 0.7k_{27}[CH_3OOH] + k_{64}[NO] + k_{65}[HONO] + \ell_{67}[M][NO_2] +$$
$$\quad k_{69}[HNO_3] + k_{71}[HNO_4] + k_{83}[HCl] + k_{86}[HOCl] + k_{102}[HBr] + k_{144}[C_2H_6] +$$
$$\quad (k_{148} + k_{149})[C_3H_8] + (k_{158} + k_{159})[PAR] + k_{168}[ALD2] + k_{172}[AONE] +$$
$$\quad k_{175}[KET] + k_{177}[MGLY] + k_{185}[NTR] + 0.51k_{189}[ROOH] + k_{190}[C_2H_4] +$$
$$\quad k_{193}[OLE] + 0.15k_{202}[ISOP] + 0.9k_{210}[C_7H_8] + G\frac{[HO_2]}{[OH]}$$

$$C = 2k_{17}[H_2][O_2(^1\Sigma)] + 2J_{18}[H_2O] + 2(k_{19}[H_2O] + k_{20}[H_2])[O(^1D)] + 2J_{26}[CH_3OOH] +$$
$$\quad 2(J_{33} + k_{31}[O] + 0.5k_{32}[Cl] + 0.5k_{108}[Br])[CH_2O] + 2J_{50}[H_2O_2] + 2J_{188}[ROOH] +$$
$$\quad k_{21}[O(^1D)][CH_4] + j_{66}[HONO] + J_{68}[HNO_3] + (J_{72} + j_{73})[HNO_4] + k_{81}[Cl][H_2] + J_{85}[HOCl] +$$
$$\quad j_{166}[ROR] + J_{169}[ALD2] + J_{176}[MGLY] + J_{184}[NTR] + J_{208}[GLY] + (0.2k_{196} + 0.88k_{197})[OLE][O_3] +$$
$$\quad 0.58k_{203}[ISOP][O_3] + 0.9k_{204}[ISOP][NO_3] + [NO]\{k_{28}[CH_3OOH] + k_{146}[C_2H_5O_2] +$$
$$\quad k_{152}[n\text{-}C_3H_7O_2] + k_{153}[i\text{-}C_3H_7O_2] + k_{160}[RO2] + k_{173}[ANO2] + (k_{191} + k_{192})[C_2H_5O_3] +$$
$$\quad k_{207}[ACO2] + 1.45k_{211}[TO2]\}$$

$$G = F + k_{25}[CH_3O_2] + k_{84}[ClO] + k_{147}[C_2H_5O_2] + k_{156}[n\text{-}C_3H_7O_2] +$$
$$\quad k_{157}[i\text{-}C_3H_7O_2] + k_{164}[RO2] + k_{165}[RO2R]$$

Box O: Wet Deposition Scheme

(O-1)   Wet Deposition Rate

$$r_i = \frac{\rho C_p Q_{conv}}{\rho_w \varDelta L_v \left\{ 1 + \left(H_i R^* T \mathcal{L} \right)^{-1} \right\}}$$

(O-2)   Liquid Water Fraction

$$\mathcal{L} = (0.18 \times 10^{-6}) \exp(-ZH/h_{LW})$$

TABLE 1
GLOBE chemical reaction mechanism.

	Reaction	Rate	Source
J_1	$O_2 + h\nu \rightarrow 2O$		De More et al. (1997), Allen & Frederick (1982)
J_2	$O_3 + h\nu \rightarrow O(^1D) + O_2$		De More et al. (1997)
J_3	$O_3 + h\nu \rightarrow O + O_2$		WMO (1986)
k_4	$O_3 + O \rightarrow 2O_2$	$8.0 \times 10^{-12}\ \exp(-2060/T)$	De More et al. (1997)
l_5	$O + O_2 + M \rightarrow O_3 + M$	$6.0 \times 10^{-34}\ (T/300)^{-2.3}$	De More et al. (1997)
l_6	$O + O + M \rightarrow O_2 + M$	$4.8 \times 10^{-34}\ \exp(900/T)$	Baulch et al. (1982)
k_7	$O(^1D) + N_2 \rightarrow O + N_2$	$1.8 \times 10^{-11}\ \exp(110/T)$	De More et al. (1997)
k_8	$O(^1D) + O_2 \rightarrow O + O_2(^1\Sigma)$	$3.2 \times 10^{-11}\ \exp(70/T)$	De More et al. (1997)
J_9	$O_2 + h\nu \rightarrow O_2(^1\Sigma)$		Bucholtz et al. (1986)
k_{10}	$O_2(^1\Sigma) + N_2 \rightarrow O_2 + N_2$	2.1×10^{-15}	De More et al. (1997)
k_{11}	$O_2(^1\Sigma) + O_2 \rightarrow 2O_2$	3.9×10^{-17}	De More et al. (1997)
k_{12}	$O_2(^1\Sigma) + CO_2 \rightarrow O_2 + CO_2$	4.2×10^{-13}	De More et al. (1997)
k_{13}	$O_2(^1\Sigma) + O_3 \rightarrow O + 2O_2$	2.2×10^{-11}	De More et al. (1997)
k_{14}	$O_2(^1\Sigma) + H_2O \rightarrow O_2 + H_2O$	5.4×10^{-12}	De More et al. (1997)
j_{15}	$O_2(^1\Sigma) \rightarrow O_2 + h\nu$	8.5×10^{-2}	Baulch et al. (1984)
k_{16}	$O_2(^1\Sigma) + N_2O \rightarrow NO + NO_2$	9.0×10^{-16}	Toumi (1993)
k_{17}	$O_2(^1\Sigma) + H_2 \rightarrow 2OH$	8.0×10^{-15}	Toumi (1993)
J_{18}	$H_2O + h\nu \rightarrow OH + H$		De More et al. (1997)
k_{19}	$H_2O + O(^1D) \rightarrow 2OH$	2.2×10^{-10}	De More et al. (1997)
k_{20}	$H_2 + O(^1D) \rightarrow H + OH$	1.1×10^{-10}	De More et al. (1997)
k_{21}	$CH_4 + O(^1D) \rightarrow CH_3 + OH$	1.5×10^{-10}	De More et al. (1997)
k_{22}	$CH_4 + O(^1D) \rightarrow CH_2O + H_2$	1.5×10^{-11}	De More et al. (1997)
k_{23}	$CH_4 + OH \rightarrow CH_3 + H_2O$	$2.45 \times 10^{-12}\ \exp(-1775/T)$	De More et al. (1997)
l_{24}	$CH_3 + O_2 + M \rightarrow CH_3O_2 + M$	∞	
k_{25}	$CH_3O_2 + HO_2 \rightarrow CH_3OOH + O_2$	$3.8 \times 10^{-13}\ \exp(800/T)$	De More et al. (1997)
J_{26}	$CH_3OOH + h\nu \rightarrow CH_3O + OH$		De More et al. (1997)
k_{27}	$CH_3OOH + OH \rightarrow 0.7CH_3O_2$ $+ 0.3CH_2O + H_2O$	$3.8 \times 10^{-12}\ \exp(200/T)$	De More et al. (1997)
k_{28}	$CH_3O_2 + NO \rightarrow CH_3O + NO_2$	$3.0 \times 10^{-12}\ \exp(280/T)$	De More et al. (1997)
k_{29}	$CH_3O + O_2 \rightarrow CH_2O + HO_2$	∞	
k_{30}	$CH_2O + OH \rightarrow CHO + H_2O$	1.0×10^{-11}	De More et al. (1997)
k_{31}	$CH_2O + O \rightarrow CHO + OH$	$3.4 \times 10^{-11}\ \exp(-1600/T)$	De More et al. (1997)
k_{32}	$CH_2O + Cl \rightarrow CHO + HCl$	$8.1 \times 10^{-11}\ \exp(-30/T)$	De More et al. (1997)
J_{33}	$CH_2O + h\nu \rightarrow CHO + H$		De More et al. (1997)
J_{34}	$CH_2O + h\nu \rightarrow CO + H_2$		De More et al. (1997)
k_{35}	$CHO + O_2 \rightarrow CO + HO_2$	∞	
k_{36}	$CO + OH \rightarrow H + CO_2$	$1.3 \times 10^{-13}\ (1+0.6P)(300/T)$	De More et al. (1997)
k_{37}	$OH + HO_2 \rightarrow H_2O + O_2$	$4.8 \times 10^{-11}\ \exp(-250/T)$	De More et al. (1997)
k_{38}	$OH + OH \rightarrow H_2O + O$	$4.2 \times 10^{-12}\ \exp(-240/T)$	De More et al. (1997)
k_{39}	$H + HO_2 \rightarrow H_2O + O$	1.6×10^{-12}	De More et al. (1997)
k_{40}	$H + HO_2 \rightarrow H_2 + O_2$	6.5×10^{-12}	De More et al. (1997)
k_{41}	$H_2 + OH \rightarrow H_2O + H$	$5.5 \times 10^{-12}\ \exp(-2000/T)$	De More et al. (1997)
k_{42}	$O_3 + OH \rightarrow HO_2 + O_2$	$1.6 \times 10^{-12}\ \exp(-940/T)$	De More et al. (1997)
k_{43}	$O + OH \rightarrow H + O_2$	$2.2 \times 10^{-11}\ \exp(120/T)$	De More et al. (1997)
k_{44}	$H + O_3 \rightarrow OH + O_2$	$1.4 \times 10^{-10}\ \exp(-470/T)$	De More et al. (1997)
l_{45}	$H + O_2 + M \rightarrow HO_2 + M$	$k_0 = 5.7 \times 10^{-32}[M](T/300)^{-1.6}$ $k_\infty = 7.5 \times 10^{-11}$ $F_c = 0.6$	De More et al. (1997)
k_{46}	$HO_2 + O \rightarrow OH + O_2$	$3.0 \times 10^{-11}\ \exp(200/T)$	De More et al. (1997)
k_{47}	$HO_2 + O_3 \rightarrow OH + 2O_2$	$1.1 \times 10^{-14}\ \exp(-500/T)$	De More et al. (1997)

k_{48}	$2HO_2 \rightarrow H_2O_2 + O_2$	$k_a = 2.3 \times 10^{-13} \exp(600/T)$ $k_b = 1.7 \times 10^{-33} \exp(1000/T)$ $k_c = 1.4 \times 10^{-21} \exp(2200/T)$ $k_{48} = (k_a + k_b[M])(1 + [H_2O]k_c)$	De More et al. (1997)
k_{49}	$H_2O_2 + OH \rightarrow H_2O + HO_2$	$2.9 \times 10^{-12} \exp(-160/T)$	De More et al. (1997)
J_{50}	$H_2O_2 + h\nu \rightarrow 2OH$		De More et al. (1997)
k_{51}	$N_2O + O(^1D) \rightarrow 2NO$	6.7×10^{-11}	De More et al. (1997)
k_{52}	$N_2O + O(^1D) \rightarrow N_2 + O_2$	4.9×10^{-11}	De More et al. (1997)
J_{53}	$N_2O + h\nu \rightarrow N_2 + O(^1D)$		De More et al. (1997)
k_{54}	$N + NO \rightarrow N_2 + O$	$2.1 \times 10^{-11} \exp(100/T)$	De More et al. (1997)
J_{55}	$NO + h\nu \rightarrow N + O$		Allen & Frederick (1982)
k_{56}	$N + O_2 \rightarrow NO + O$	$1.5 \times 10^{-11} \exp(-3600/T)$	De More et al. (1997)
k_{57}	$NO + HO_2 \rightarrow NO_2 + OH$	$3.5 \times 10^{-12} \exp(250/T)$	De More et al. (1997)
J_{58}	$NO_2 + h\nu \rightarrow NO + O$		De More et al. (1997)
k_{59}	$NO + O_3 \rightarrow NO_2 + O_2$	$2.0 \times 10^{-12} \exp(-1400/T)$	De More et al. (1997)
k_{60}	$NO_2 + O \rightarrow NO + O_2$	$6.5 \times 10^{-12} \exp(120/T)$	De More et al. (1997)
k_{61}	$NO_2 + O_3 \rightarrow NO_3 + O_2$	$1.2 \times 10^{-13} \exp(-2450/T)$	De More et al. (1997)
J_{62}	$NO_3 + h\nu \rightarrow NO_2 + O$		Wayne et al. (1991)
J_{63}	$NO_3 + h\nu \rightarrow NO + O_2$		Wayne et al. (1991)
l_{64}	$NO + OH + M \rightarrow HONO + M$	$k_0 = 7.0 \times 10^{-31}[M](T/300)^{-2.6}$ $k_\infty = 3.6 \times 10^{-11}(T/300)^{-0.1}$ $F_c = 0.6$	De More et al. (1997)
k_{65}	$HONO + OH \rightarrow NO_2 + H_2O$	$1.8 \times 10^{-11} \exp(-390/T)$	De More et al. (1997)
J_{66}	$HONO + h\nu \rightarrow OH + NO$		De More et al. (1997)
l_{67}	$OH + NO_2 + M \rightarrow HNO_3 + M$	$k_0 = 3.39 \times 10^{-30}[M](T/300)^{-4.4}$ $k_\infty = 4.77 \times 10^{-11}(T/300)^{-1.7}$ $F_c = 0.3$	Donahue et al. (1997)
J_{68}	$HNO_3 + h\nu \rightarrow OH + NO_2$		De More et al. (1997)
k_{69}	$HNO_3 + OH \rightarrow NO_3 + H_2O$	$k_a = 7.1 \times 10^{-15} \exp(785/T)$ $k_b = 4.1 \times 10^{-16} \exp(1440/T)$ $k_c = 1.9 \times 10^{-33} \exp(725/T)$ $k_{48} = (k_a + k_b[M])(1 + [M]k_c/k_b)$	De More et al. (1997)
l_{70}	$HO_2 + NO_2 + M \rightarrow HNO_4 + M$	$k_0 = 1.8 \times 10^{-31}[M](T/300)^{-3.2}$ $k_\infty = 4.7 \times 10^{-12}(T/300)^{-1.4}$ $F_c = 0.6$	De More et al. (1997)
k_{71}	$HNO_4 + OH \rightarrow H_2O + NO_2$	$1.3 \times 10^{-12} \exp(380/T)$	De More et al. (1997)
J_{72}	$HNO_4 + h\nu \rightarrow HO_2 + NO_2$		De More et al. (1997)
j_{73}	$HNO_4 \rightarrow HO_2 + NO_2$	$k_0 = 4.1 \times 10^{-5}[M]\exp(-10{,}650/T)$ $k_\infty = 5.7 \times 10^{15} \exp(-11{,}170/T)$ $F_c = 0.5$	Atkinson et al. (1997b)
l_{74}	$NO_3 + NO_2 + M \rightarrow N_2O_5 + M$	$k_0 = 2.2 \times 10^{-30}[M](T/300)^{-3.9}$ $k_\infty = 1.5 \times 10^{-12}(T/300)^{-0.7}$ $F_c = 0.6$	De More et al. (1997)
J_{75}	$N_2O_5 + h\nu \rightarrow NO_3 + NO_2$		De More et al. (1997)
j_{76}	$N_2O_5 \rightarrow NO_3 + NO_2$	$k_0 = 1.0 \times 10^{-3}[M]\exp(-11{,}000/T)$ $\times (T/300)^{-3.5}$ $k_\infty = 9.7 \times 10^{14} \exp(-11{,}080/T)$ $\times (T/300)^{-0.1}$ $F_c = 2.5 \exp(-1950/T)$ $+ 0.9 \exp(-T/430)$	Atkinson et al. (1997b)
k_{77}	$Cl + O_3 \rightarrow ClO + O_2$	$2.9 \times 10^{-11} \exp(-260/T)$	De More et al. (1997)
k_{78}	$ClO + O \rightarrow Cl + O_2$	$3.0 \times 10^{-11} \exp(70/T)$	De More et al. (1997)
k_{79}	$ClO + NO \rightarrow Cl + NO_2$	$6.4 \times 10^{-12} \exp(290/T)$	De More et al. (1997)
k_{80}	$Cl + CH_4 \rightarrow CH_3 + HCl$	$1.1 \times 10^{-11} \exp(-1400/T) \cdot$	De More et al. (1997)
k_{81}	$Cl + H_2 \rightarrow H + HCl$	$3.7 \times 10^{-11} \exp(-2300/T)$	De More et al. (1997)
k_{82}	$Cl + HO_2 \rightarrow O_2 + HCl$	$1.8 \times 10^{-11} \exp(170/T)$	De More et al. (1997)

k_{83}	$HCl + OH \rightarrow H_2O + Cl$	$2.6\times10^{-12}\ \exp(-350/T)$	De More et al. (1997)
k_{84}	$HO_2 + ClO \rightarrow HOCl + O_2$	$4.8\times10^{-13}\ \exp(700/T)$	De More et al. (1997)
J_{85}	$HOCl + h\nu \rightarrow OH + Cl$		De More et al. (1997)
k_{86}	$HOCl + OH \rightarrow H_2O + ClO$	$3.0\times10^{-12}\ \exp(-500/T)$	De More et al. (1997)
l_{87}	$ClO + NO_2 + M \rightarrow ClONO_2 + M$	$k_0=1.8\times10^{-31}[M](T/300)^{-3.4}$ $k_\infty=1.5\times10^{-11}\ (T/300)^{-1.9}$ $F_c=0.6$	De More et al. (1997)
J_{88}	$ClONO_2 + h\nu \rightarrow Cl + NO_3$		De More et al. (1997)
J_{89}	$ClONO_2 + h\nu \rightarrow ClO + NO_2$		De More et al. (1997)
l_{90}	$ClO + ClO + M \rightarrow 2Cl + M + O_2$	$k_0=2.2\times10^{-32}[M](T/300)^{-3.1}$ $k_\infty=3.5\times10^{-12}\ (T/300)$ $F_c=0.6$	De More et al. (1997)
k_{91}	$ClO + OH \rightarrow HO_2 + Cl$	$(0.93)\ 1.1\times10^{-11}\ \exp(120/T)$	De More et al. (1997), Lipson et al. (1997)
k_{92}	$ClO + OH \rightarrow HCl + O_2$	$(0.07)\ 1.1\times10^{-11}\ \exp(120/T)$	De More et al. (1997), Lipson et al. (1997)
k_{93}	$Br + O_3 \rightarrow BrO + O_2$	$1.7\times10^{-11}\ \exp(-800/T)$	De More et al. (1997)
J_{94}	$BrO + h\nu \rightarrow Br + O$		De More et al. (1997)
k_{95}	$BrO + O \rightarrow Br + O_2$	$1.9\times10^{-11}\ \exp(230/T)$	De More et al. (1997)
k_{96}	$BrO + NO \rightarrow Br + NO_2$	$8.8\times10^{-12}\ \exp(260/T)$	De More et al. (1997)
k_{97}	$BrO + OH \rightarrow Br + HO_2$	7.5×10^{-11}	De More et al. (1997)
k_{98}	$BrO + ClO \rightarrow Br + Cl + O_2$	$2.9\times10^{-12}\ \exp(220/T)$	De More et al. (1997)
k_{99}	$BrO + ClO \rightarrow Br + ClO + O$	$1.6\times10^{-12}\ \exp(430/T)$	De More et al. (1997)
k_{100}	$Br + NO_3 \rightarrow BrO + NO_2$	1.6×10^{-11}	De More et al. (1997)
k_{101}	$Br + HO_2 \rightarrow HBr + O_2$	$1.5\times10^{-11}\ \exp(-600/T)$	De More et al. (1997)
k_{102}	$HBr + OH \rightarrow Br + H_2O$	1.1×10^{-11}	De More et al. (1997)
l_{103}	$BrO + NO_2 + M \rightarrow BrONO_2 + M$	$k_0=5.2\times10^{-31}[M](T/300)^{-3.2}$ $k_\infty=6.9\times10^{-12}\ (T/300)^{-2.9}$ $F_c=0.6$	De More et al. (1997)
J_{104}	$BrONO_2 + h\nu \rightarrow Br + NO_3$		De More et al. (1997)
J_{105}	$BrONO_2 + h\nu \rightarrow BrO + NO_2$		De More et al. (1997)
k_{106}	$BrO + HO_2 \rightarrow HOBr + O_2$	$3.4\times10^{-12}\ \exp(540/T)$	De More et al. (1997)
J_{107}	$HOBr + h\nu \rightarrow Br + OH$		De More et al. (1997)
k_{108}	$Br + CH_2O \rightarrow HBr + CHO$	$1.7\times10^{-11}\ \exp(-800/T)$	De More et al. (1997)
J_{109}	$COCl_2 + h\nu \rightarrow 2Cl$		De More et al. (1997)
k_{110}	$COCl_2 + O(^1D) \rightarrow 2Cl$	3.6×10^{-10}	De More et al. (1997)
J_{111}	$CCl_4 + h\nu \rightarrow 2Cl + COCl_2$		De More et al. (1997)
J_{112}	$CFCl_3 + h\nu \rightarrow 3Cl$		De More et al. (1997)
J_{113}	$CF_2Cl_2 + h\nu \rightarrow 2Cl$		De More et al. (1997)
k_{114}	$CF_3Cl + O(^1D) \rightarrow Cl$	8.7×10^{-11}	De More et al. (1997)
J_{115}	$CH_3CCl_3 + h\nu \rightarrow Cl + COCl_2$		De More et al. (1997)
k_{116}	$CH_3CCl_3 + OH \rightarrow Cl + COCl_2$	$1.8\times10^{-12}\ \exp(-1550/T)$	De More et al. (1997)
J_{117}	$CH_3Cl + h\nu \rightarrow Cl$		De More et al. (1997)
k_{118}	$CH_3Cl + OH \rightarrow Cl$	$4.0\times10^{-12}\ \exp(-1400/T)$	De More et al. (1997)
J_{119}	$CF_2ClCFCl_2 + h\nu \rightarrow Cl + COCl_2$		De More et al. (1997)
J_{120}	$CF_2ClCF_2Cl + h\nu \rightarrow 2Cl$		De More et al. (1997)
J_{121}	$CF_3CF_2Cl + h\nu \rightarrow Cl$		De More et al. (1997)
J_{122}	$CHClF_2 + h\nu \rightarrow Cl$		De More et al. (1997)
k_{123}	$CHClF_2 + OH \rightarrow Cl$	$1.0\times10^{-12}\ \exp(-1600/T)$	De More et al. (1997)
J_{124}	$CF_3CHCl_2 + h\nu \rightarrow 2Cl$		De More et al. (1997)
k_{125}	$CF_3CHCl_2 + OH \rightarrow 2Cl$	$7.0\times10^{-13}\ \exp(-900/T)$	De More et al. (1997)
J_{126}	$CF_3CHFCl + h\nu \rightarrow Cl$		De More et al. (1997)
k_{127}	$CF_3CHFCl + OH \rightarrow Cl$	$8.0\times10^{-13}\ \exp(-1350/T)$	De More et al. (1997)
J_{128}	$CH_3CFCl_2 + h\nu \rightarrow 2Cl$		De More et al. (1997)
k_{129}	$CH_3CFCl_2 + OH \rightarrow 2Cl$	$1.7\times10^{-12}\ \exp(-1700/T)$	De More et al. (1997)
J_{130}	$CH_3CF_2Cl + h\nu \rightarrow Cl$		De More et al. (1997)
k_{131}	$CH_3CF_2Cl + OH \rightarrow Cl$	$1.3\times10^{-12}\ \exp(-1800/T)$	De More et al. (1997)

k_{132}	$CF_3CF_2CHCl_2 + OH \rightarrow 2Cl$	$1.0\times10^{-12}\ \exp(-1100/T)$	De More et al. (1997)
k_{133}	$CF_2ClCF_2CHClF + OH \rightarrow 2Cl$	$5.5\times10^{-13}\ \exp(-1250/T)$	De More et al. (1997)
J_{134}	$CH_3Br + h\nu \rightarrow Br$		De More et al. (1997)
k_{135}	$CH_3Br + OH \rightarrow Br$	$4.0\times10^{-12}\ \exp(-1470/T)$	De More et al. (1997)
J_{136}	$CF_3Br + h\nu \rightarrow Br$		De More et al. (1997)
k_{137}	$CF_3Br + OH \rightarrow Br$	1.2×10^{-16}	De More et al. (1997)
k_{138}	$CF_2ClBr + OH \rightarrow Cl + Br$	1.5×10^{-16}	De More et al. (1997)
J_{139}	$CF_2BrCF_2Br + h\nu \rightarrow 2Br$		De More et al. (1997)
J_{140}	$CF_2Br_2 + h\nu \rightarrow 2Br$		De More et al. (1997)
k_{141}	$CF_2HBr + OH \rightarrow Br$	$1.1\times10^{-12}\ \exp(-1400/T)$	De More et al. (1997)
k_{142}	$CF_3CHFBr + OH \rightarrow Br$	$7.2\times10^{-13}\ \exp(-1110/T)$	De More et al. (1997)
k_{143}	$CF_3CHClBr + OH \rightarrow Cl + Br$	$1.3\times10^{-12}\ \exp(-995/T)$	De More et al. (1997)
k_{144}	$C_2H_6 + OH \rightarrow C_2H_5O_2 + H_2O$	$7.9\times10^{-12}\ \exp(-1030/T)$	Atkinson et al. (1997a)
k_{145}	$C_2H_6 + Cl \rightarrow C_2H_5O_2 + HCl$	$8.1\times10^{-11}\ \exp(-90/T)$	Atkinson et al. (1997a)
k_{146}	$C_2H_5O_2 + NO \rightarrow NO_2 + HO_2 + ALD2$	8.7×10^{-12}	Atkinson et al. (1997a)
k_{147}	$C_2H_5O_2 + HO_2 \rightarrow ROOH$	$7.5\times10^{-13}\ \exp(700/T)$	De More et al. (1997)
k_{148}	$C_3H_8 + OH \rightarrow n\text{-}C_3H_7O_2 + H_2O$	$(0.28)8.0\times10^{-12}\ \exp(-590/T)$	Atkinson et al. (1997a)
k_{149}	$C_3H_8 + OH \rightarrow i\text{-}C_3H_7O_2 + H_2O$	$(0.72)8.0\times10^{-12}\ \exp(-590/T)$	Atkinson et al. (1997a)
k_{150}	$C_3H_8 + Cl \rightarrow n\text{-}C_3H_7O_2 + HCl$	$(0.28)1.2\times10^{-10}\ \exp(-40/T)$	Atkinson et al. (1997a)
k_{151}	$C_3H_8 + Cl \rightarrow i\text{-}C_3H_7O_2 + HCl$	$(0.72)1.2\times10^{-10}\ \exp(-40/T)$	Atkinson et al. (1997a)
k_{152}	$n\text{-}C_3H_7O_2 + NO \rightarrow NO_2 + HO_2 + ALD2$	4.9×10^{-12}	Atkinson et al. (1997a)
k_{153}	$i\text{-}C_3H_7O_2 + NO \rightarrow NO_2 + HO_2 + AONE$	4.8×10^{-12}	Atkinson et al. (1997a)
k_{154}	$n\text{-}C_3H_7O_2 + NO \rightarrow NTR$	1.0×10^{-13}	Atkinson et al. (1997a)
k_{155}	$i\text{-}C_3H_7O_2 + NO \rightarrow NTR$	2.1×10^{-13}	Atkinson et al. (1997a)
k_{156}	$n\text{-}C_3H_7O_2 + HO_2 \rightarrow ROOH$	$1.66\times10^{-13}\ \exp(1300/T)$	Stockwell et al. (1997)
k_{157}	$i\text{-}C_3H_7O_2 + HO_2 \rightarrow ROOH$	$1.66\times10^{-13}\ \exp(1300/T)$	Stockwell et al. (1997)
k_{158}	$PAR + OH \rightarrow RO2$	9.2×10^{-14}	Gery et al. (1989)
k_{159}	$PAR + OH \rightarrow RO2R$	7.2×10^{-13}	Gery et al. (1989)
k_{160}	$RO2 + NO \rightarrow NO_2 + HO_2$ $+ ALD2 - PAR$	7.7×10^{-12}	Gery et al. (1989)
k_{161}	$RO2R + NO \rightarrow NO_2 + ROR$	7.0×10^{-12}	Gery et al. (1989)
k_{162}	$RO2 + NO \rightarrow NTR$	$4.4\times10^{-11}\ \exp(-1400/T)$	Gery et al. (1989)
k_{163}	$RO2R + NO \rightarrow NTR$	$1.2\times10^{-10}\ \exp(-1400/T)$	Gery et al. (1989)
k_{164}	$RO2 + HO_2 \rightarrow ROOH$	$1.66\times10^{-13}\ \exp(1300/T)$	Stockwell et al. (1997)
k_{165}	$RO2R + HO_2 \rightarrow ROOH$	$1.66\times10^{-13}\ \exp(1300/T)$	Stockwell et al. (1997)
j_{166}	$ROR \rightarrow HO_2 + KET$	1.6×10^{3}	Gery et al. (1989)
k_{167}	$ROR + NO_2 \rightarrow NTR$	1.5×10^{-11}	Gery et al. (1989)
k_{168}	$ALD2 + OH \rightarrow C2O3 + H_2O$	$5.6\times10^{-12}\ \exp(310/T)$	Atkinson et al. (1997a)
J_{169}	$ALD2 + h\nu \rightarrow CH_3O_2 + HO_2 + CO$		Martinez et al. (1992)
J_{170}	$AONE + h\nu \rightarrow 2CH_3O_2 + CO$		Martinez et al. (1992), McKeen et al. (1997)
J_{171}	$AONE + h\nu \rightarrow CH_3O_2 + C2O3$		Martinez et al. (1992), McKeen et al. (1997)
k_{172}	$AONE + OH \rightarrow ANO2 + H_2O$	$2.8\times10^{-12}\ \exp(-760/T)$	Atkinson et al. (1997a)
k_{173}	$ANO2 + NO \rightarrow NO_2 + HO_2 + MGLY$	8.1×10^{-12}	Gery et al. (1989)
J_{174}	$KET + h\nu \rightarrow C2O3 + RO2 - 2PAR$		Martinez et al. (1992)
k_{175}	$KET + OH \rightarrow XO2$	$5.68\times10^{-18}\ T^2\ \exp(-760/T)$	Atkinson (1994)
J_{176}	$MGLY + h\nu \rightarrow C2O3 + HO_2 + CO$		Staffelbach et al. (1995)
k_{177}	$MGLY + OH \rightarrow C2O3 + H_2O + CO$	1.5×10^{-11}	Atkinson et al. (1997a)
k_{178}	$C2O3 + NO \rightarrow NO_2 + CH_3O_2 + CO_2$	2.0×10^{-11}	Atkinson et al. (1997a)
k_{179}	$C2O3 + HO_2 \rightarrow CH_3O_2 + OH + O_2$	$(0.74)4.3\times10^{-13}\ \exp(1040/T)$	Atkinson et al. (1997a)
k_{180}	$C2O3 + HO_2 \rightarrow ROOH$	$(0.26)4.3\times10^{-13}\ \exp(1040/T)$	Atkinson et al. (1997a)
l_{181}	$C2O3 + NO_2 + M \rightarrow PAN$	$k_0 = 2.7\times10^{-28}[M](T/300)^{-7.1}$ $k_\infty = 1.2\times10^{-11}\ (T/300)^{-0.9}$ $F_c = 0.3$	Atkinson et al. (1997a)

J_{182}	$PAN + h\nu \to C2O3 + NO_2$		De More et al. (1997)
j_{183}	$PAN \to C2O3 + NO_2$	$k_0=4.9\times10^{-3}[M]\exp(-12,100/T)$ $k_\infty=5.4\times10^{16}\exp(-13,830/T)$ $F_c=0.3$	Atkinson et al. (1997a)
J_{184}	$NTR + h\nu \to NO_2 + HO_2$		Roberts & Fajer (1989)
k_{185}	$NTR + OH \to NO_2 + XO2$	$5.31\times10^{-12}\exp(-260/T)$	Stockwell et al. (1997)
k_{186}	$XO2 + NO \to NO_2$	1.78×10^{-12}	Atkinson (1994)
k_{187}	$XO2 + HO_2 \to ROOH$	$3.5\times10^{-13}\exp(1000/T)$	Atkinson (1994)
J_{188}	$ROOH + h\nu \to HO_2 + OH + ALD2$	same as J_{26}	
k_{189}	$ROOH + OH \to 0.51XO2 + 0.49OH$	$3.4\times10^{-12}\exp(190/T)$	Stockwell et al. (1997)
l_{190}	$C_2H_4 + OH + M \to C_2H_5O_3 + M$	$k_0=7.0\times10^{-29}[M](T/300)^{-3.1}$ $k_\infty=9.0\times10^{-12}$ $F_c=0.7$	Atkinson et al. (1997a)
k_{191}	$C_2H_5O_3+NO \to NO_2+HO_2+2CH_2O$	6.3×10^{-12}	Gery et al. (1989)
k_{192}	$C_2H_5O_3+NO \to NO_2+HO_2+ALD2$	1.8×10^{-12}	Gery et al. (1989)
k_{193}	$OLE+OH \to CH_3O_2+ALD2-PAR$	$5.2\times10^{-12}\exp(504/T)$	Gery et al. (1989)
k_{194}	$OLE+O_3 \to ALD2+CRIG-PAR$	$2.8\times10^{-15}\exp(-2105/T)$	Gery et al. (1989)
k_{195}	$OLE+O_3 \to CH_2O+MCRG-PAR$	$2.8\times10^{-15}\exp(-2105/T)$	Gery et al. (1989)
k_{196}	$OLE+O_3 \to ALD2+0.2HO_2+0.7CO$	$4.3\times10^{-15}\exp(-2105/T)$	Gery et al. (1989)
k_{197}	$OLE+O_3 \to 1.08CH_2O+0.72CH_3O_2$ $+ 0.56HO_2+0.32OH$ $+ 0.4CO$	$4.3\times10^{-15}\exp(-2105/T)$	Gery et al. (1989)
k_{198}	$CRIG + NO \to NO_2 + CH_2O$	7.0×10^{-12}	Gery et al. (1989)
k_{199}	$MCRG + NO \to NO_2 + ALD2$	7.0×10^{-12}	Gery et al. (1989)
k_{200}	$CRIG + H_2O \to products$	4.0×10^{-16}	Gery et al. (1989)
k_{201}	$MCRG + H_2O \to products$	4.0×10^{-16}	Gery et al. (1989)
k_{202}	$ISOP+OH \to 0.85XO2+0.61CH_2O$ $+ 0.85HO_2+0.03MGLY$ $+ 0.58OLE+0.15XO2N$ $+ 0.63PAR$	$2.54\times10^{-11}\exp(410/T)$	Atkinson (1994)
k_{203}	$ISOP+O_3 \to 0.9CH_2O+0.55OLE$ $+ 0.18XO2+0.36CO$ $+ 0.15C2O3+0.03MGLY$ $+ 0.63PAR+0.30HO_2$ $+ 0.28OH$	$1.23\times10^{-14}\exp(-2013/T)$	Atkinson (1994)
k_{204}	$ISOP+NO_3 \to 0.9NTR+0.9HO_2$ $+ 0.03CH_2O+0.45OLE$ $+ 0.12ALD2+0.1NO_2$ $+ 0.08MGLY$	7.8×10^{-13}	Willie et al. (1991)
k_{205}	$XO2N + NO \to NTR$	6.8×10^{-12}	Gery et al. (1989)
l_{206}	$C_2H_2 + OH + M \to HOC_2H_2O_2$	$k_0=5.0\times10^{-30}[M](T/300)^{-1.5}$ $k_\infty=9.0\times10^{-13}(T/300)^{2.0}$ $F_c=0.62$	Atkinson et al. (1997a)
k_{207}	$HOC_2H_2O_2 + NO \to HO_2 + GLY$ $+ NO_2$	same as k_{28}	
J_{208}	$GLY + h\nu \to HO_2 + 0.13CH_2O$ $+ 0.87H_2 + 1.87CO$	∞	
k_{209}	$GLY + OH \to 2CO + HO_2 + H_2O$	1.1×10^{-11}	Atkinson et al. (1997a), Atkinson (1990)
k_{210}	$C_7H_8 + OH \to 0.1XO2 + 0.1HO_2$ $+ 0.9TO2$	$1.81\times10^{-12}\exp(355/T)$	Stockwell et al. (1997)
k_{211}	$TO2 + NO \to 0.95NO_2 + 0.95HO_2$ $+ 0.65GLY+ 1.20GLY$ $+ 0.5DCB + 0.05NTR$	4.0×10^{-12}	Stockwell et al. (1997)
J_{212}	$DCB + h\nu \to C2O3 + HO_2$	∞	
m_{213}	$N_2O_5 + H_2O(s) \to 2HNO_3$		
m_{214}	$ClONO_2+H_2O(s) \to HNO_3+HOCl$		
m_{215}	$N_2O_5 + H_2O(s) \to 2HNO_3(s)$		

TABLE 2

Photolysis rates (s^{-1}) for photocomp scenarios.

Level	$O_3 + h\nu \rightarrow$ $O(^1D) + O_2$	$NO_2 + h\nu \rightarrow$ $NO + O$	$H_2O_2 + h\nu \rightarrow$ $2OH$	$CH_2O + h\nu \rightarrow$ $CHO + H$
	Method A **Diurnal**			
Surface	0.88×10^{-5} $(+23.9\%)$	0.57×10^{-2} $(+46.2\%)$	3.06×10^{-6} $(+14.6\%)$	1.20×10^{-5} $(+15.4\%)$
4 km	1.09×10^{-5} $(+15.6\%)$	0.66×10^{-2} $(+32.0\%)$	4.11×10^{-6} $(+12.6\%)$	1.82×10^{-5} $(+19.7\%)$
8 km	1.02×10^{-5} $(+8.5\%)$	0.68×10^{-2} $(+19.3\%)$	4.36×10^{-6} $(+4.3\%)$	2.12×10^{-5} $(+15.8\%)$
12 km	0.99×10^{-5} $(+15.1\%)$	0.68×10^{-2} $(+11.5\%)$	4.50×10^{-6} $(+0.2\%)$	2.32×10^{-5} $(+16.0\%)$
	Photocomp **Diurnal**			
Surface	0.71×10^{-5} $(\pm7.2\%)$	0.39×10^{-2} $(\pm7.1\%)$	2.67×10^{-6} $(\pm8.3\%)$	1.04×10^{-5} $(\pm10.8\%)$
4 km	0.94×10^{-5} $(\pm8.0\%)$	0.50×10^{-2} $(\pm7.8\%)$	3.65×10^{-6} $(\pm10.6\%)$	1.52×10^{-5} $(\pm14.0\%)$
8 km	0.94×10^{-5} $(\pm8.6\%)$	0.57×10^{-2} $(\pm7.2\%)$	4.18×10^{-6} $(\pm12.0\%)$	1.83×10^{-5} $(\pm14.0\%)$
12 km	0.86×10^{-5} $(\pm8.0\%)$	0.61×10^{-2} $(\pm7.4\%)$	4.49×10^{-6} $(\pm13.0\%)$	2.00×10^{-5} $(\pm13.5\%)$
	Method B **Noon**			
Surface	2.80×10^{-5} $(+3.7\%)$	0.99×10^{-2} $(+8.8\%)$	6.39×10^{-6} (-13.4%)	2.71×10^{-5} (-9.7%)
4 km	3.51×10^{-5} $(+1.2\%)$	1.11×10^{-2} $(+2.8\%)$	8.33×10^{-6} (-12.0%)	3.96×10^{-5} (-3.6%)
8 km	3.31×10^{-5} (-2.1%)	1.09×10^{-2} (-6.0%)	8.49×10^{-6} (-16.8%)	4.41×10^{-5} (-5.0%)
12 km	3.22×10^{-5} $(+4.5\%)$	1.07×10^{-2} (-9.3%)	8.54×10^{-6} (-17.7%)	4.67×10^{-5} (-2.9%)
	Photocomp **Noon**			
Surface	2.70×10^{-5} $(\pm6.7\%)$	0.91×10^{-2} $(\pm6.1\%)$	7.38×10^{-6} $(\pm7.4\%)$	3.00×10^{-5} $(\pm9.7\%)$
4 km	3.47×10^{-5} $(\pm6.4\%)$	1.08×10^{-2} $(\pm7.2\%)$	9.47×10^{-6} $(\pm9.8\%)$	4.11×10^{-5} $(\pm12.6\%)$
8 km	3.38×10^{-5} $(\pm6.8\%)$	1.16×10^{-2} $(\pm7.2\%)$	10.20×10^{-6} $(\pm12.1\%)$	4.64×10^{-5} $(\pm13.2\%)$
12 km	3.08×10^{-5} $(\pm6.4\%)$	1.18×10^{-2} $(\pm7.4\%)$	10.38×10^{-6} $(\pm13.7\%)$	4.81×10^{-5} $(\pm12.6\%)$

TABLE 3

Surface dry deposition velocities (cm/s).

Chemical species	Land	Ocean and sea ice
Ozone (O_3)	0.80	0.10
Hydrogen peroxide (H_2O_2)	0.50	0.50
Nitrogen dioxide (NO_2)	0.25	0.10
Nitric acid (HNO_3)	2.00	0.80
Peroxyacyl nitrate (PAN)	0.25	0.10
Organic nitrate (NTR)	0.25	0.10

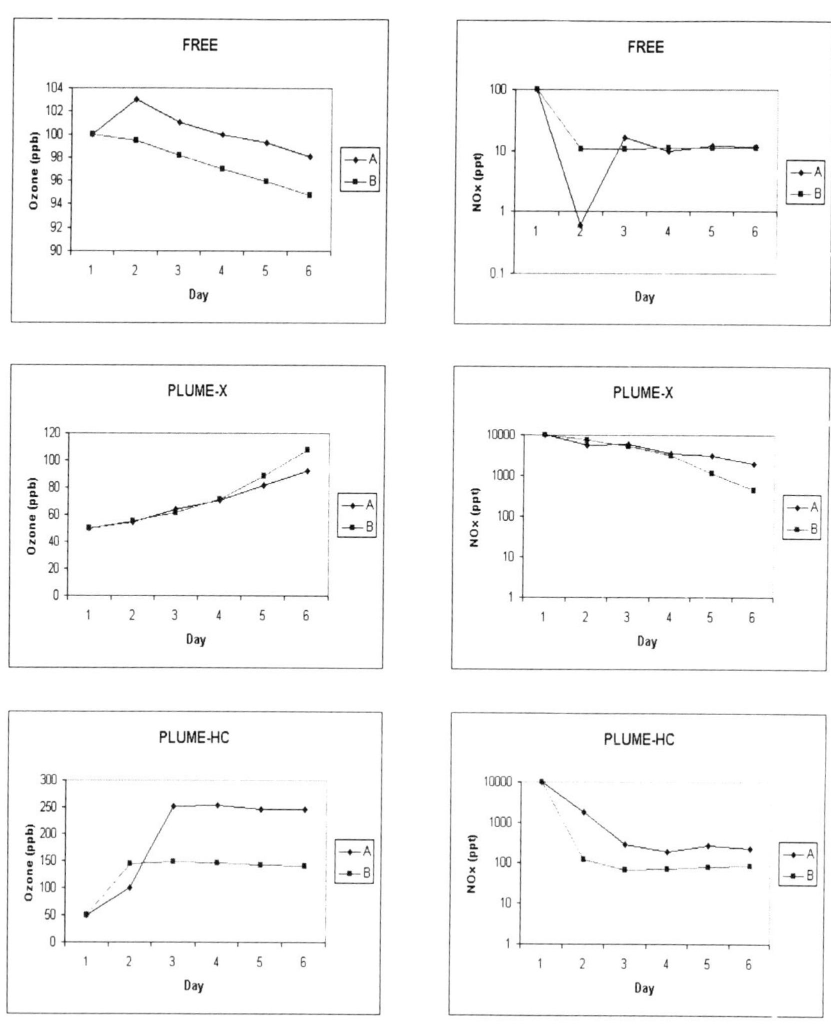

Fig. 1. *GLOBE results for Photocomp cases FREE, PLUME-X and PLUME-HC.*

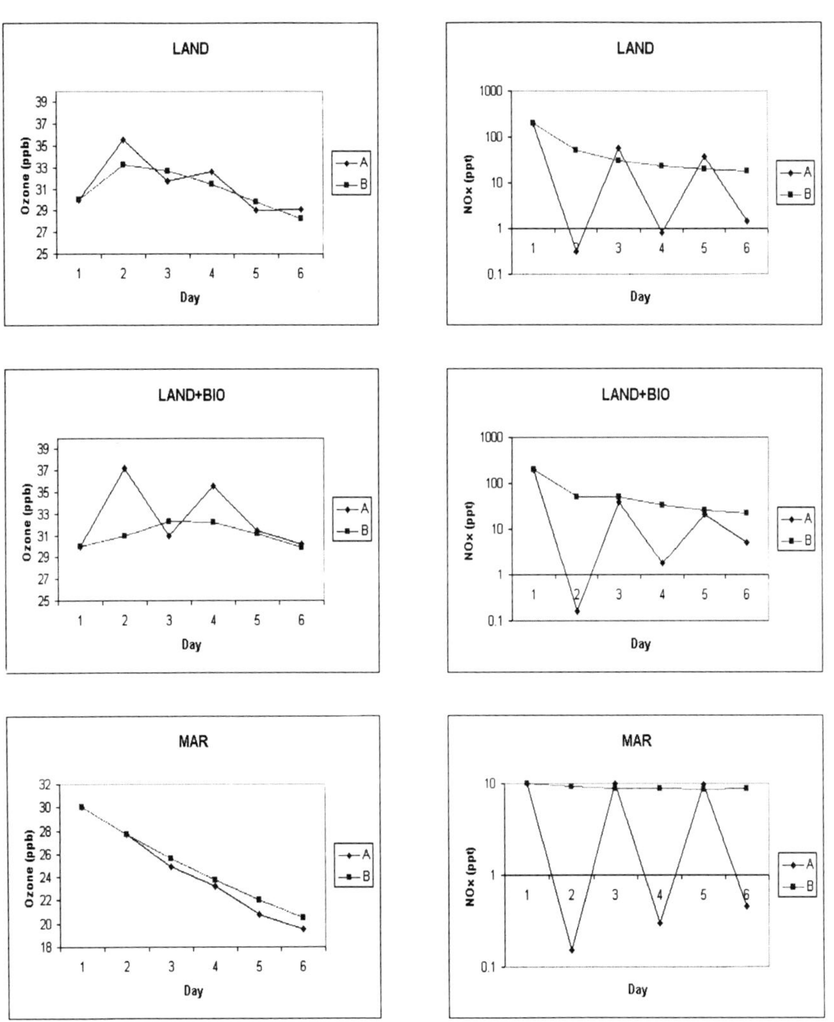

FIG. 2. *GLOBE results for Photocomp cases LAND, LAND+BIO and MAR.*

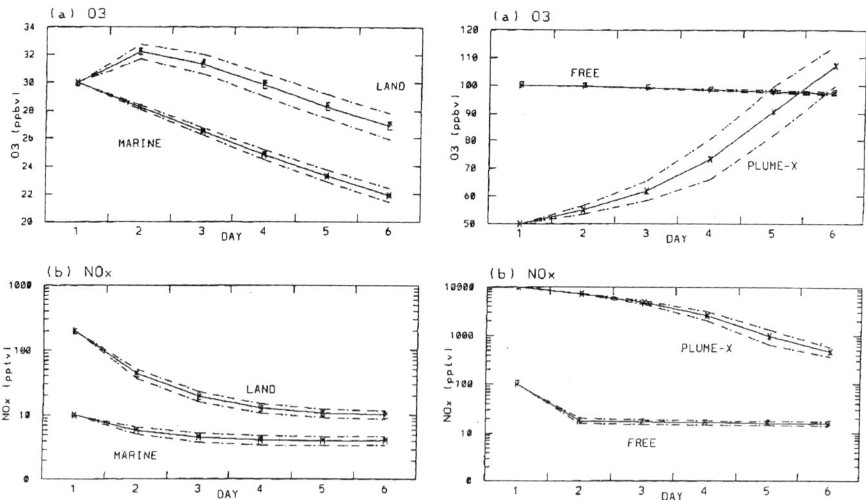

FIG. 3. *Photocomp results for LAND and MAR (left), and FREE and PLUME-X (right) cases (after Olson et al., 1997).*

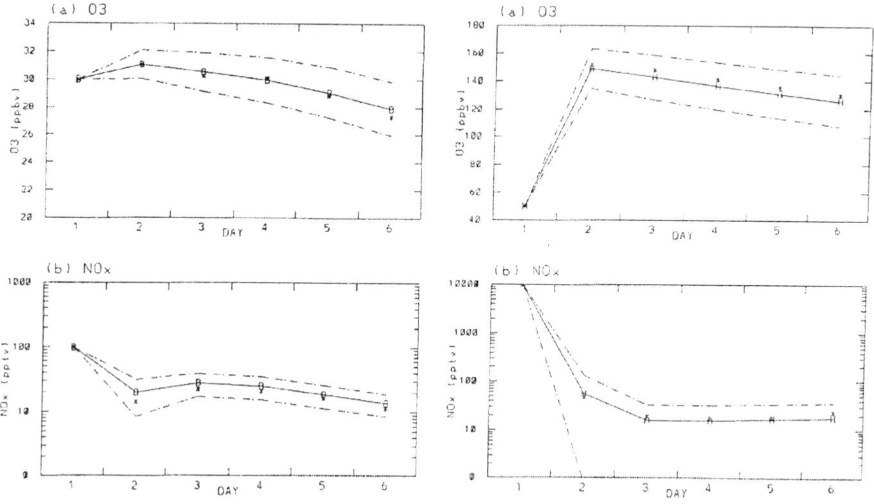

FIG. 4. *Photocomp results for LAND+BIO (left) and PLUME-HC (right) cases (after Olson et al., 1997).*

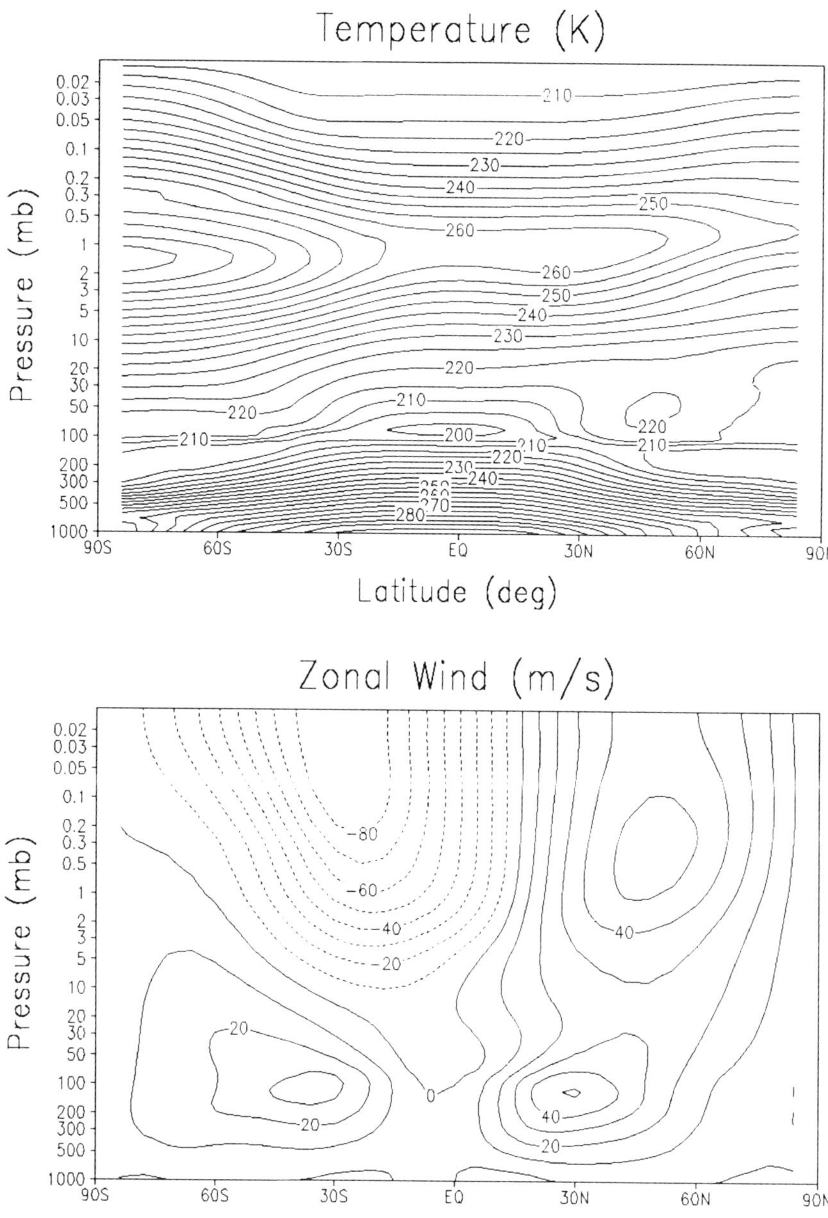

FIG. 5. *Simulated January mean temperature and zonal wind for Model Year 5.*

 EDUARDO P. OLAGUER

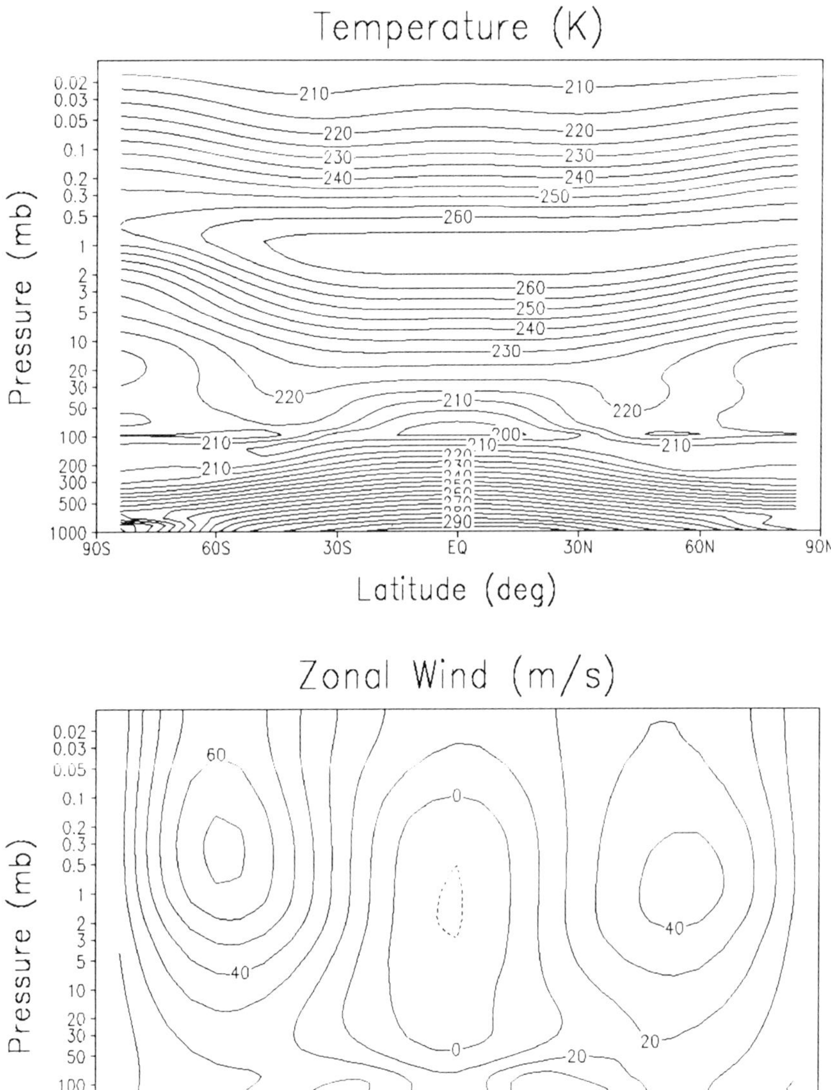

FIG. 6. *Simulated April mean temperature and zonal wind for Model Year 5.*

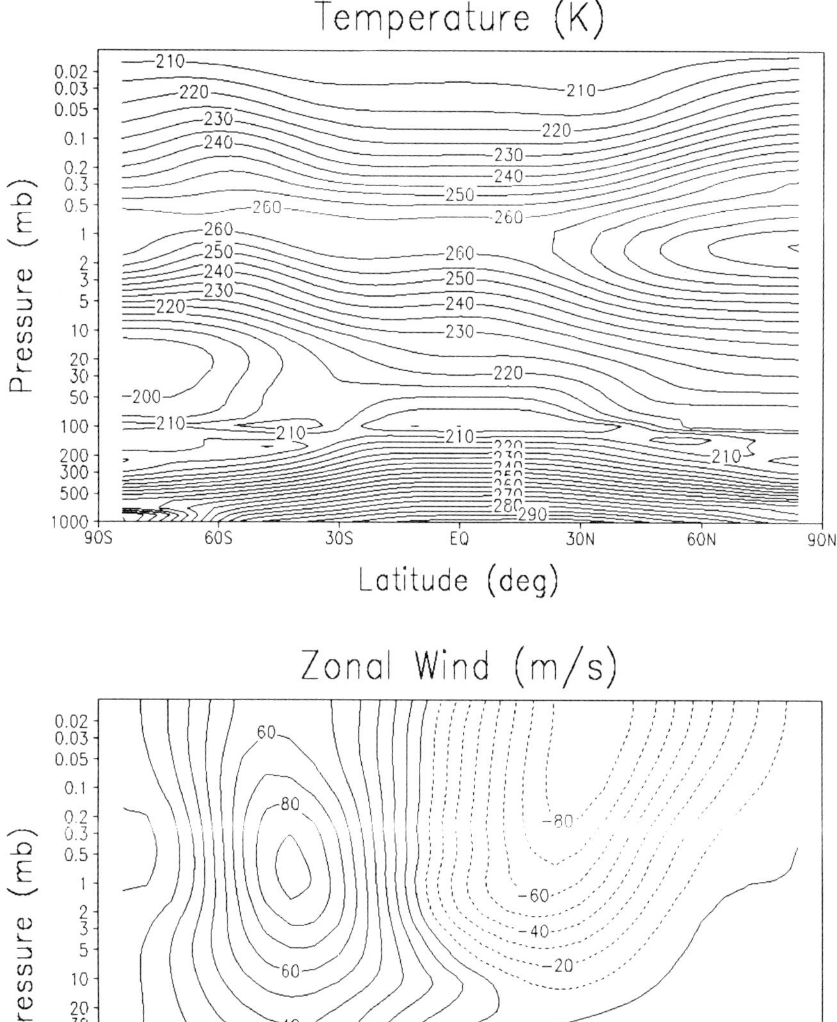

FIG. 7. *Simulated July mean temperature and zonal wind for Model Year 5.*

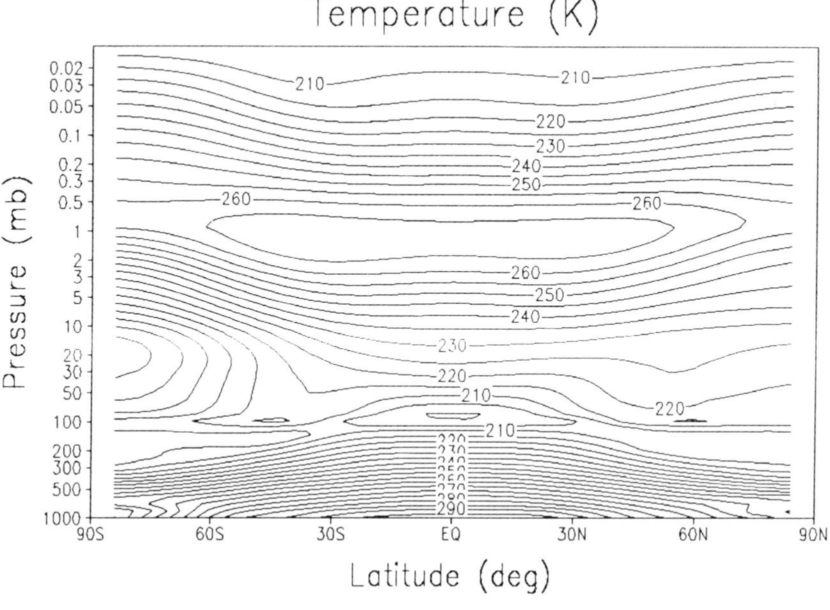

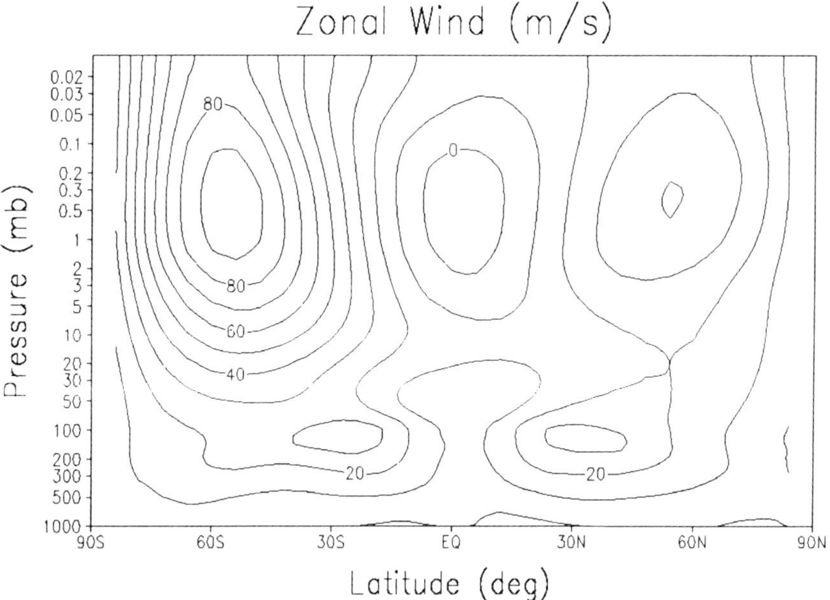

FIG. 8. *Simulated October mean temperature and zonal wind for Model Year 5.*

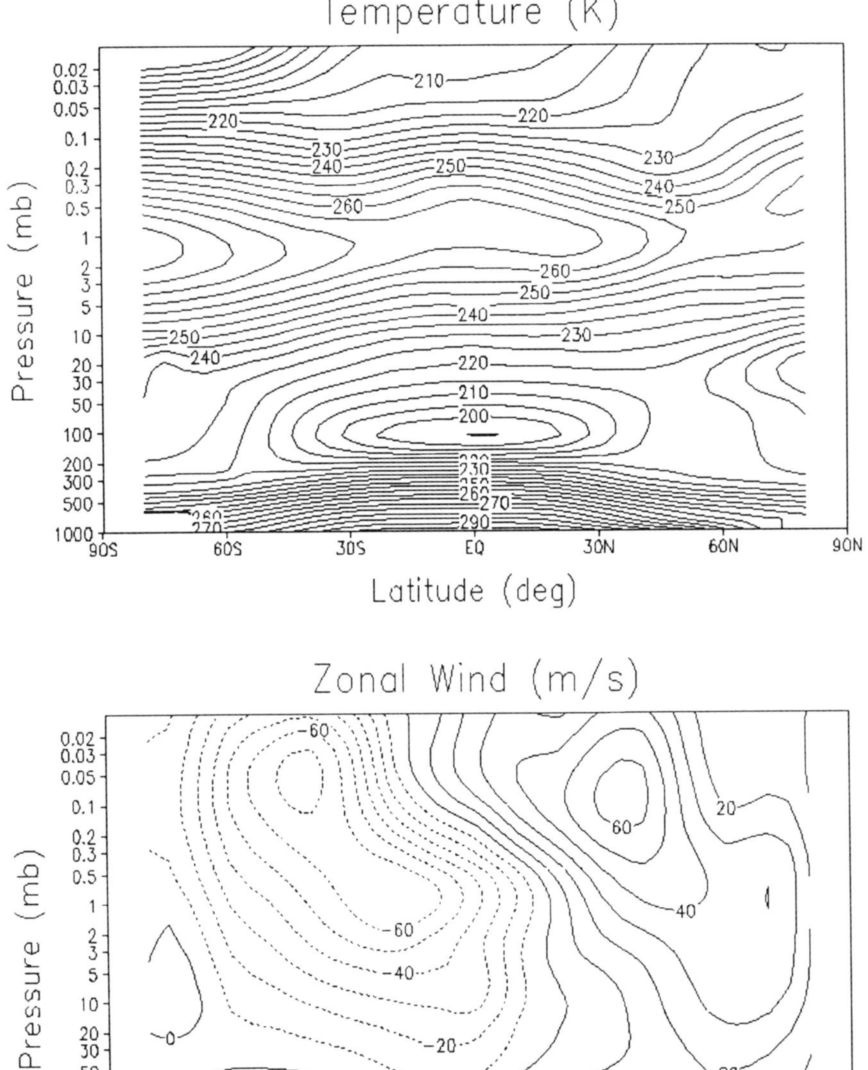

FIG. 9. *Observed January mean temperature and zonal wind from CIRA-86.*

Temperature (K)

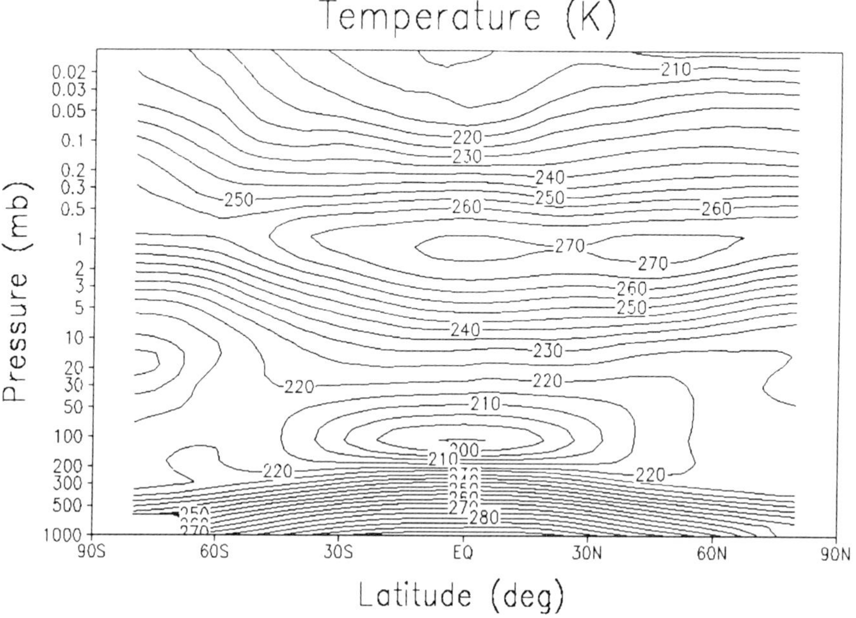

Zonal Wind (m/s)

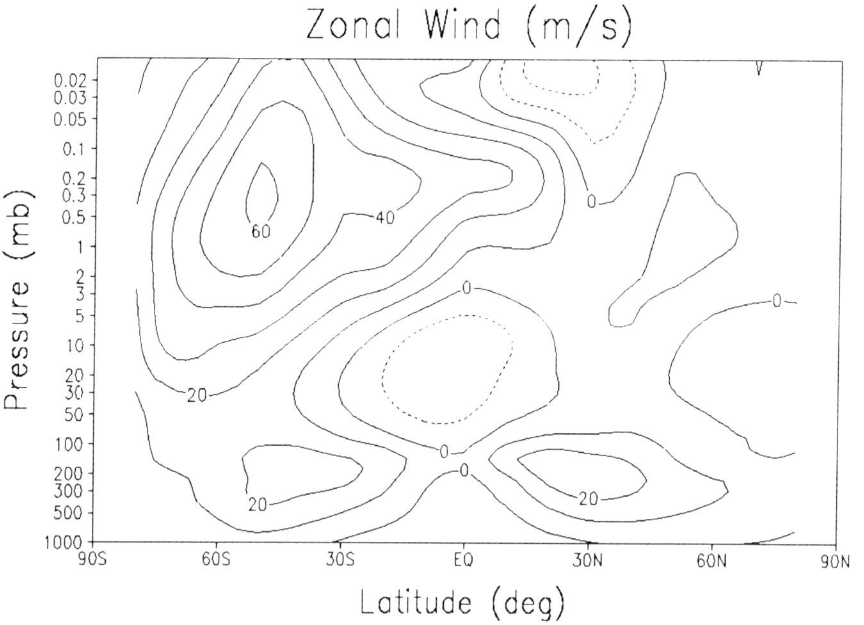

FIG. 10. *Observed April mean temperature and zonal wind from CIRA-86.*

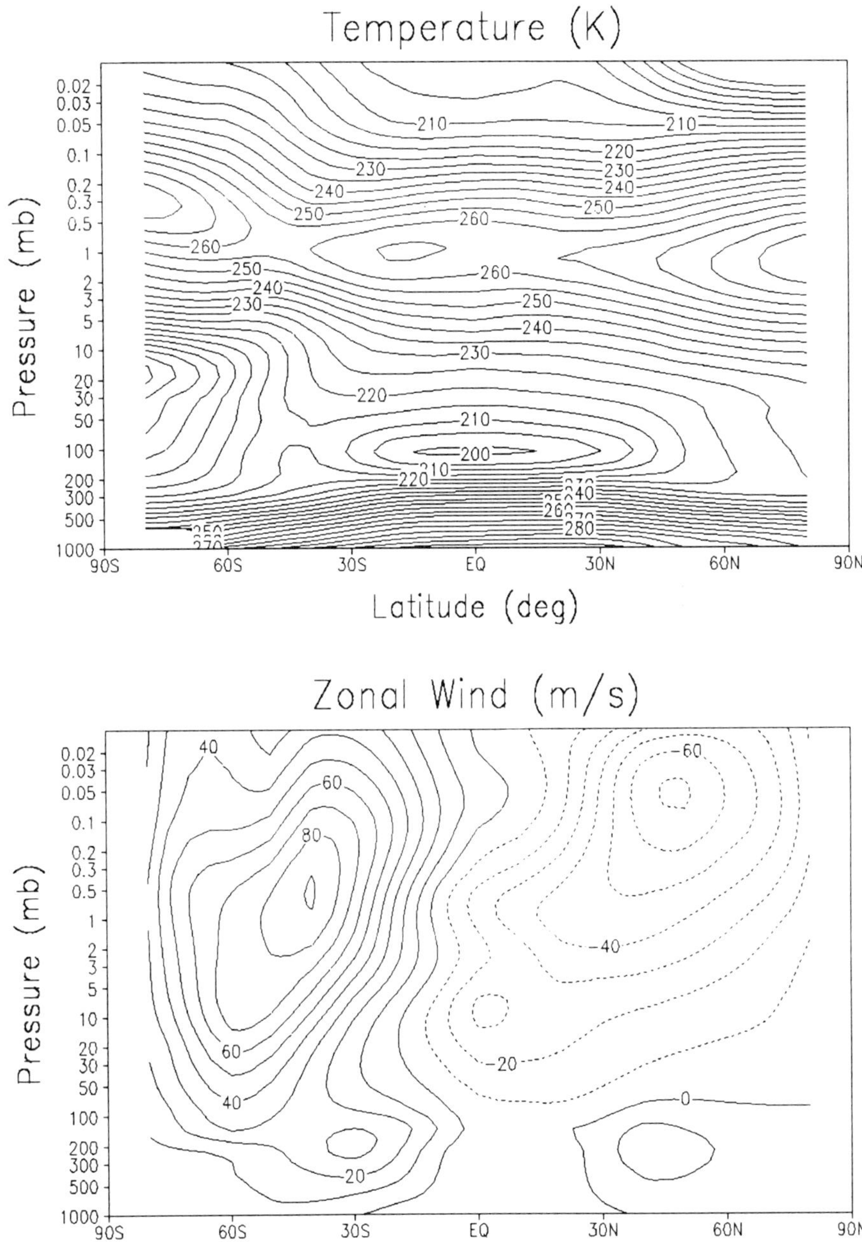

FIG. 11. *Observed July mean temperature and zonal wind from CIRA-86.*

Temperature (K)

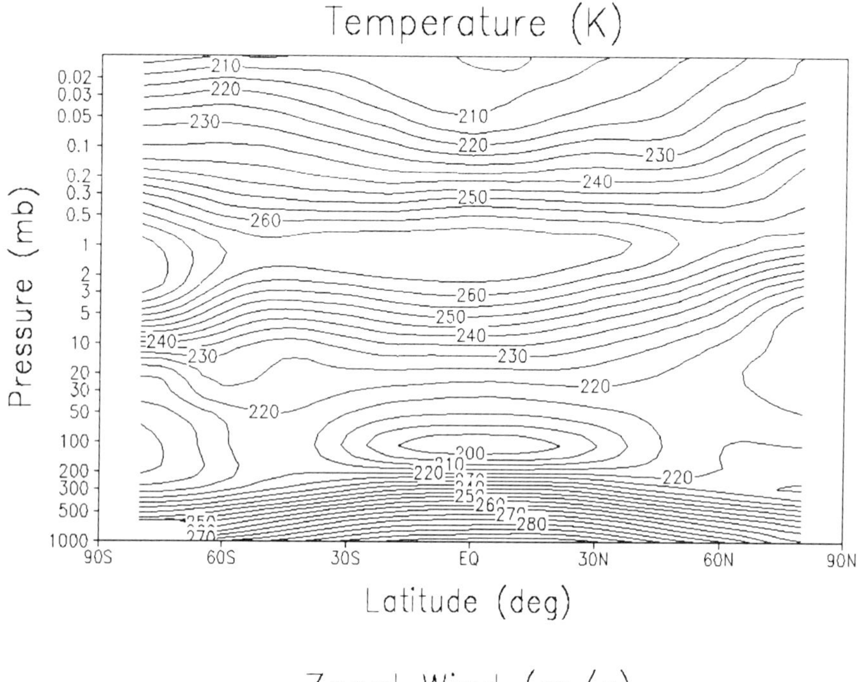

Zonal Wind (m/s)

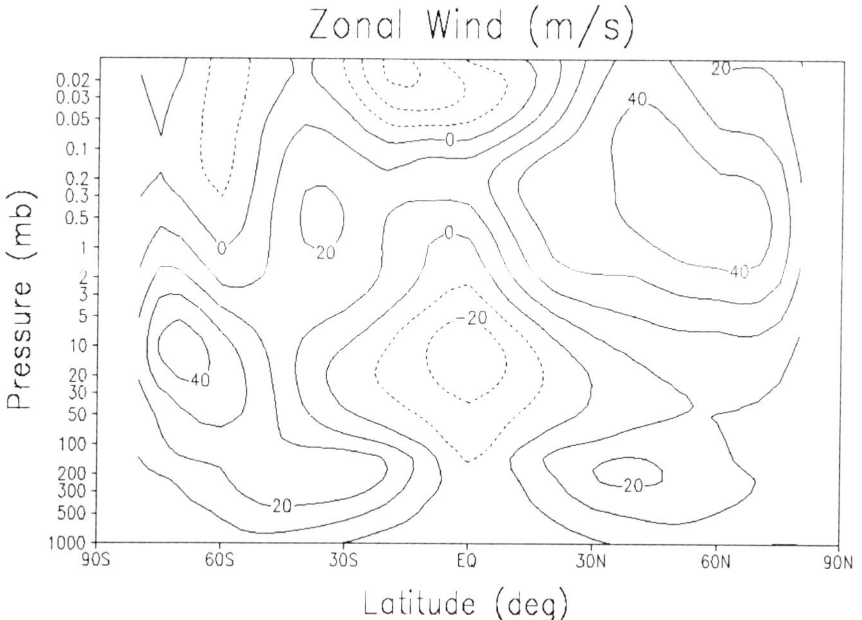

FIG. 12. *Observed October mean temperature and zonal wind from CIRA-86.*

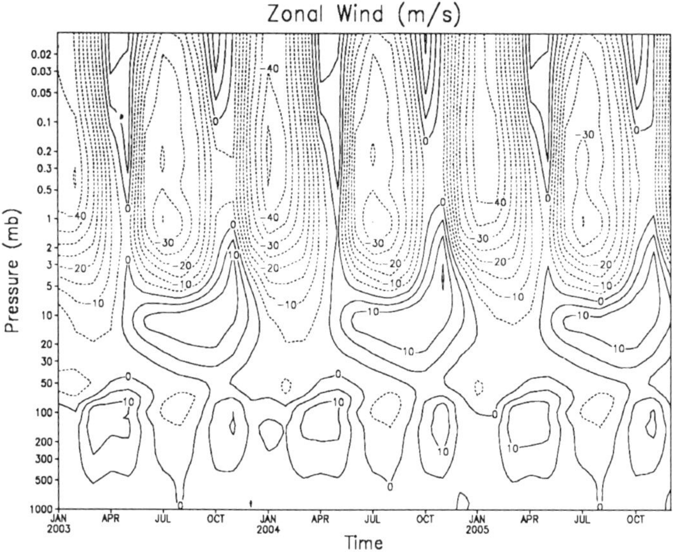

FIG. 13. *Time development of equatorial zonal wind for Model Years 3–5.*

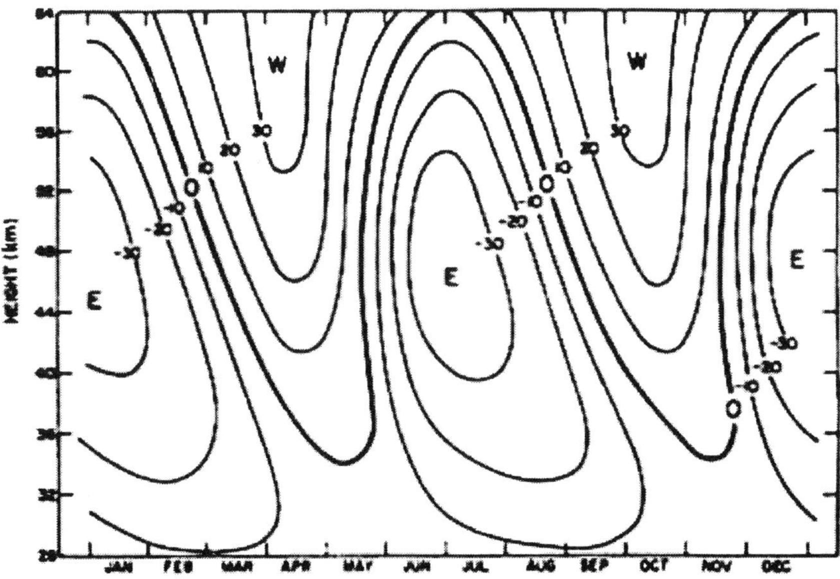

FIG. 14. *Time-height section of the semiannual zonal wind oscillation (m/s) at the equator (after Reed, 1966).*

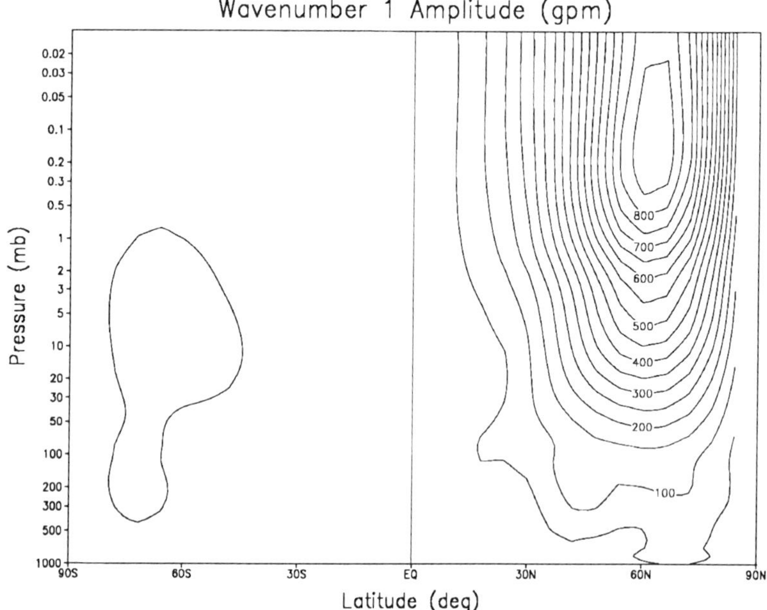

FIG. 15. *Approximate geopotential (fψ) wavenumber 1 amplitude for January, Model Year 5.*

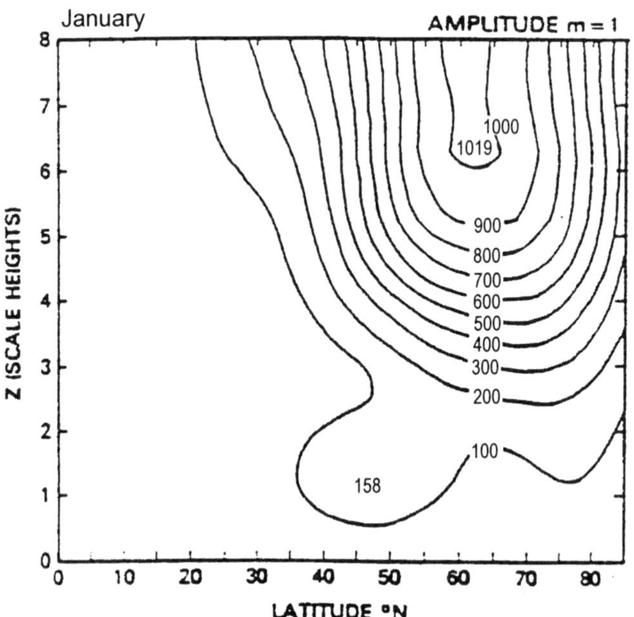

FIG. 16. *Observed geopotential wavenumber 1 amplitude (m) for January in the northern hemisphere (after Geller et al., 1983).*

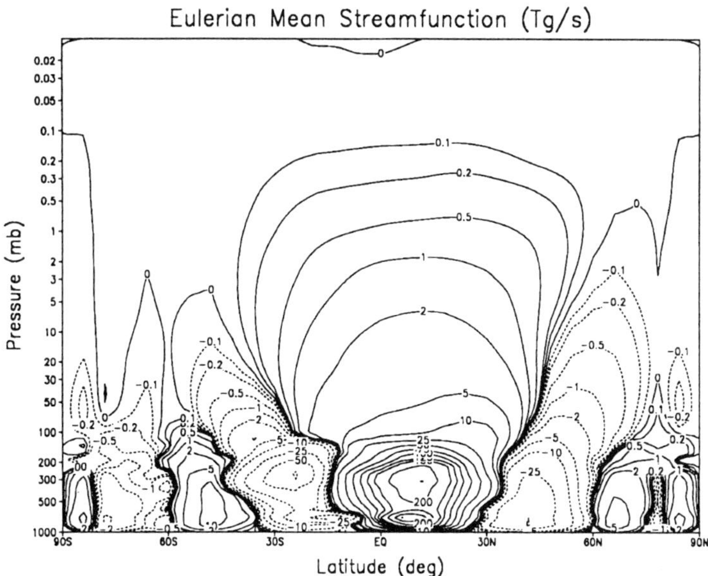

FIG. 17. *Simulated Eulerian mean meridional circulation for December, Model Year 4 to February, Model Year 5.*

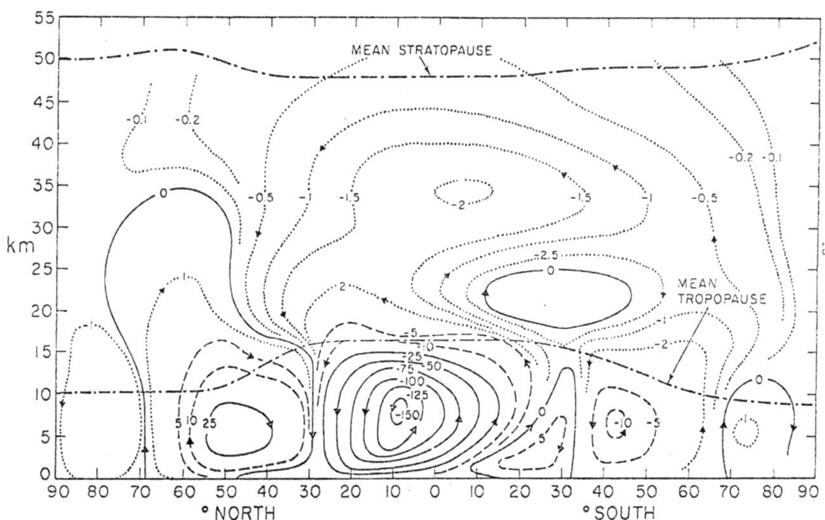

FIG. 18. *Inferred winter mean meridional circulation (Tg/s) after Louis (1975). Note reversal of poles and of sign convention relative to Figure 13.*

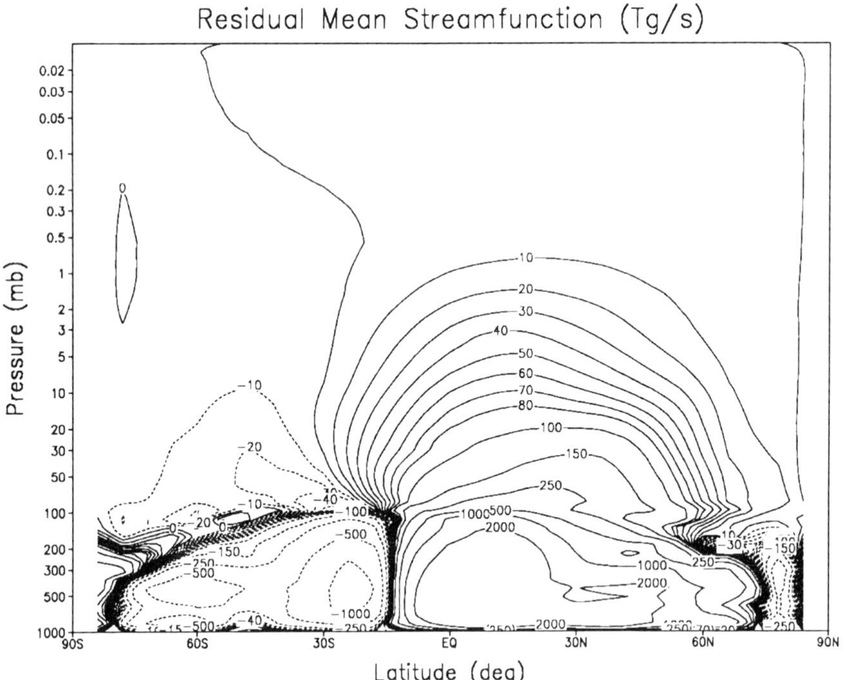

FIG. 19. *January residual streamfunction (kg/m/s) for Model Year 5. Contour levels are ±2000, ±1000, ±500, ±250, ±150, ±100, ±80, ±70, ±60, ±50, ±40, ±30, ±20, ±10, and 0.*

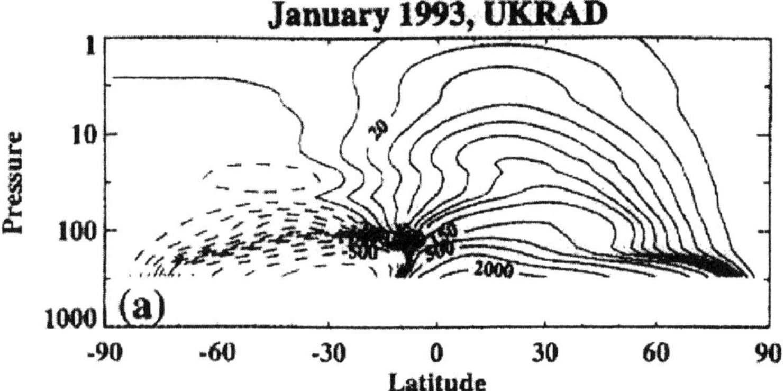

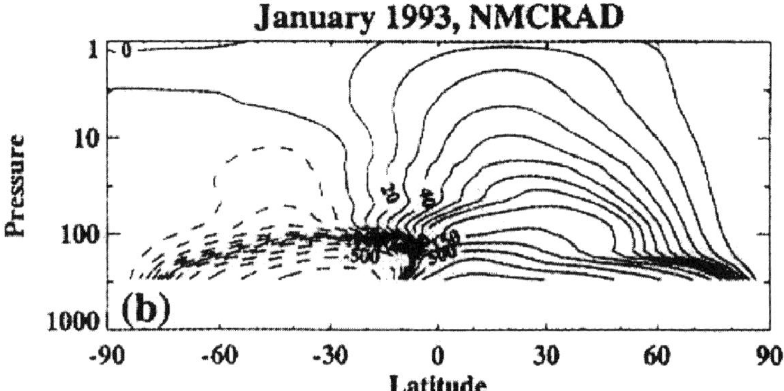

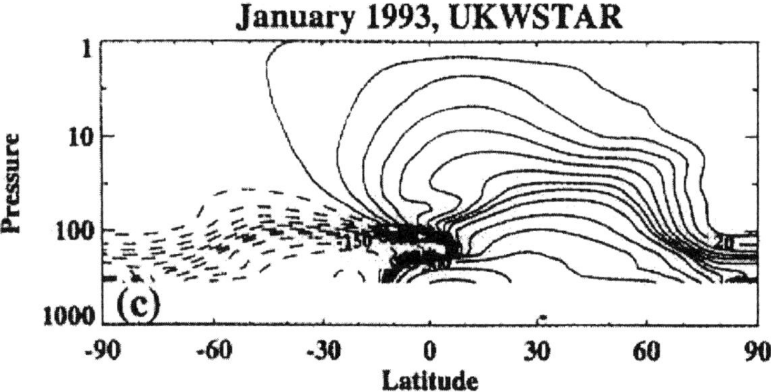

FIG. 20. *Residual streamfunction (kg/m/s) for January 1993 calculated from (a) radiative model heating rates using UKMO-assimilated temperatures, (b) radiative model heating rates using NMC temperatures, (c) UKMO-assimilated residual mean vertical velocities. Contour levels are ±2000, ±1000, ±500, ±250, ±150, ±100, ±80, ±70, ±60, ±50, ±40, ±30, ±20, ±10, and 0 (after Rosenlof, 1995).*

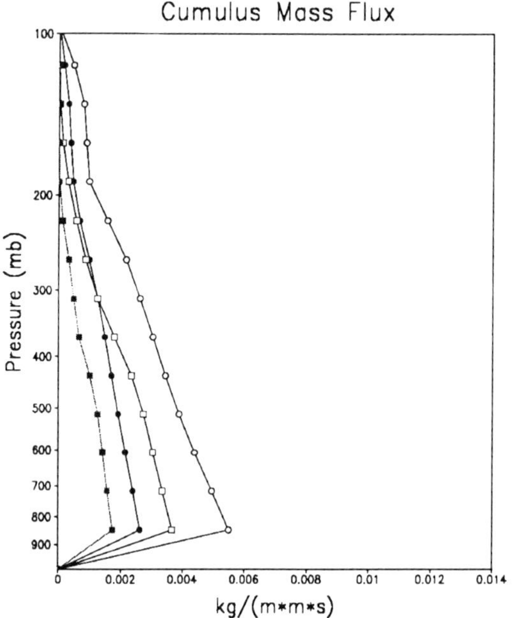

FIG. 21. *July mean cumulus mass flux for Model Year 5 at 0-20N (open circles), 20N-40N (filled circles), 40N-60N (open squares), and 60N-80N (filled squares).*

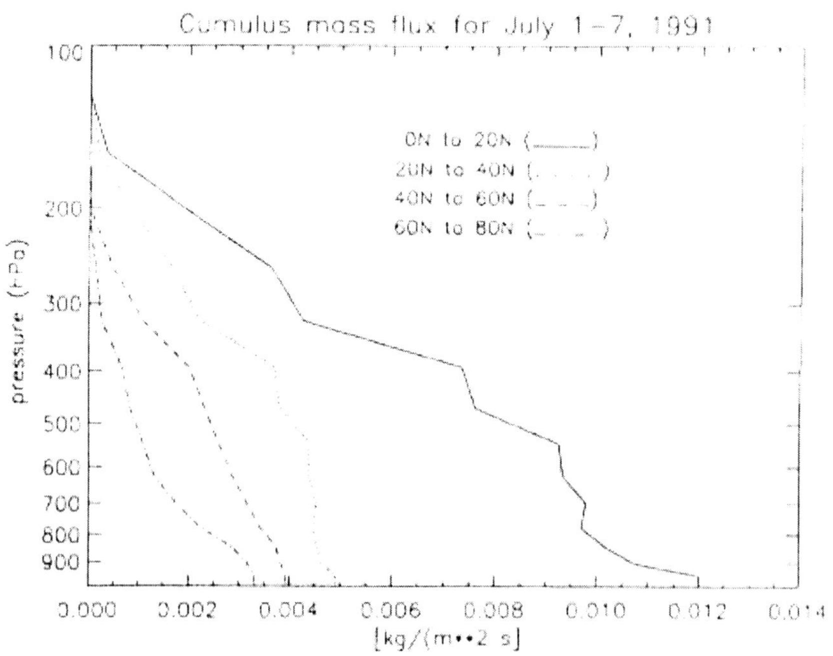

FIG. 22. *GEOS-1 DAS cumulus mass flux after Allen et al. (1996).*

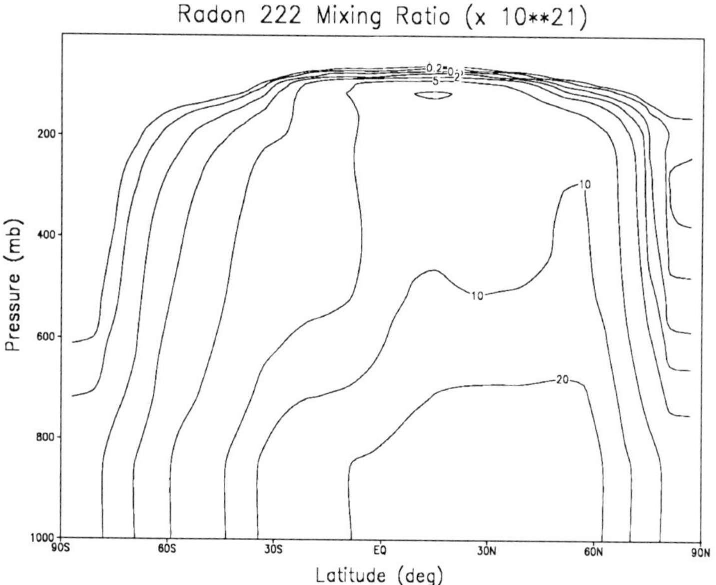

FIG. 23. *Zonal mean* ^{222}Rn *mixing ratio for June–August of Model Year 5.*

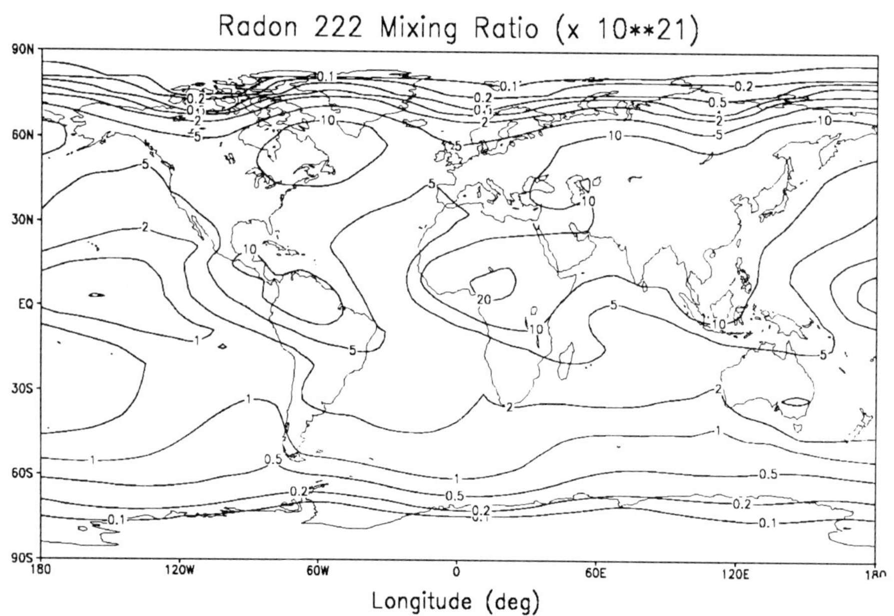

FIG. 24. ^{222}Rn *mixing ratio at 300 mb for June–August of Model Year 5.*

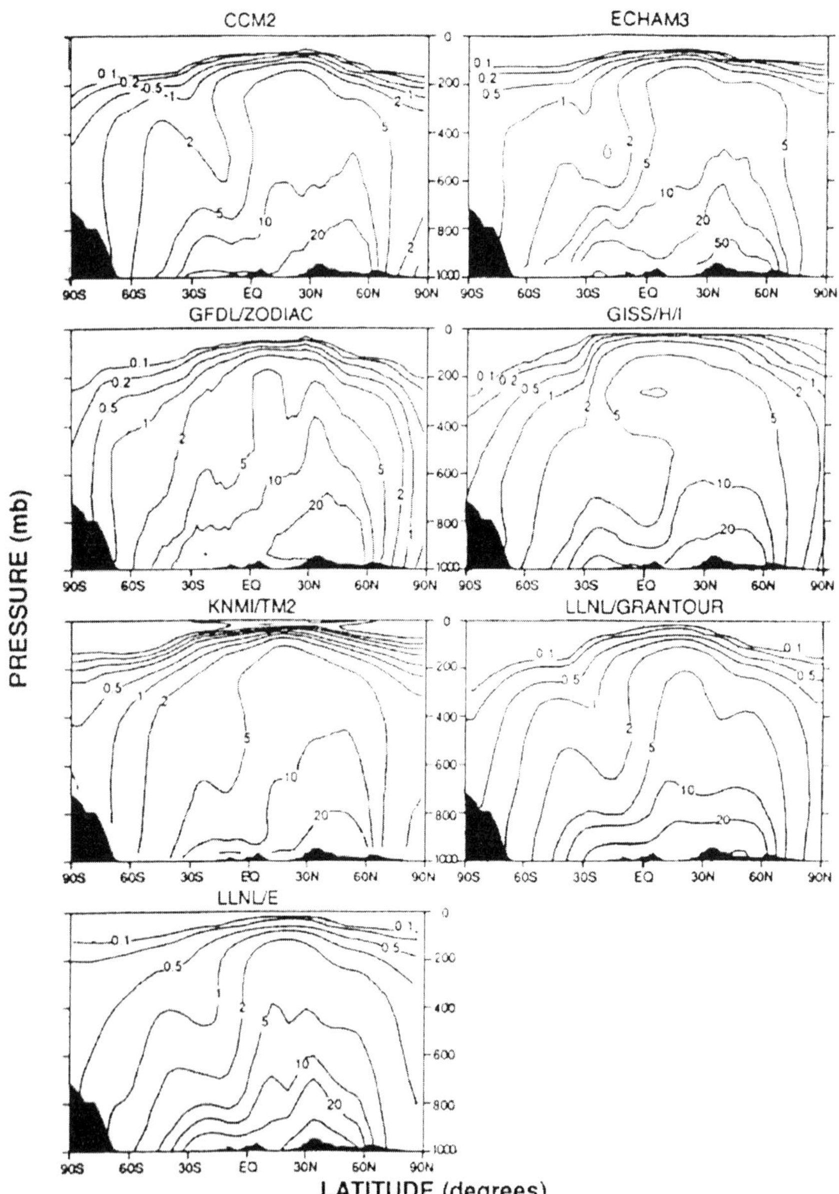

FIG. 25. *Zonal mean* ^{222}Rn *mixing ratio (times 10^{21}) as simulated by various 3-D models for June–August (after Jacob et al., 1997).*

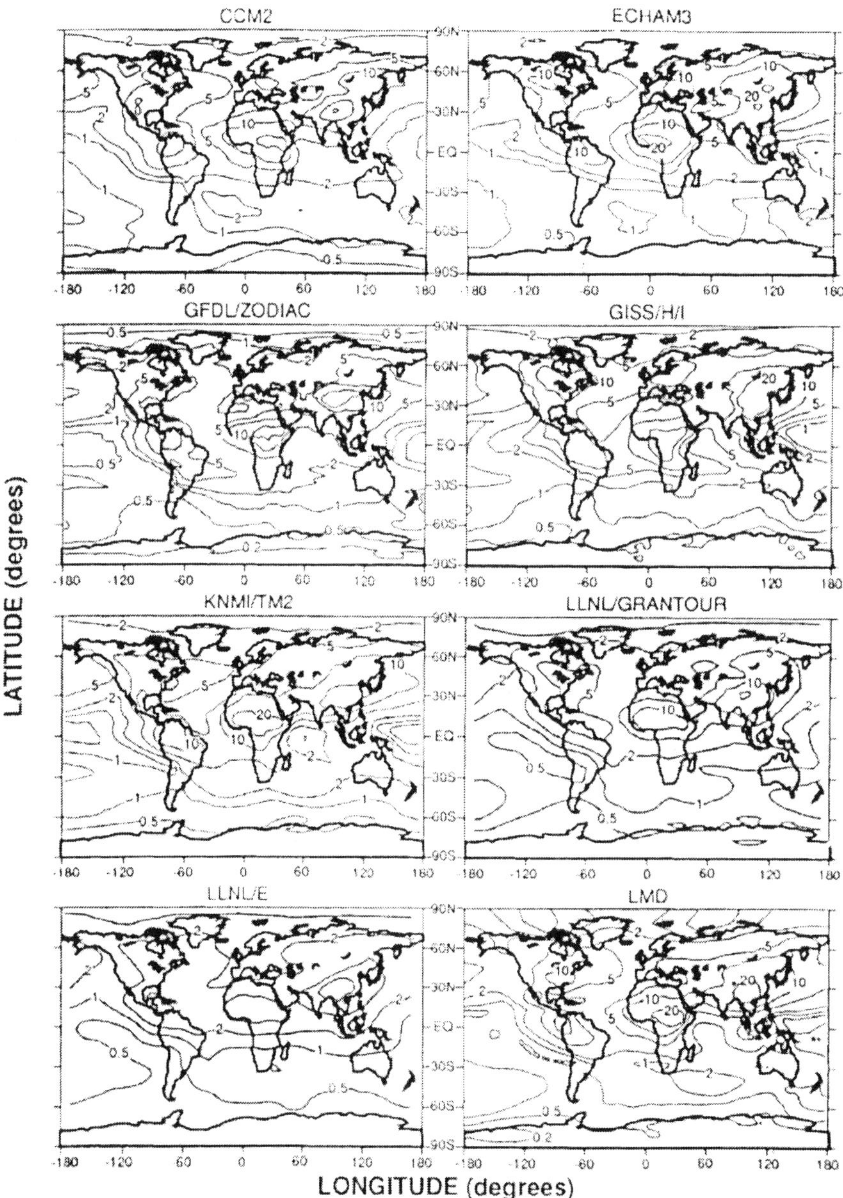

FIG. 26. ^{222}Rn mixing ratio (times 10^{21}) at 300 mb as simulated by various 3-D models for June–August (after Jacob et al., 1997).

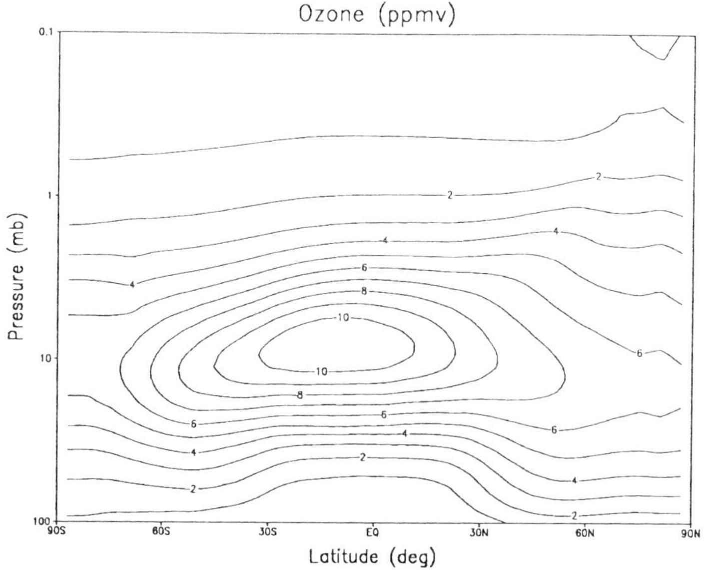

FIG. 27. *Simulated January daytime stratospheric ozone for Model Year 5.*

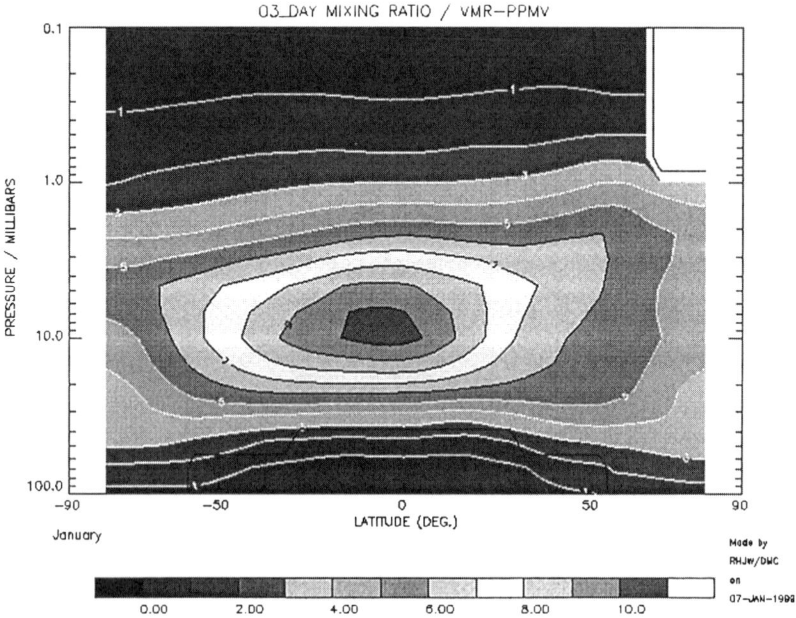

FIG. 28. *URAP 1998 January daytime stratospheric ozone (baseline standard).*

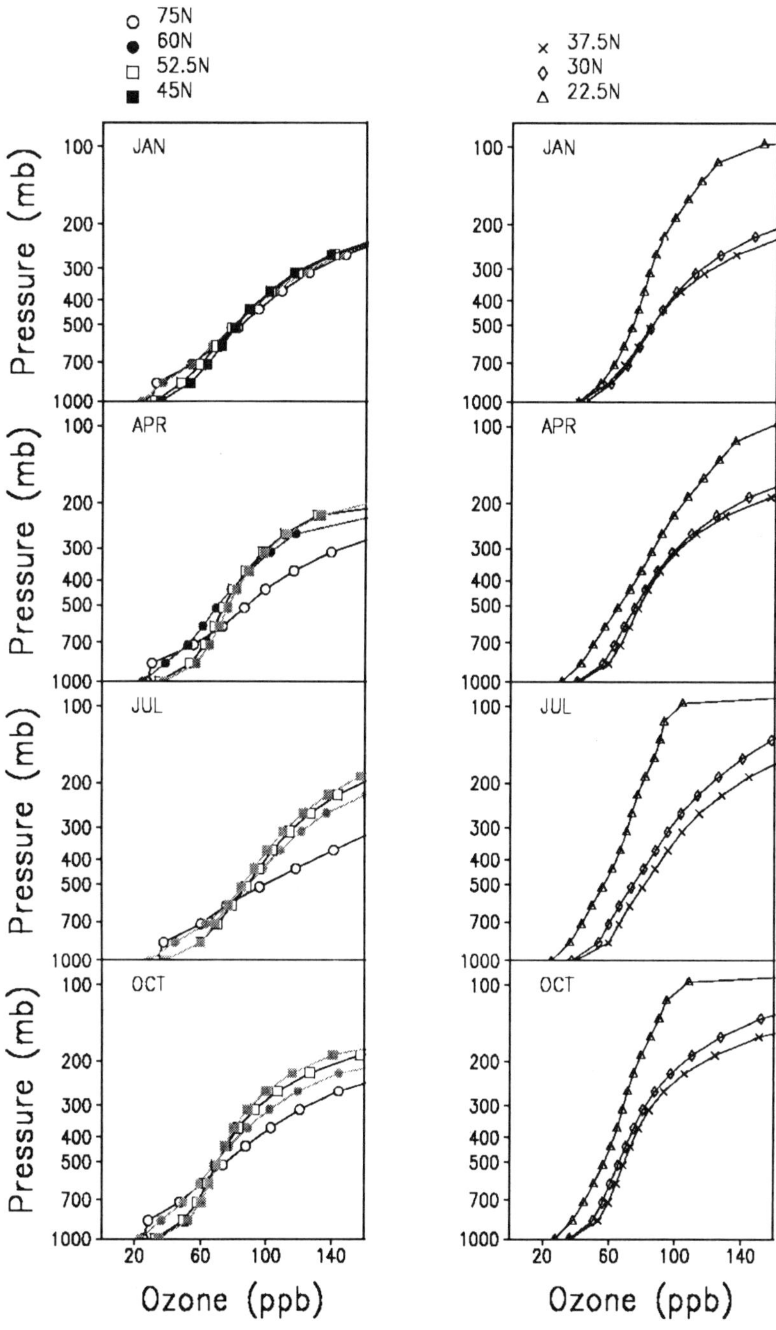

FIG. 29. *Simulated vertical profile of tropospheric ozone for Model Year 5.*

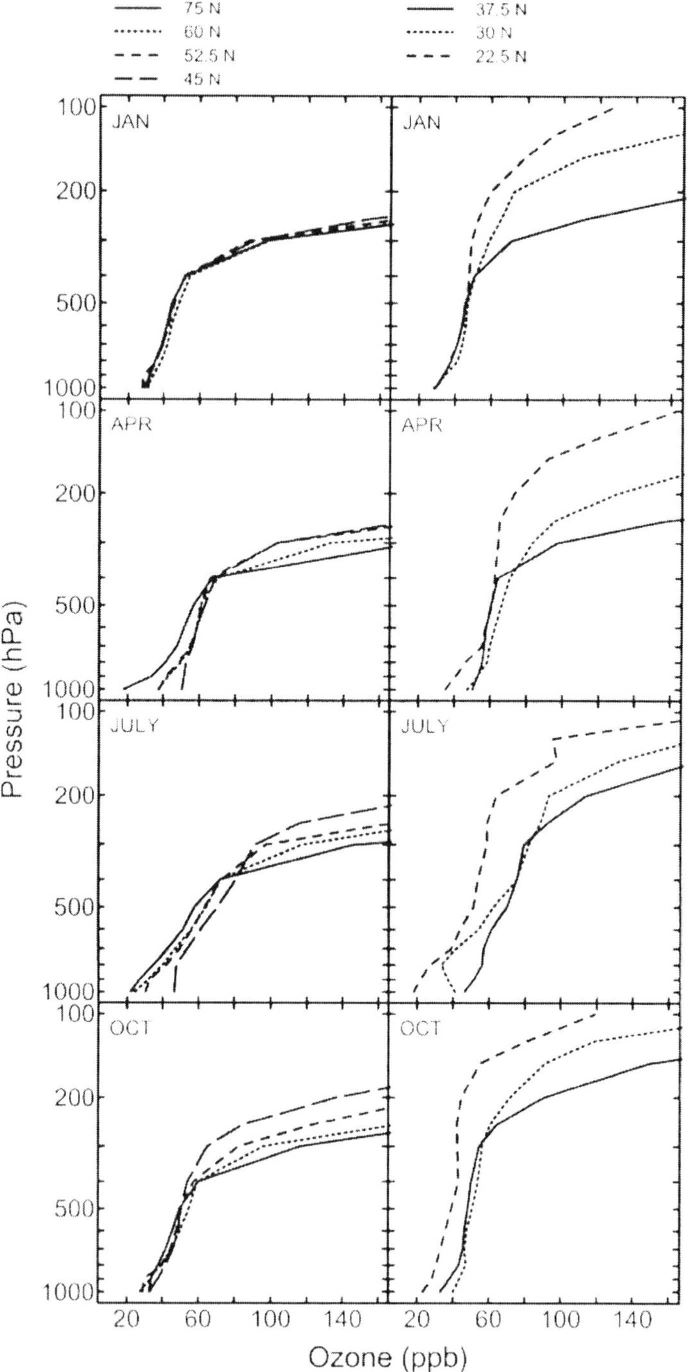

FIG. 30 *Tropospheric ozone climatology (vertical profile) after Logan (1999).*

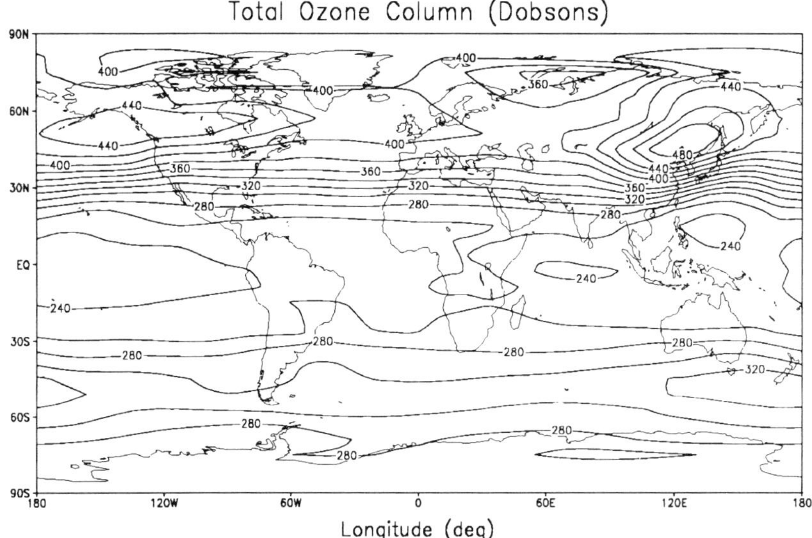

FIG. 31. *Simulated January mean total ozone column for Model Year 5.*

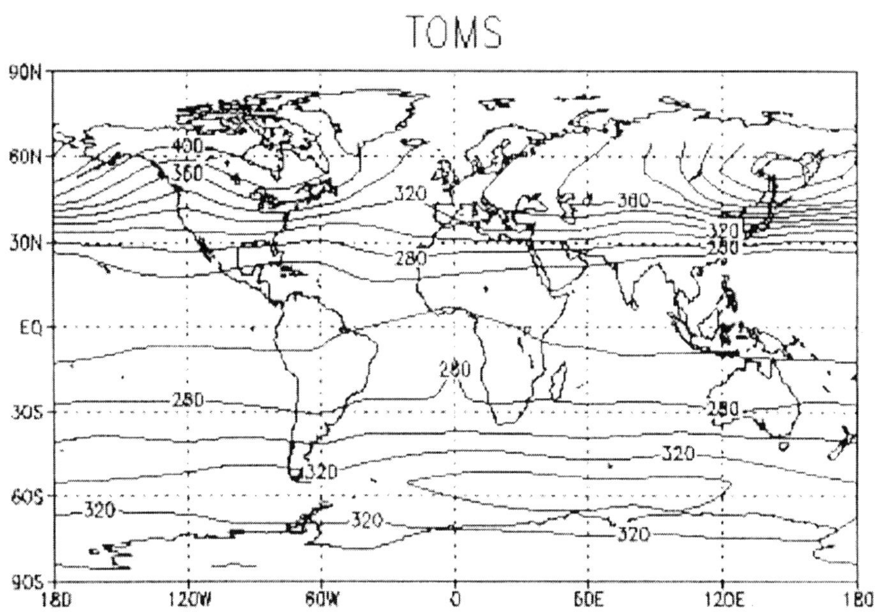

FIG. 32. *Observed January mean total ozone column (Dobsons) for 1979–1991.*

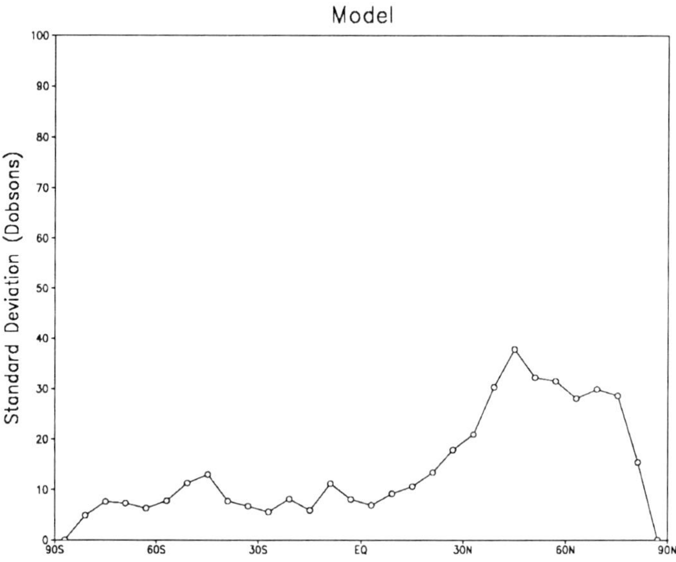

FIG. 33. *Standard deviation of simulated total ozone column for January, Year 5.*

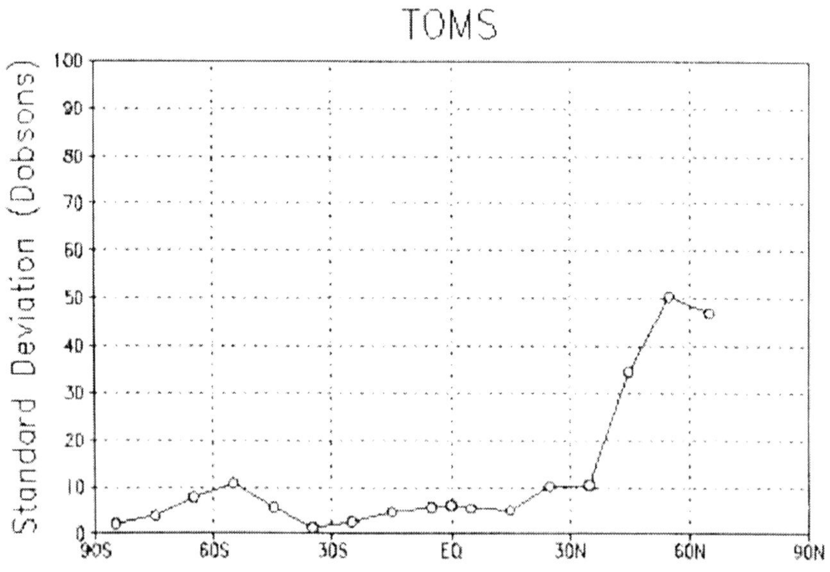

FIG. 34. *Standard deviation of observed total ozone column for January, 1979– 1991 from TOMS (Stanford et al., 1995).*

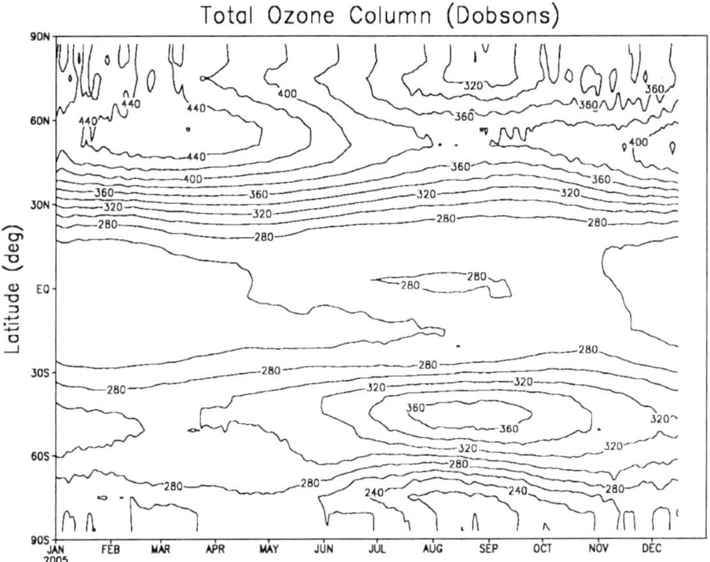

FIG. 35. *Seasonal variation of total ozone column (Dobsons) for Model Year 5.*

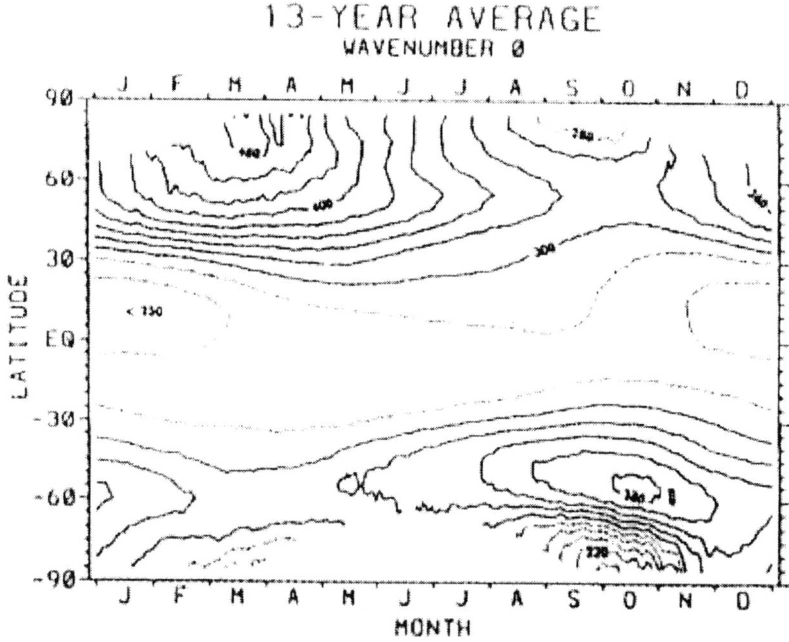

FIG. 36. *Seasonal variation of observed total ozone column (Dobsons) from TOMS averaged over 1979 to 1991 (after Stanford et al., 1995).*

REFERENCES

Allen, M. and J.E. Frederick, Effective photodissociation cross sections for molecular oxygen and nitric oxide in the Schumann-Runge bands, *J. Atmos. Sci.*, **39**, 2066–2075, 1982.

Allen, D.J., R.R. Rood, A.M. Thompson, and R.D. Hudson, Three-dimensional radon 222 calculations using assimilated meteorological data and a convective mixing algorithm, *J. Geophys. Res.*, **101**, 6871–6881, 1996.

Andrews, D.G. and M.E. McIntyre, Planetary waves in horizontal and vertical sheer: The Generalized Eliassen-Palm relation and the mean zonal circulation, *J. Atmos. Sci.*, **33**, 2031–2048, 1976.

Atkinson, R., Gas-phase tropospheric chemistry of organic compounds: A review, *Atmos. Environ.*, **24**A, 1–41, 1990.

Atkinson, R., Gas-phase tropospheric chemistry of organic compounds, *J. Phys. Chem. Ref. Data,* Monogr. **2**, 1994.

Atkinson, R., D.L. Baulch, R.A.Cox, R.F. Hampson, Jr., J.A. Kerr, M.J. Rossi, and J. Troe, Evaluated kinetic and photochemical data for atmospheric chemistry: Supplement V, *J. Phys. Chem. Ref. Data,* **26**, 521–1011, 1997a.

Atkinson, R., D.L. Baulch, R.A.Cox, R.F. Hampson, Jr., J.A. Kerr, M.J. Rossi, and J. Troe, Evaluated kinetic and photochemical data for atmospheric chemistry: Supplement VI, *J. Phys. Chem. Ref. Data,* **26**, 1329–1499, 1997b.

Baulch, D.L., R.A.Cox, P.J. Crutzen, R.F. Hampson, Jr., J.A. Kerr, J. Troe, and R.T. Watson, Evaluated kinetic and photochemical data for atmospheric chemistry: Supplement I, *J. Phys. Chem. Ref. Data,* **11**, 327–496, 1982.

Baulch, D.L., R.A.Cox, R.F. Hampson, Jr., J.A. Kerr, J. Troe, and R.T. Watson, Evaluated kinetic and photochemical data for atmospheric chemistry: Supplement II, *J. Phys. Chem. Ref. Data,* **13**, 1259–1380, 1984.

Bott, A., A positive definite advection scheme obtained by nonlinear renormalization of the advective fluxes, *Mon. Wea. Rev.*, **117**, 1006, 1989.

Boville, B.A., Middle atmosphere version of CCM2 (MACCM2): Annual cycle and interannual variability, *J. Geophys. Res.*, **100**, 9017–9039, 1995.

Brasseur, G. and S. Solomon, *Aeronomy of the Middle Atmosphere,* D. Reidel Publ. Co., Dordrecht, Holland, p. 452, 1986.

Briegleb, B.P., Delta-Eddington approximation for solar radiation in the NCAR community climate model, *J. Geophys. Res.*, **97**, 7603–7612, 1992.

Bucholtz, A., W.R. Skinner, J. Abreu, and P.B. Hays, The dayglow of the O_2 atmospheric band system, *Planet. Space Sci.*, **34**, 1031–1035, 1986.

Carmichael, G.R., I. Uno, M.J. Phadnis, Y. Zhang, and Y. Sunwoo, Tropospheric ozone production and transport in the springtime in east Asia, *J. Geophys. Res.*, **103**, 10649–10672, 1998.

Chatfield, R.B. and A.C. Delany, Convection links biomass burning to increased tropical ozone: However, models will tend to overpredict O_3, *J. Geophys. Res.*, **95**, 18473, 1990.

Cunnold, D.M., F.N. Alyea, N. Phillips, and R. Prinn, A three-dimensional dynamical-chemical model of atmospheric ozone, *J. Atmos. Sci.*, **32**, 170–194, 1975.

Cunnold, D.M. M.J. Newchurch, L.E. Floyd, H.J. Wang, J.M. Russell III, R. McPeters, J. Zawodny, and L. Froideveaux, Uncertainties in upper stratospheric ozone trends from 1979 to 1996, *J. Geophys. Res.*, **105**, 4427–4444, 2000.

DeMore, W.B., S.P. Sander, D.M. Golden, R.F. Hampson, M.J. Kurylo, C.J. Howard, A.R. Ravishankara, C.E. Kolb, and M.J. Molina, *Chemical Kinetics and Photochemical Data for Use in Stratospheric Modeling, Evaluation*, No. 12, Jet Propulsion Laboratory, Publication 97-4, Pasadena, CA, 1997.

Donahue, N.M., M.K. Dubey, R. Mohrschlatdt, K.L. Demerjian, and J.G. Anderson, High-pressure flow Study of the reaction $NO_x + OH \longrightarrow HONO_x$: Errors in the falloff region, *J. Geophvs. Res.*, **102**, 6159–6168, 1997.

Dunkerton, T.J., The role of gravity waves in the quasi-biennial oscillation, *J. Geophys. Res.*, **102**, 26053–26076, 1997.

Fels, S.B., A parameterization of scale-dependent radiative damping rates in the middle atmosphere, *J. Atmos. Sci.*, **39**, 1141–1152, 1982.

Fleming, E.L., S. Chandra, M.R. Shoeberl, and J.J. Barnett, Monthly mean global climatology of temperature, wind, geopotential height and pressure for 0–120 km, National Aeronautics and Space Administration, Technical Memorandum 100697, Washington, D.C., 1988.

Fomichev, V.I., A.A. Kutepov, R.A. Akmaev, and G.M. Shved, Parameterization of the 15 μ CO_2 band cooling in the middle atmosphere (15–115 kin), *J. Atmos. Terr. Phys.*, **55**, 7–18, 1993.

Fuglestvedt, J.S., I.S.A. Isaksen, and W.-C. Wang, Estimates of indirect global warming potentials for CH_4, CO and NO_x, *Climatic Change*, **34**, 405–437, 1996.

Fung, I., J. John, J. Lerner, E. Matthews, M. Prather, L.P. Steele, and P.J. Fraser, Three-dimensional Model synthesis of the global methane cycle, *J. Geophvs. Res.*, **96**, 13033–13065, 1991.

Garcia, R.R. and S. Solomon, A possible relationship between interannual variability in Antarctic ozone and the quasi-biennial oscillation, *Geophys. Res. Lett.*, **14**, 848–851, 1987.

Garcia, R.R., Parameterization of planetary wave breaking in the middle atmosphere, *J. Atmos. Sci.*, **48**, 1405–1419, 1991.

Geller, M.A., M.-F. Wu, and M.E. Gelman, Troposphere-stratosphere (surface-55 km) monthly winter general circulation statistics for the northern hemisphere-four year averages, *J. Atmos. Sci.*, **40**, 1334–1352, 1983.

Gery, M.W., G.Z. Whitten, J.P. Killus, and M.C. Dodge, A photochemical mechanism for urban and regional scale computer modeling, *J. Geophys. Res.*, **94**, 12925–12956, 1989.

Granier, C. and G. Brasseur, Impact of heterogeneous chemistry on model predictions of ozone changes, *J. Geophys. Res.*, **97**, 18015–18033, 1992.

Gray, L.J. and J.A. Pyle, The semi-annual oscillation and equatorial tracer distributions, *Quart. J. Roy. Meteor. Soc.*, **112**, 387–407, 1986.

Gray, L.J. and J.A. Pyle, Two-dimensional model studies of equatorial dynamics and tracer distributions, *Quart. J. Roy. Meteor. Soc.*, **113**, 635–651, 1987.

Holton, J.R. and R.S. Lindzen, An updated theory for the quasi-biennial cycle of the tropical stratosphere, *J. Atmos. Sci.*, **29**, 1076–1080, 1972.

Holton, J.R., The role of gravity wave induced drag and diffusion in the momentum budget of the mesosphere, *J. Atmos. Sci.*, **39**, 791–799, 1982.

Holton, J.R., The influence of gravity wave breaking on the general circulation of the middle atmosphere, *J. Atmos. Sci.*, **40**, 2497–2507, 1983.

Holton, J.R., P.H. Haynes, M.E. McIntyre, A.R. Douglass, R.B. Rood, and L. Pfister, Stratosphere-troposphere exchange, *Rev. Geophvs.*, **33**, 403–439, 1995.

Houweling, S., F. Dentener, and J. Lelieveld, The impact of nonmethane hydrocarbon compounds on tropospheric photochemistry, *J. Geophys. Res.*, **103**, 10673–10696, 1998.

Jacob, D.J., M.J. Prather, P.J. Rasch, R.-L. Shia, Y.J. Balkanski, S.R. Beagley, D.J. Bergmann, W.T. Blackshear, M. Brown, M. Chiba, M.P. Chipperfield, J. de Grand-pre, J.E. Dignon, J. Feichter, C. Genthon, W.L. Grose, P.S. Khasibhatla, I. Kohler, M.A. Kritz, K. Law, J.E. Penner, M. Ramonet, C.E. Reeves, D.A. Rotman, D.Z. Stockwell, P.F.J. Van Velthoven, G. Verver, O. Wild, H. Yang, and P. Zimmerman, Evaluation and intercomparison of global atmospheric transport models using ^{222}Rn and other short-lived tracers, *J. Geophys. Res.*, **102**, 5953–5970, 1997.

Jiang, Y., Y.L. Yung, A.R. Douglass, and K.K. Tung, The standard deviation of column ozone from the zonal mean, *Geophys. Res. Lett.*, **25**, 911–914, 1998.

Johnson, C.E. and R.G. Derwent, Relative radiative forcing consequences of global emissions of hydrocarbons, carbon monoxide, and NO_x from human activities estimated with a zonally-averaged two-dimensional model, *Climatic Change*, **34**, 439–462, 1996.

Kiehl, J.T., J.J. Hack, and B.P. Briegleb, The simulated Earth radiation budget of the National Center for Atmospheric Research community climate model CCM2 and comparisons with the Earth Radiation Budget Experiment (ERBE), *J. Geophys. Res.*, **99**, 20815–20828, 1994.

Kuo, H.L., On formation and intensification of tropical cyclones through latent heat release by cumulus convection, *J. Atmos. Sci.*, **22**, 40–63, 1965.

Legates, D.R. and C.J. Willmott, Mean seasonal and spatial variability in global surface air temperature, *Theor. and Appl. Clim.*, **41**, 11–21, 1990a.

Legates, D.R. and C.J. Willmott, Mean seasonal and spatial variability in gauge-corrected global precipitation, *International Journal of Climatology.*, **10**, 111–127, 1990b.

Lelieveld, J. and P.J. Crutzen, Role of deep cloud convection in the ozone budget of the troposphere, *Science*, **264**, 1759–1761, 1994.

Leovy, C.B., C.-R. Sun, M.H. Hitchman, E.E. Remsberg, J.M. Russell, L.L. Girdley, J.C. Gille, and L.V. Lyjak, Transport of ozone in the middle stratosphere: evidence for planetary wave breaking, *J. Atmos. Sci.*, **42**, 230–244, 1985.

Li, Y. and J.S. Chang, A mass-conservative, positive-definite, and efficient Eulerian advection scheme in spherical geometry and on a nonuniform grid system, *J. Appl. Met.*, **35**, 1897–1913, 1996.

Lindzen, R.S., Turbulence and stress owing to gravity wave and tidal breakdown, *J. Geophys. Res.*, **86**, 9707–9714, 1981.

Logan, J.A., An analysis of ozonesonde data for the troposphere: Recommendations for testing 3-D models and development of a gridded climatology for tropospheric ozone, *J. Geophys. Res.*, **104**, 16115–16149, 1999.

Lorenz, E.N., Energy and numerical weather prediction, *Tellus*, **12**, 304–373, 1960.

Louis, J.F., Chap. 6, CIAP Monograph No. 1, E.R. Reiter, ed., U.S. Dept. of Transportation, Washington, D.C., 1975.

Manabe, S. and R.T. Wetherald, Thermal equilibrium of the atmosphere with a given distribution of relative humidity, *J. Atmos. Sci.*, **24**, 241–259, 1967.

Martinez, R.D., A.A. Buttrago, N.W. Howell, C.H. Hearn, and J.A. Joens, The near U.V. absorption spectra of several aliphatic aldehydes and ketones at 300 K. *Atmos. Environ.*, **26A**, 785–792, 1992.

Matsuno, T., Vertical propagation of stationary planetary waves in the winter northern hemisphere, *J. Atmos. Sci.*, **27**, 871–883, 1970.

McFarlane, N.A., The effect of orographically excited gravity wave drag on the general circulation of the lower stratosphere and troposphere, *J. Atmos. Sci.*, **44**, 1775–1800, 1987.

McIntyre, M.E., and T.N. Palmer, Breaking planetary waves in the stratosphere, *Nature*, **305**, 593–600, 1983.

McKeen, S.A., T. Gierczak, J.B. Burkholder, P.O. Wennberg, T.F. Hanisco, E.R. Keim, R.-S. Gao, S.C. Liu, A.R. Ravishankara, and D.W. Fahey, The photochemistry of acetone in the upper troposphere: source of odd hydrogen radicals, *Geophys. Res. Lett.*, **24**, 3177–3180, 1997.

Moura, A.D., The eigensolution of linearized balance equations over a sphere, *J. Atmos.Sci.*, **33**, 877–907, 1976.

Newell, R.E., J.W. Kidson, D.G. Vincent, and G.J. Boer, *The General Circulation of the Tropical Atmosphere and Interactions with Extratropical Latitudes*, Vol. **2**, MIT Press, p. 370, 1974.

Olaguer, E.P., H. Yang and K.K. Tung, A reexamination of the radiative balance of the stratosphere, *J. Atmos. Sci.*, **49**, 1242–1263, 1992.

Olivier, J.G.J., A.F. Bouwman, C.M.W. van der Maas, J.J.M. Berdowski, C. Veldt, J.P.J. Bloos, A.J.H. Visschedijk, P.Y.J. Zandveld, and J.L. Haverlag, Description of EDGAR Version 2.0: A set of global emission inverntories of greenhouse gases and ozone-depleting substances for all Anthropogenic and most natural sources on a per country basis and on a $1°$ by $1°$ grid, RIVM Report No. 771060-002, National Institute of Public Health and the Environment, Bilthoven, the Netherlands, 1996.

Olson, J., M. Prather, T. Berntsen, G. Carmichael, R. Chatfield, P. Connell, R. Derwent, L. Horowitz, S. Jin, M. Kanakidou, P. Kasibhatla, R. Kotamarthi, M. Kuhn, K. Law, J. Penner, L. Perliski, S. Sillman, F. Stordal, A. Thompson, and O. Wild, Results from the Intergovernmental Panel on Climate Change Photochemical Model Intercomparison (PhotoComp), *J. Geophys. Res.*, **102**, 5979–5991, 1997.

Oort, A.H., *Global Atmospheric Circulation Statistics 1958–1973*, NOAA Professional Paper 14, U.S. Dept. of Commerce, Rockville, MD, 1983.

Pandis, S.N. and J.H. Seinfeld, Sensitivity analysis of a chemical mechanism for aquaeous-phase atmospheric chemistry, *J. Geophys. Res.*, **94**, 1105–1126, 1989.

Peixoto, J.P., and A.H. Oort, *Physics of Climate*, American Institute of Physics, New York, NY, p. 520, 1992.

Pitari, G., S. Palermi, G. Visconti, and R.G. Prinn, Ozone response to a CO_2 doubling: results from a stratospheric circulation model with heterogeneous chemistry, *J. Geophys. Res.*, **97**, 5953–5962, 1992.

Prados, A.I., R.R. Dickerson, B.G. Doddridge, P.A. Milne, J.L. Moody, and J.T. Merrill, Transport of ozone and pollutants from North America to the North Atlantic Ocean during the 1996 Atmosphere/Ocean Chemistry Experiment (AEROCE) intensive, *J. Geophys. Res.*, **104**, 26219–26233, 1999.

Prospero, J.M., Long-term measurements of the transport of African mineral dust to the southeastern United states: Implications for regional air quality, *J. Geophys. Res.*, **104**, 15917–15927, 1999.

Reed, R.J., Zonal wind behavior in the equatorial stratosphere and lower mesosphere, *J. Geophys. Res.*, **71**, 4223–4233, 1966.

Randel, W.J., *Global Atmospheric Circulation Statistics*, National Center for Atmospheric Research, Boulder, Colorado, NCAR/TN-366+STR, 1992.

Roberts, J.M., and R.W. Fajer, UV absorption cross sections of organic nitrates of potential atmospheric importance and estimation of atmospheric lifetimes, *Environ. Sci. Technol.*, **23**, 945–951, 1989.

Rosenlof, K.H., Seasonal cycle of the residual mean meridional circulation in the stratosphere, *J. Geophys. Res.*, **100**, 5173–5192, 1995.

Rudolph, J. and F.J. Johnen, Measurements of light atmospheric hydrocarbons over the Atlantic in regions of low biological activity, *J. Geophys. Res.*, **95**, 20583–20591, 1990.

Schultz, M.G., D.J. Jacob, Y. Wang, J.A. Logan, E.L. Atlas, D.R. Blake, N.J. Blake, J.D. Bradshaw, E.V. Browell, M.A. Fenn, F. Flocke, G.L. Gregory, B.G. Heikes, G.W. Sachse, S.T. Sandholm, R.E. Shetter, H.B. Singh, and R.W. Talbot, Chemical NO_x budget in the upper troposphere over the tropical South Pacific, *J. Geophys. Res.*, **104**, 5829–5843, 1999.

Singh, H.B., D. Herlth, D. O'Hara, K. Zahnle, J.D. Bradshaw, S.T. Sandholm, R. Talbot, P.J. Crutzen, and, M. Kanakidou, Relationship of peroxyacetyl nitrate to active and total odd nitrogen at northern high latitudes: Influence of reservoir species on NO_x and O_3, *J. Geophys. Res.*, **97**, 16523–16530, 1992.

Singh, H., D. O'Hara, D. Herlth, W. Sachse, D.R. Blake, J.D. Bradshaw, M. Kanakidou, and P.J. Crutzen, Acetone in the atmosphere: Distribution, sources, and sinks, *J. Geophys. Res.*, **99**, 1805–1819, 1994.

Singh, H., Y. Chen, A. Tabazadeh, Y. Fukui, I. Bey, R. Yantosca, D. Jacob, F. Arnold, K. Wohlfrom, A. Atlas, F. Flocke, D. Blake, N. Blake, B. Heikes, J. Snow, R. Talbot, G. Gregory, G. Sachse, S. Vay, and Y. Kondo, Distribution and fate of select oxygenated organic species in the troposphere and lower stratosphere over the Atlantic, *J. Geophys. Res.*, **105**, 3795–3805, 2000.

Slingo, A., A GCM parameterization for the shortwave radiative properties of water clouds, *J. Atmos. Sci.*, **46**, 1419–1427, 1989.

Spivakovsky, C.M., S.C. Wofsy, and M.J. Prather, A numerical method for parameterization of atmospheric chemistry: Computation of tropospheric OH, *J. Geophys. Res.*, **95**, 18433–18440, 1990.

Staffelbach, T.A., J.J. Orlando, G.S. Tyndall, and J.G. Calvert, The UV-visible absorption spectrum and photolysis quantum yields of methylglyoxal, *J. Geophys. Res.*, **100**, 14189–14198, 1995.

Stanford, J.L., J.R. Ziemke, R.D. McPeters, A.J. Krueger, and P.K. Bhartia, Spectral analyses, climatology, and interannual variability of Nimbus-7 TOMS Version 6 total column ozone, NASA Ref. Publ. 1360, 1995.

Stockwell, W.R., F. Kirchner, and M. Kuhn, A new mechanism for regional atmospheric chemistry modeling, *J. Geophys. Res.*, **102**, 25847–25879, 1997.

Stohl, A. and T. Trickl, A textbook example of long-range transport: Simultaneous observation of ozone maxima of stratospheric and North American origin in the free troposphere ever Europe, *J. Geophys. Res.*, **104**, 30445–30462, 1999.

Toumi, R., A potential new source of OH and odd-nitrogen in the atmosphere, *Geophys. Res. Lett.*, **20**, 25–28, 1993.

Turco, R.P., O.B. Toon, and P. Hamill, Heterogeneous physicochemistry of the polar ozone hole, *J. Geophys. Res.*, **94**, 16493–16510, 1989.

Wang, Y., J. A. Logan, and D.J. Jacob, Global simulation of tropospheric O_3-NO_x - hydrocarbon chemistry, 2. Model evaluation and global ozone budget, *J. Geophys. Res.*, **103**, 10727–10755, 1998a.

Wang, Y., J.A. Logan, and D.J. Jacob, Global simulation of tropospheric O_3-NO_x - hydrocarbon chemistry, 3. Origin of tropospheric ozone and effects of nonmethane hydrocarbons, *J. Geophys. Res.*, **103**, 10757–10767, 1998b.

Wang, Y., D.J. Jacob, and J.A. Logan, Global simulation of tropospheric O_3-NO_x - hydrocarbon chemistry, 3. Origin of tropospheric ozone and effects of nonmethane hydrocarbons, *J. Geophys. Res.*, **103**, 10757–10767, 1998c.

Wayne, R.P., I. Barnes, J.P.Burrows, C.E. Canova-Mas, J. Hjorth, G. Le Bras, G.K. Moortgat, D. Perner, G. Poulet, G. Restelli, and H. Sidebottom, The nitrate radical: physics, chemistry, and the atmosphere, *Atmos. Environ.*, **25**A, 1–203, 1991.

Willie, U.E. Becker, R.N. Schindler, I.T. Lancer, G. Poulet, and G. LeBras, A discharge flow mass-spectrometric study of the reaction between the NO_3 radical and isoprene, *J. Atmos. Chem.*, **13**, 183–193, 1991.

World Meteorological Organization, Atmospheric ozone 1985: Assessment of our understanding of the processes controlling its present distribution and change. Global Ozone Research and Monitoring Project, Report No. 16, Geneva, Switzerland, 1986.

World Meteorological Organization, Scientific assessment of ozone depletion: 1992, Global Ozone Research and Monitoring Project, Report No. 25, Geneva, Switzerland, 1992.

Wuebbles, D.J., J.S. Tamaresis, and K.O. Patten, *Effects of Increasing Trace Gas* Emissions on Global Atmospheric Chemistry and Climate: An Interim Report, U.S. Environmental Protection Agency, Research Triangle Park, NC, 1993.

Wuebbles, D.J., R. Kotamarthi, and K.O. Patten, Updated evaluation of ozone depletion potentials for chlorobromomethane (CH_2CIBr) and 1-bromo-propane ($CH_2BrCH_2CH_2$), *Atmos. Environ.*, **33**, 1352–2510, 1999.

Zhong, W. and J.D. Haigh, Improved broadband emissivity parameterization for water vapor cooling rate calculations, *J. Atmos. Sci.*, **52**, 124–138, 1995.

DEVELOPMENT OF A GLOBAL-THROUGH-URBAN SCALE NESTED AND COUPLED AIR POLLUTION AND WEATHER FORECAST MODEL AND APPLICATION TO THE SARMAP FIELD CAMPAIGN

MARK Z. JACOBSON*

Abstract. A global-through-urban scale nested air pollution and weather forecast model was developed and applied to the SARMAP (San Joaquin Valley Air Quality Study [SJVAQS] and Atmospheric Utility Signatures Predictions and Experiments Study [AUSPEX] Regional Modeling Adaptation Project) field campaign. The model, GATOR-GCMM (Gas, Aerosol, Transport, Radiation, General Circulation, and Mesoscale Meteorological) model, treats nesting of all important air quality and meteorological parameters simultaneously, from the global-through-urban scales (down to < 5 km grid spacing). The model does not use nudging or data assimilation; it is prognostic, except for the emission inventory. The model was applied with five global-through-urban scale nested grids to the August 3–6, 1990 SARMAP field campaign. Parameters compared with observations included air temperatures, air pressures, relative humidities, wind speeds and directions, and 20 gases. In the inner-nested domain, the gross error in near-surface temperatures, normalized over all 8617 day and night observations in the model domain, was 1.02% . The gross error in near-surface ozone, normalized over all 2866 measurements above 50 ppbv during the four days was 22.5%.

1. Introduction. Limited-area models (LAMs) are used for predicting weather and air quality on urban and regional scales. Nesting is a common method of providing lateral boundary conditions for LAMs. Unfortunately, unless the largest LAM scale is the global scale, boundary conditions are still needed for the largest LAM domain. Thus, an ideal nested model uses a global model for boundary conditions around the largest LAM domain.

Two widely-used limited-area meteorological models with nesting features are the Penn State University (PSU)/National Center for Atmospheric Research (NCAR) Mesoscale Model [MM4, *Anthes et al.*, 1987; MM5, *Grell et al.*, 1993] and the Colorado State University Regional Atmospheric Modeling System (CSU/RAMS) [*Pielke et al.*, 1992; *Copeland et al.*, 1996]. Limited-area meteorological models have been nested with global models or observations to produce a variety of regional climate models (RCMs) [e.g., *Dickinson et al.*, 1989; *Giorgi and Bates*, 1989; *McGregor and Walsh*, 1993; *Juang and Kanamitsu*, 1994; *Marinucci et al.*, 1995; *Leung and Ghan*, 1995; *Podzun et al*, 1995; *Giorgi and Marinucci*, 1996; *Dudek et al.*, 1996; *Giorgi and Mearns*, 1999; *Qian et al.*, 1999; *Leung and Ghan*, 1999; *McGregor*, 1997; *Jones et al.*, 1997; *Laprise et al.*, 1998].

Nested limited-area meteorological models have also been used to provide boundary conditions for limited-area chemical transport models (CTMs). In some cases, the CTM and mesoscale model have both been

---

*Department of Civil & Environmental Engineering, Stanford University, Stanford, CA 94305-4020.

nested (although not necessarily occupying the same domains) [e.g., *Pleim et al.*, 1991; *Jakobs et al.*, 1995; *Odman and Ingram*, 1996; *Byun and Ching*, 1999; *McHenry et al.*, 1999]. In other cases, the meteorological model has been nested, but the CTM has occupied only the innermost meteorological-model domain [e.g., *DaMassa et al.*, 1996; *Qian and Giorgi* , 1999; *Umeda and Martien*, 2000]. In other cases, the CTM has been nested [*Morris et al.*, 1992; *Kumar et al.*, 1994] or configured with heterogeneous grid spacing [*Mathur et al.*, 1992], and winds were obtained offline from meteorological observations or model predictions.

Although RCMs in use today treat meteorology from the global scale to resolutions of 20 km [*Marinucci et al.*, 1995], 45 km [*Laprise et al.*, 1998]; or 50 km [*Jones et al.*, 1997; *Giorgi and Marinucci*, 1996], no RCM has treated meteorology, gas chemistry, or aerosols from the global to urban scale (< 5 km). Similarly, whereas all nested mesoscale LAMs coupled with CTMs account for gas chemistry and some [e.g., EPA Models-3, *Byun and Ching*, 1999; EDSS, *McHenry et al.*, 1999] account for aerosols, none treats nesting of aerosols, chemistry, or meteorology up to the global scale. Current nested models are also limited by computer memory. They either need separate memory for each grid (limiting the number of grids on a machine) or require that simulations for a grid be completed and the code recompiled for the next grid to be solved.

For this work, a one-way nested model was designed to treat gases, size/composition-resolved aerosols, radiation, and meteorology from the global through urban (< 5 km) scales. The model runs over a single, continuous simulation, and, regardless of the number of domains, the central memory required never exceeds about 1.5 and 2.1 times that of the largest domain for gas simulations and gas/aerosol simulations, respectively. The model accounts for radiative feedback from photochemically-active gases, size-resolved aerosols, and size-resolved liquid water and ice particles to meteorology on all scales. The model and some results are briefly described below.

2. Description of the model. GATOR-GCMM derives from a global model, GATORG (Gas, Aerosol, Transport, Radiation, and General Circulation model) [*Jacobson*, 2001a,b], and a regional model, GATORM (Gas, Aerosol, Transport, Radiation, and Mesoscale Meteorological model) [*Jacobson et al.*, 1996; *Jacobson*, 1997a,b; 1998a, 1999b,c]. The gas, aerosol, and radiative parts of the two models are the same, but the meteorological and transport parts differ.

GATOR-GCMM is a single, unified model where all processes are run online. It uses one common block and has switches to run in global mode, regional mode, nested mode, and with/without gases, aerosols, radiation, meteorology, transport, deposition, cloud physics, surface processes, etc. The model allows any number (limited only by computer time) of one-way nested layers between the global and urban scales, and in each layer, any number of domains for both air quality and meteorological calculations. In all domains in all layers, the air quality and meteorological domains occupy the same space and have the same grid spacing, except that, in all nonglobal-domains, each meteorological domain contains two more rows and columns than each air quality domain. In this "buffer zone," topographic gradients are eliminated to dampen gravity-wave reflections at boundaries. The horizontal and vertical coordinates in all domains are the spherical and sigma-pressure coordinates, respectively.

The model is presently run without spinup or data assimilation so that it is prognostic, except for the emissions inventory. Several components of the model are discussed next. More details, including treatment of surface processes, are given in *Jacobson* [2001c]. Aerosol processes are discussed in a paper in preparation.

2.1. Gas processes. Gas processes in GATOR-GCMM include emissions, photochemistry, heterogeneous chemistry, nucleation, condensation/ evaporation, dissolution/evaporation, and dry deposition. These processes are solved in all domains. Area- and mobile-source emissions are placed in the bottom model layer of all domains. Point-source emissions are placed in a layer determined from the stack height and estimated plume rise of the emissions. The method of determining plume rise height is obtained from *Briggs* [1975].

Gas photochemistry and heterogeneous chemistry are solved with SMVGEAR II [*Jacobson*, 1995; 1998b]. In all domains, 115 gases, 220 kinetic reactions, and 27 photolysis reactions are treated, except that 3 additional photolysis reactions are solved for in the global domain, since the top of the global domain extends to 55 km whereas the tops of the regional domains are limited here to 16 km.

Dry deposition velocities for gases are calculated as the inverse sum of three resistances [Equations 20.10*ff* of *Jacobson*, 1999a]. The surface resistance treatment [*Wesely*, 1989; *Walmsley and Wesely*, 1996] depends on landcover type. Landcover categories at 1 km resolution were obtained from *USGS* [1999]. The 24 USGS landcover classes are compressed into each of Wesely's 11 landcover categories in each cell in each domain for dry deposition calculations. Predicted deposition velocities over each category in each cell are weighted by the fractional landcover of the category in the cell.

2.2. Radiative transfer. Radiative transfer is used to determine diabatic heating rates for temperature calculations and actinic fluxes for photolysis calculations. Radiative calculations in the model depend on scattering and absorption by gases, aerosols, cloud liquid, and cloud ice. UV and visible gas absorption by all photolyzing gases, solar-IR absorption by H_2O, CO_2, O_3, and O_2 and thermal-IR absorption by H_2O, CO_2, O_3, CH_4, N_2O, $CFCl_3$, CF_2Cl_2, $CFCl_3$, and CCl_4 [with absorption coefficients from *Mlawer et al.*, 1997] are included.

Cloud liquid and ice affect optics in each domain as follows: Bulk cloud liquid and ice contents from the cumulus parameterization are combined with modified-gamma size distribution data from *Welch et al.* [1980] and power-law/Marshall-Palmer distribution data from *Platt* [1997] for different cloud types to estimate cloud drop and ice crystal number and cross-sectional area distributions. Mie calculations are then combined with the distributions to give spectral cloud liquid and ice optical depths, single-scattering, albedos, and asymmetry parameters, which are input into the radiative calculation. Irradiances for diabatic heating rate calculations are determined for 86, 67, and 256 wavelength intervals in the regions < 0.8 μm, 0.8–4.5 μm, and 4.5–1000 μm, respectively, with the radiative transfer algorithm described in *Toon et al.* [1989]. Actinic fluxes are determined for the 86 intervals in the region < 0.8 μm.

2.3. Global dynamics. The global dynamics module integrates equations for momentum (under the hydrostatic assumption), thermodynamic energy, and total water. The solution schemes originate from a 1994 version of the UCLA GCM [*Arakawa and Lamb*, 1977; 1981; *Arakawa and Suarez*, 1983]. The solution to the momentum equations is a 4th-order scheme that conserves potential enstrophy and energy [*Arakawa and Lamb*, 1981], designed to improve nonlinear aspects of flow over steep topography. The advection of thermodynamic energy and water are solved with a 13-point 4th-order scheme [*Arakawa*, 1995] that conserves the global mass integral of the square of the transported variable.

2.4. Global transport. Transport of gases other than water vapor and transport of all aerosol components are performed with a code written here that uses the 13-point 4th-order scheme of *Arakawa* [1995]. The code was written to minimize redundant calculations when applied to different trace species. The scheme exactly conserves the constant mixing ratio of a species. Air density changes during advection of each species are also exactly consistent with air density changes during advection of momentum.

2.5. Regional dynamics. The regional dynamics module integrates equations for momentum (under the hydrostatic assumption), thermodynamic energy, and total water. The module was developed by *Lu and Turco* [1994, 1995] from finite-difference solutions to the momentum, thermodynamic energy, and water continuity equations by *Arakawa and Lamb* [1977] and *Arakawa and Suarez* [1983]. The regional dynamics module contains a hybrid boundary-layer turbulence scheme [*Lu and Turco*, 1994] in which three turbulence regimes are considered. Under strongly stable conditions, turbulence is suppressed and flow is assumed to be laminar. Under strongly unstable conditions, a convective plume model that simulates rising convective turbulence and subsidence, is used to calculate turbulent fluxes and the height to which thermals accelerate buoyantly. Between these two limits, a first-order closure technique is used, and diffusion coefficients are calculated from *Blackadar* [1976].

2.6. Regional transport. Transport of gases and aerosols in regional-scale domains is carried out with the scheme of *Walcek and Aleksic* [1998]. The scheme uses operator-splitting and is absolutely monotonic for all dimensional flows, and constant mixing ratios are guaranteed to be maintained through a treatment of the density changes in each dimension of a multi-dimensional calculation. The scheme preserves peaks and odd shapes extremely well.

2.7. Cumulus parameterization. The cumulus parameterization used for all nested domains is a modified *Arakawa-Schubert* [1974] algorithm developed by *Ding and Randall* [1998]. The algorithm allows cloud bases at any altitude; the original scheme allowed cloud bases in the boundary layer only. The modified scheme also accounts for downdrafts. The scheme predicts cumulus precipitation, liquid water and ice contents, and adjustments to large-scale potential temperature, momentum, and water vapor due to cumulus convection. The original UCLA GCM also included algorithms that estimated large-scale precipitation (large-scale plus cumulus precipitation is total precipitation), stratus cloud occurrence, and adjustments to large-scale potential temperature and water vapor due to stratus. These algorithms were retained for the global domain and added to the regional domains.

3. Description of simulations. The model was applied to study weather and air pollution during the August 3–6 (Friday–Monday), 1990 SARMAP field campaign in Northern and Central California [*Lagarias and Sylte*, 1991; *Blumenthal*, 1993; *Ranzieri and Thuillier*, 1994; *Solomon and Thuillier*, 1995]. During the SARMAP study, measurements of over 40 parameters were taken at over 326 sites. The measurement locations ranged from 34.3°N to 42.4°N and −123.1°W to −117.6°W.

A nested simulation that included five nested grids inclusive of global and Northern/Central California grids was run from Friday, August 3, 03:30 Pacific Standard Time (PST) (11:30 Greenwich Mean Time) to Monday, August 6, 23:30 PST (92-hour simulations). The simulation was run without model spinup or data assimilation. Simulations were started with initial interpolated meteorological and gas fields.

The processes accounted for in all domains from the global through urban scale were emissions, gas-phase chemistry, radiative transfer, meteorology, surface processes, species transport, and cloud formation. Aerosol processes were excluded due the lack of a size-segregated aerosol emissions inventory and detailed aerosol observations during this simulation period, but such aerosol processes are included in the model on all scales.

4. Results and analysis. Results from the inner grid of a nested simulation are discussed below. Only near-surface results and results for a few parameters are discussed here. Results for more parameters, for elevated parameters, and for outer nested grids are discussed in *Jacobson* [2001d].

4.1. Temperatures, relative humidities, and dew points. Figure 1 shows time series predictions versus observations of near-surface temperatures and RHs for the 92-hour nested simulation. Table 1 shows that gross errors, normalized over all observations in the innermost domain during the 92-h period, were 1.02% for Kelvin temperatures and 26.2% for RHs. The RH error is the percent error of the RH value itself; thus, a 25% percent error of an observed RH of 50% is 12.5% of the RH. Normalized biases indicate slight underpredictions for both parameters.

TABLE 1

Normalized Gross Errors (NGE) and Normalized Biases (NB) for Several Near-Surface Meteorological and Air Quality Parameters From the Innermost Nested Grid of the Nested Simulation.

Parameter	Cutoff[1]	No. obs.[2]	NGE (%)	NB (%)
Temperature	0 K	8617	1.02	−0.17
Relative humidity	0 %	5597	26.2	−9.2
Wind speed	1.5 m s^{-1}	3534	43.6	−8.5
Wind direction[3]	$\parallel$	3534	16.9	0.6
Air pressure	0 mb	307	0.15	0.10
Ozone (O_3)	50 ppbv	2866	22.5	−8.9
Nitric oxide (NO)	0 ppbv	1277	80.2	−32.7
Nitrogen dioxide (NO_2)	0 ppbv	3707	65.9	−13.3
Carbon monoxide (CO)	0 ppbv	2119	49.9	−44.7
Sulfur dioxide (SO_2)	0 ppbv	182	64.3	−63.4
NMOC	0 ppbv	275	48.6	24.6
Methane (CH_4)	0 ppbv	1371	6.4	−2.6
Ethane (C_2H_6)	0 ppbv	275	40.3	4.5
Propane (C_3H_8)	0 ppbv	275	49.1	−18.7
Paraffins (PAR)	0 ppbv	275	52.3	25.6
Ethene (C_2H_4)	3 ppbv	45	47.6	7.6
Olefins (OLE)	0 ppbv	275	53.0	−43.1
Formaldehyde (HCHO)	3 ppbv	33	54.4	37.1
Acetaldehyde (CH_3CHO)	2 ppbv	147	37.7	−1.2
Acetone (CH_3COCH_3)	1 ppbv	243	33.3	3.4
Ketones (KET)	2 ppbv	17	32.4	10.0
Toluene ($C_6H_5CH_3$)	0 ppbv	275	52.7	−34.2
Xylene ($C_6H_4[CH_3]_2$)	0 ppbv	275	67.8	−62.4
Benzaldehyde ($C_6H_5CH_2O_2$)	0 ppbv	178	47.4	−32.9
Isoprene (C_5H_8)	3 ppbv	13	38.5	31.7

[1] A cutoff value is the lowest values of the observed parameter considered in the comparison. No. obs. is the number of observations above the cutoff value. For wind direction, statistics are taken only when the wind speed cutoff is exceeded.

[2] No. obs. is the number of observations in the innermost domain. Comparisons were made only in this domain.

[3] Wind direction normalized gross errors are percentages of 360 degrees.

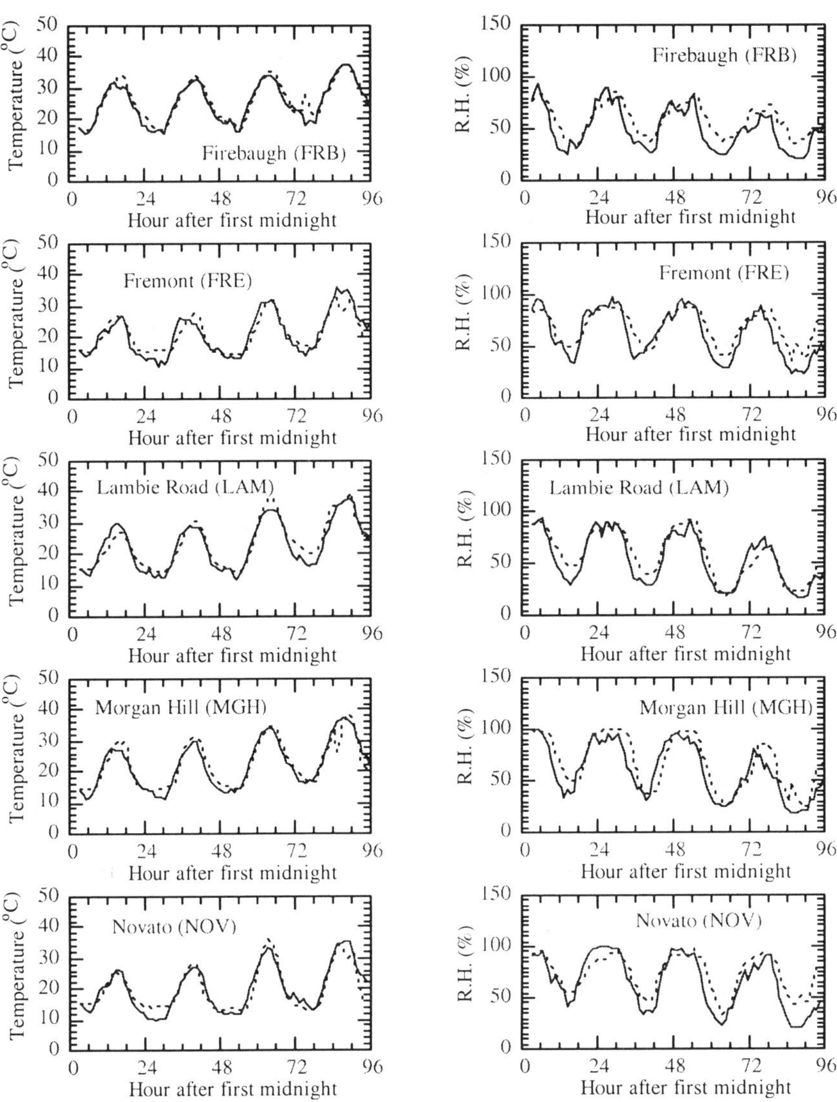

FIG. 1. *Time series plots, over the 92-hour simulation, of predicted versus observed temperatures and RHs at several locations in the innermost domain during a nested simulation. Solid lines are predictions, dashed lines and circles are observations. These and subsequent figures were originally published in Jacobson [2001d].*

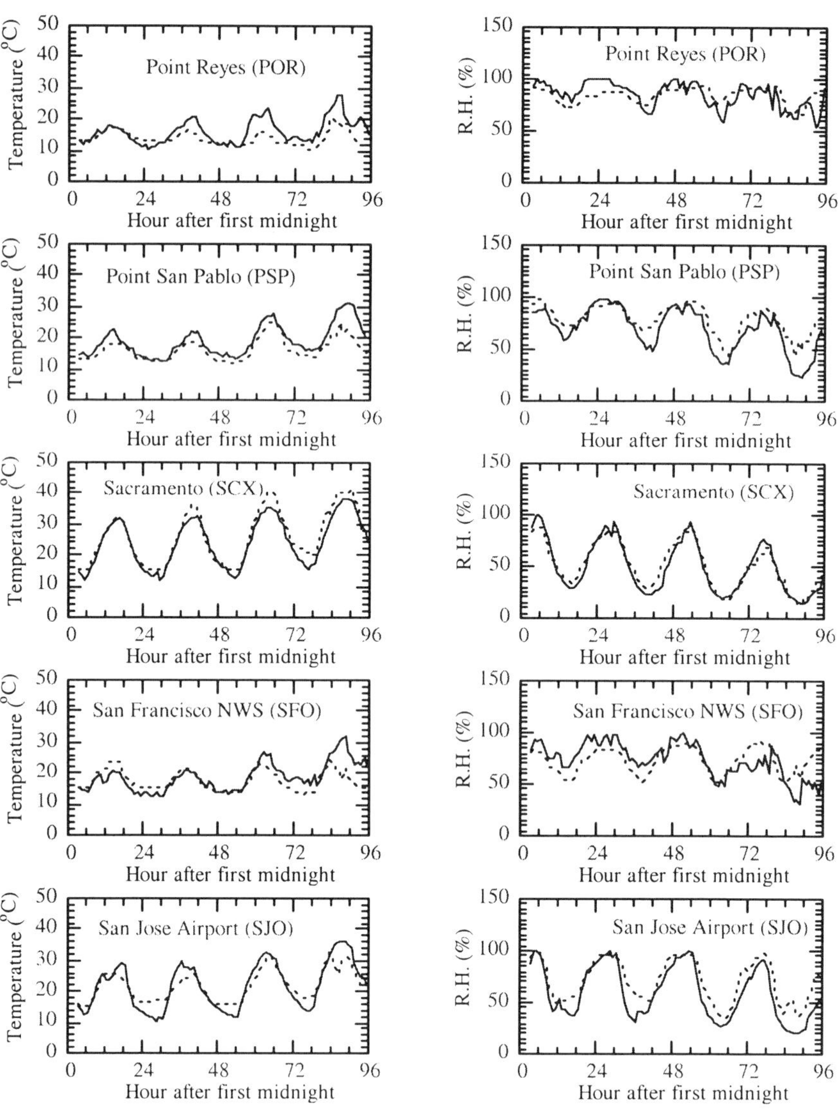

FIG. 1. *Continued.*

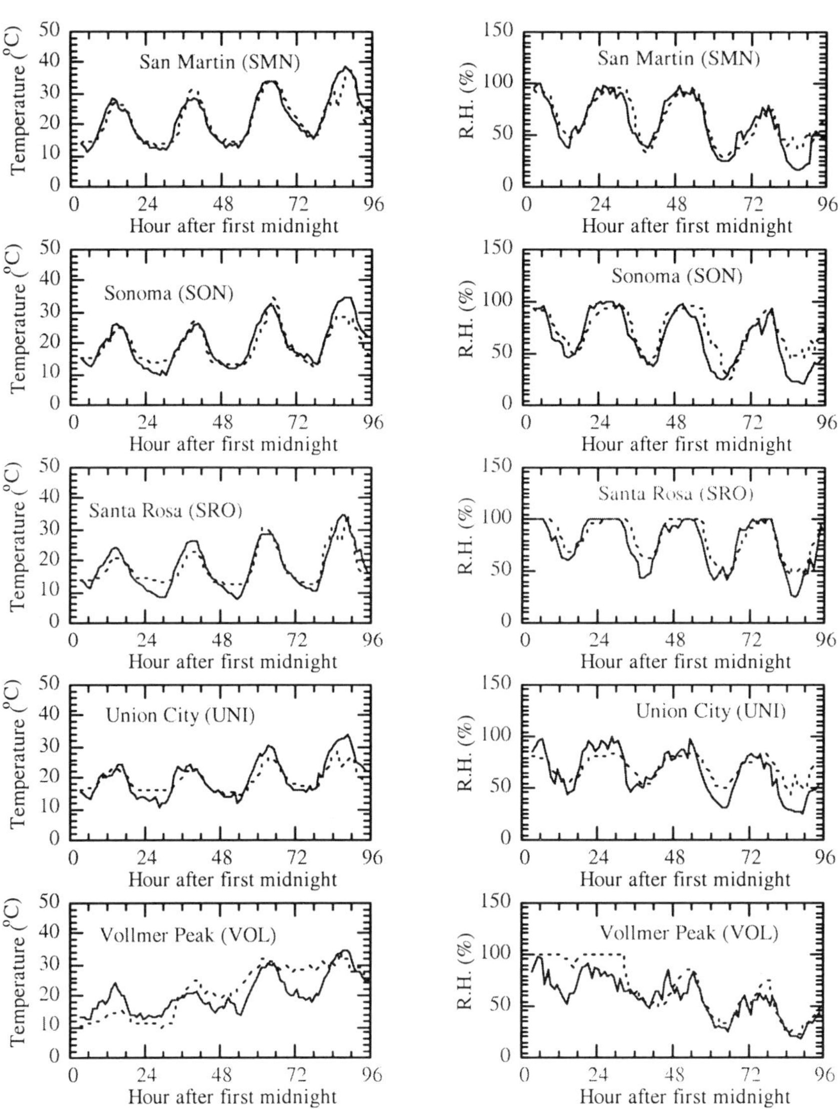

FIG. 1. *Continued.*

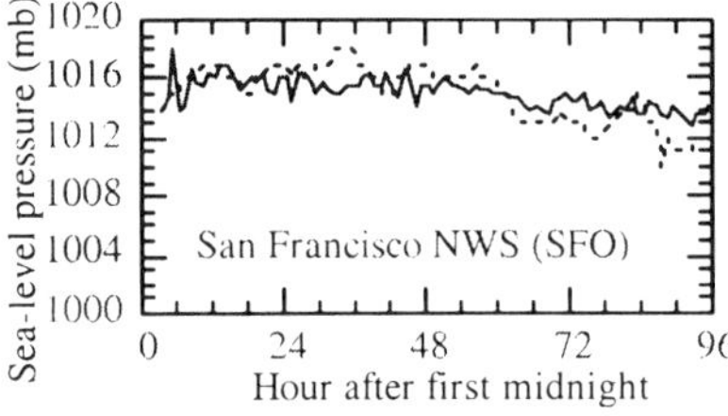

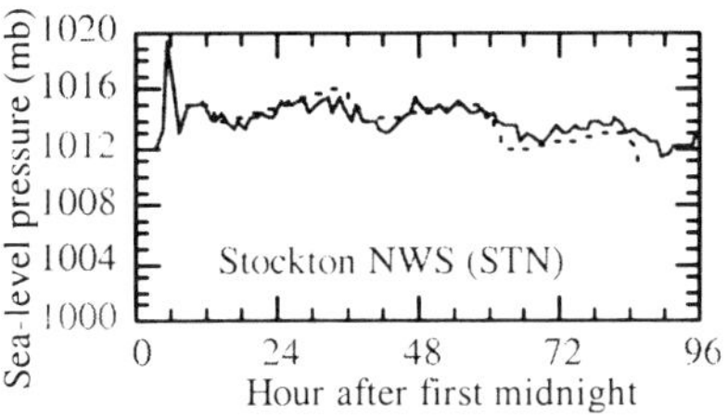

FIG. 2. *Same as Figure 1, but for air pressure.*

Figure 1 indicates that the model was able to capture the gradual increases in temperatures and decreases in RHs that occurred each successive day. Temperature variations near the coast are generally weaker than those inland, since coastal areas are exposed to an influx of relatively constant-temperature ocean air, and soil moisture contents near the coast are generally high. RH variations are similarly weaker near the coast than inland due to the weaker temperature variations and constant source of water vapor near the coast. As shown in Figure 1, the model reproduced weak diurnal temperature and RH variations well at near coastal sites, such as Mount Pise (MTP), Mount Tamalpais (MTT), Point Reyes (POR), Point San Pablo (PSP), and San Francisco (SFO). It also reproduced strong diurnal variations of both parameters at inland locations, such as Bellota (BEL), Firebaugh (FRB), Lambie Road (LAM), Lodi (LOD), Manteca (MAN), Modesto (MDT), Morgan Hill (MGH), Novato (NOV), Sacramento (SCX), and San Martin (SMN). The model also captured moderate variations at intermediate locations, such as Fremont (FRE), San Jose (SJO), and Union City (UNI). At three mountainside stations, Mount Tamalpais (MTT), Mount Pise (MTP), and Vollmer Peak (VOL), observed peak temperatures increased by 10–15 °C and observed peak RHs (at Vollmer Peak) decreased by about 40% between the first and fourth days of simulation. The model captured these trends at all three locations.

4.2. Wind speed, wind direction, and air pressure. Wind speeds and directions in northern California depend a lot on the locations of the Pacific High and Aleutian low. During the simulation period, the NCEP reanalysis 500 mb height over the SARMAP domain increased and surface air pressures decreased due to enhanced thermal low conditions (e.g., higher surface temperatures). By August 6 at 16:00 PST, the 500 mb height maximum had moved just to the northwest of the innermost domain, and had increased to 5975 m from its initial value of about 5900 m. In the largest regional model domain, a 500 mb height maximum was predicted near this location at the same time. Figure 2 shows that observed surface air pressures decreased by about 4 mb during the simulation period, and the model (in the innermost domain) was able to predict these decreases at both Stockton and San Francisco.

Winds in northern California also depend on local factors, such as sea and bay breezes, low altitude nocturnal jets and eddies in the San Joaquin Valley [particularly near Fresno and on the east side of the valley, e.g., *Smith et al.*, 1981; *Roberts et al*, 1990], nighttime drainage flows from the Sierra Nevada Mountains, and convergence/divergence through the Carquinez Straits, Altamont Pass, Cholame Pass, and Pacheco Pass.

Figure 3 shows predicted versus observed near-surface wind speeds and directions. The model predicted the observed strong diurnal wind speed and direction variations at Altamont Pass (ALT) and the strong diurnal wind speed variation at Dixon (DIX). The model captured the observed day/night reversal in wind direction due to nighttime drainage flow at North Fork (NFK), at the base of the Sierra-Nevada Mountain Range. It also captured the one-time strong variations in wind direction at Pacheco Pass (PAC), a thoroughfare for pollutants from the Bay Area to the San Joaquin Valley. The model captured the four strong variations in wind direction at El Patio (EPT), in the southern Santa Clara Valley.

4.3. Near-surface ozone. Figure 4 shows time series plots of predicted versus observed near-surface ozone at a variety of sites. The gross error in ozone predictions, normalized over all hourly observations in which ozone mixing ratios exceeded 50 ppbv, was 22.5% (Table 1).

In order to predict observed daytime ozone peaks, it was necessary to initialize the model with observed (from aircraft) elevated ozone layers. When such initialization was not performed, daytime peaks in near-surface ozone were not captured. Although the effect of initialization on accuracy decreased each day, it was still noticeable by the fourth day.

Ozone predictions compared well with observations under a variety of conditions, including conditions of strong diurnal variations [e.g., Altamont (ALT), Collegeville (COL), Modesto (M14), Sacramento (S13), and Stockton-Lodi (STO)], weak diurnal variations but high mixing ratios [e.g., Yosemite (YOS)], and low mixing ratios [e.g., Oakland (OKA), Point Reyes (POR), San Francisco (SFA)]. In particular, observed near-surface ozone mixing ratios at coastal sites in Northern California were quite low, as seen in Figure 4 for Point Reyes (consistently 30 ppbv), San Francisco (consistently 20 ppbv), and Oakland (consistently 20–30 ppbv). One reason for these low mixing ratios is that winds originating from the ocean carry relatively clean, low-ozone air into these coastal locations. In addition, the same winds advect emissions of ozone precursors from coastal locations inland, reducing *in situ* ozone production along the coast. The model was able to reproduce the low mixing ratios at all the coastal locations.

During the simulation period, observed ozone mixing ratios tended to increase. These increases were predicted, for the most part, by the model, except for a few cases (e.g., day 3 at Livermore, Fremont, and San Jose), where abnormally high observed ozone mixing ratios were not replicated.

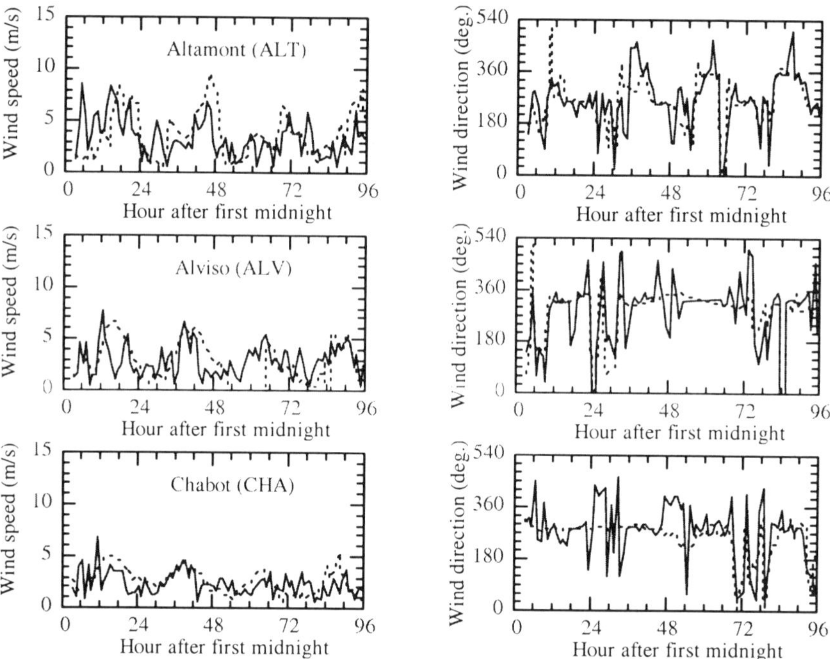

FIG. 3. *Same as Figure 1, but for wind speed and direction. A wind direction of zero degrees is a northerly wind (from the north). Directions between 360° and 540° are the same as those between 0 and 180°, but used to avoid discontinuities at 360°. Modeled winds are presented at 10 m altitude, the height of the observations.*

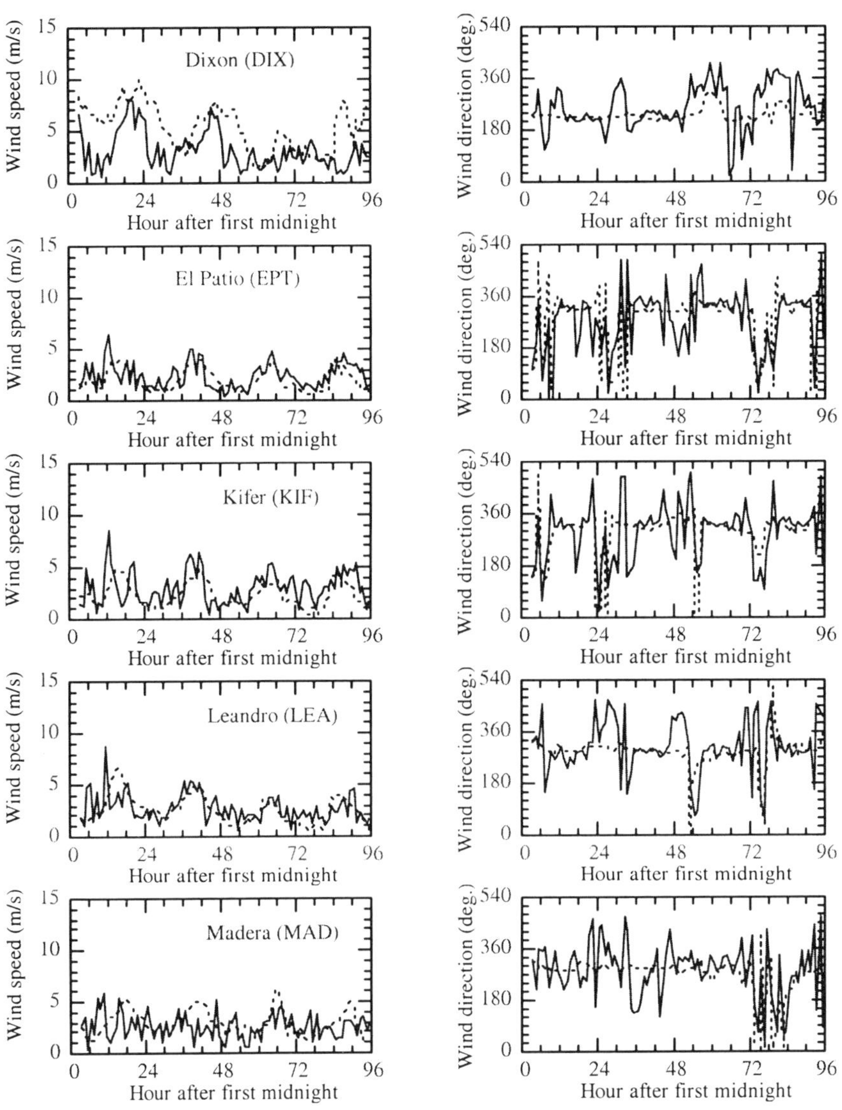

FIG. 3. *Continued.*

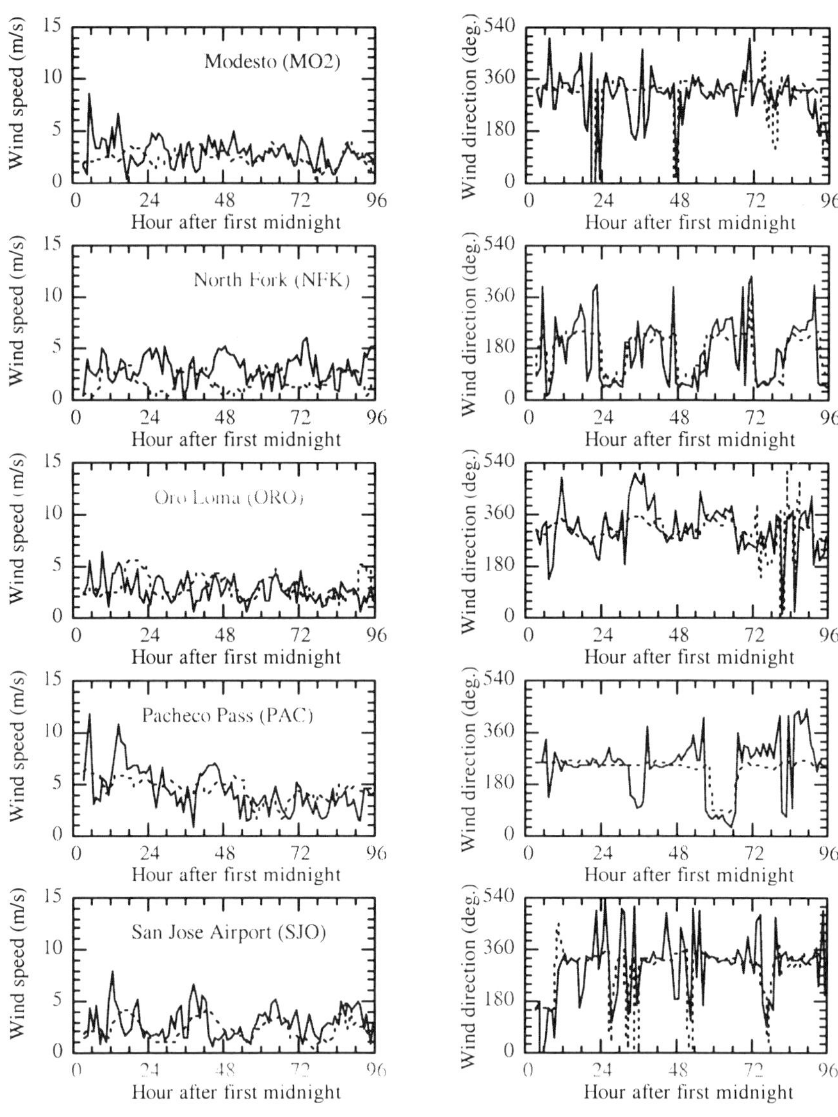

FIG. 3. *Continued.*

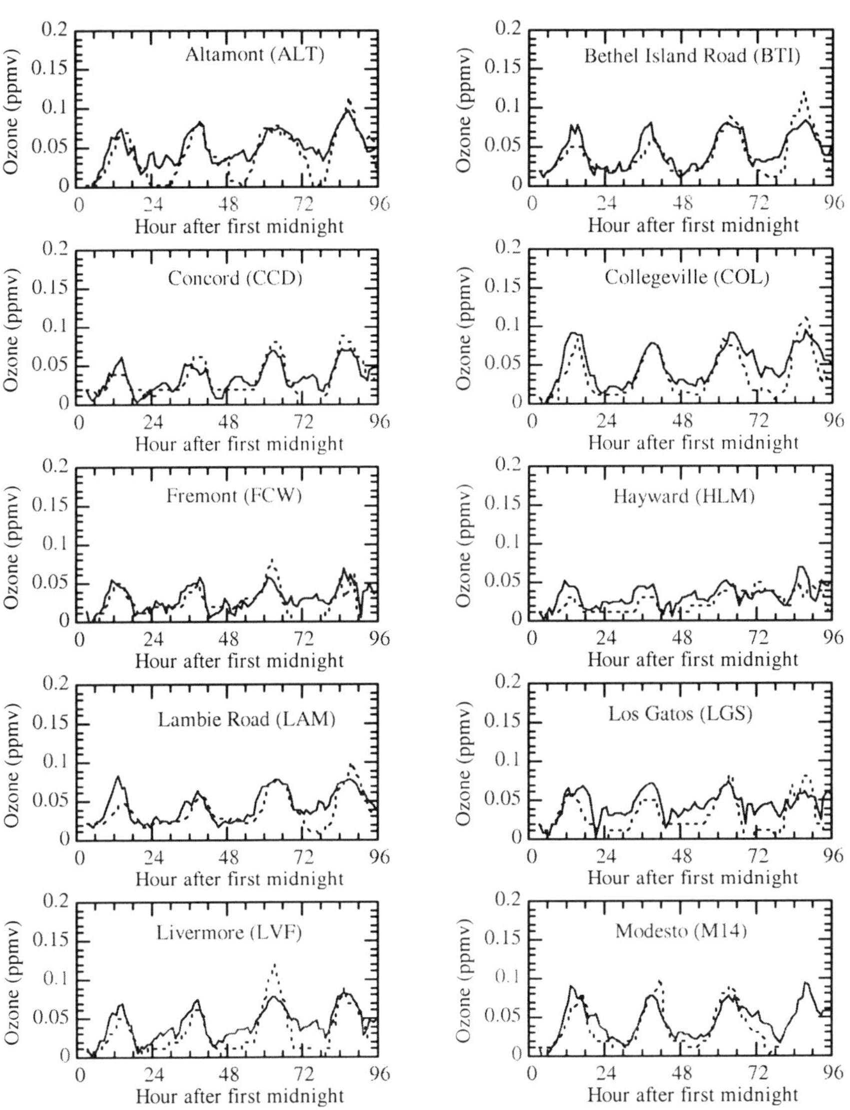

F𝖨𝖦. 4. *Same as Figure 1, but for ozone.*

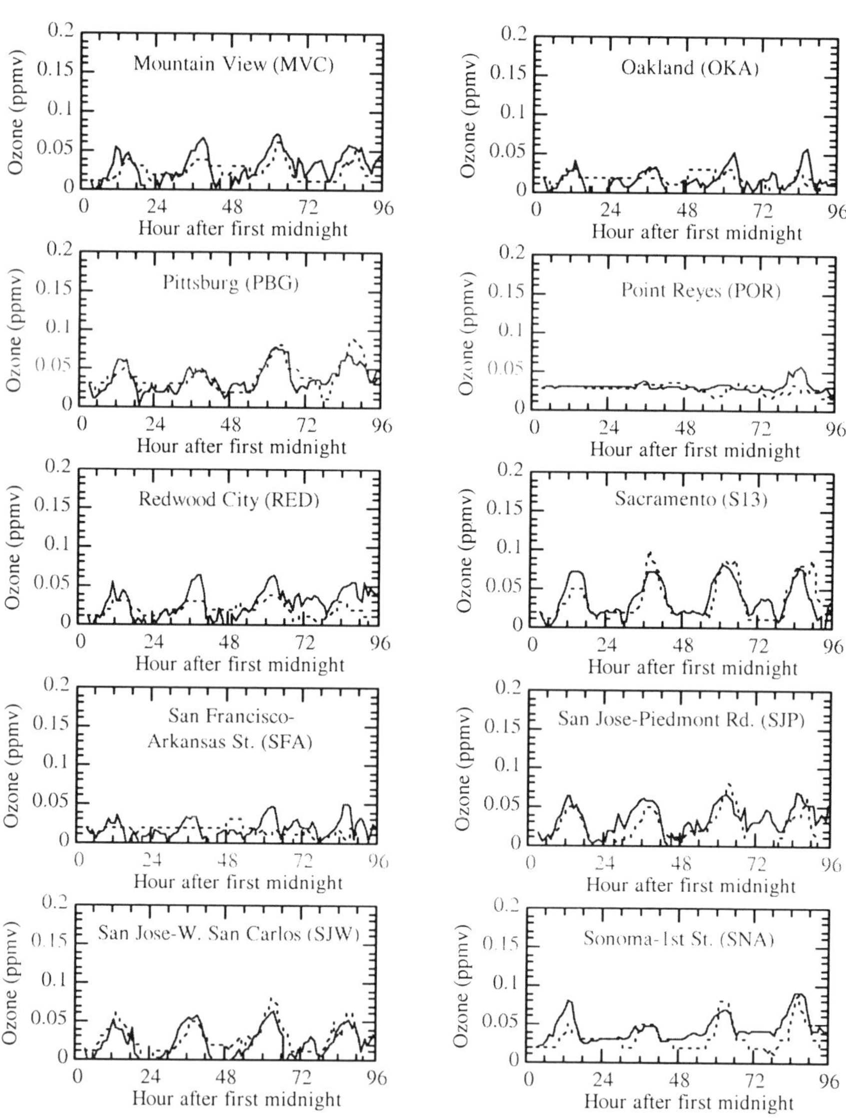

FIG. 4. *Continued.*

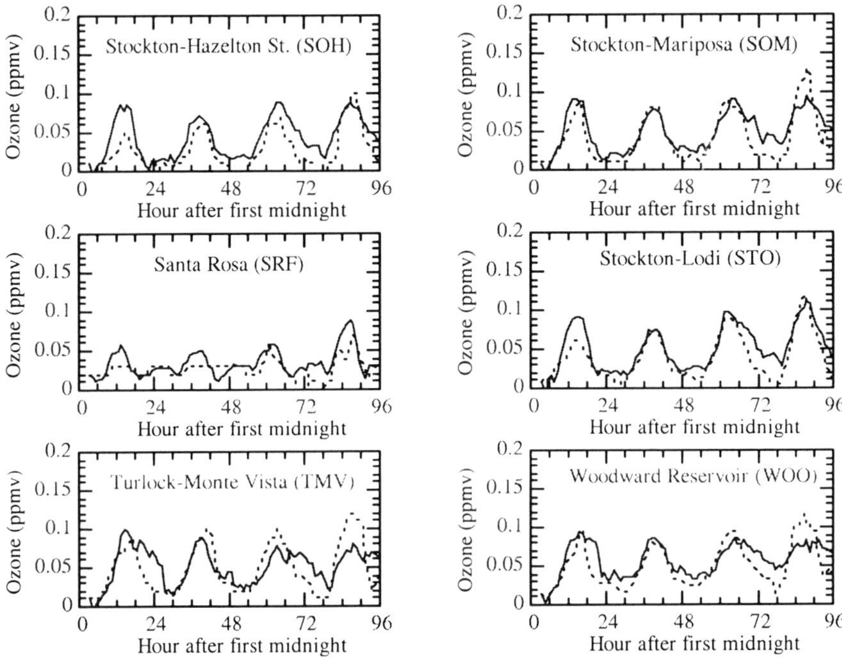

FIG. 4. *Continued.*

Ozone increases during the period coincided with increases in near-surface temperatures (Figure 1), decreases in surface pressures (Figure 2), and decreases in near-surface wind speeds (Figure 3). The enhanced stagnation of air may have led to the gradual increase in ozone. Since the simulation period ran from a Friday through Monday, it is unlikely that enhanced emissions caused the ozone buildup, particularly on day 3 (Sunday, August 5).

5. Conclusion. A global- through urban-scale air pollution/weather forecast model was developed and applied to the SARMAP field campaign of August 3–6, 1991. The model, treats nesting of all important air quality and meteorological parameters simultaneously, from the global scale through the urban scale (< 5 km grid spacing). It treats any number of nested layers and any number of meteorological and air quality grids in each layer between the global and urban scales. The model does not use nudging or data assimilation; it is entirely prognostic, except for the emission inventory. The model was applied with five global-through-urban scale nested grids to the August 3–6, 1990 SARMAP field campaign. Parameters compared with observations included air temperatures, air pressures, relative humidities, wind speeds and directions, and the following 20 gases and carbon bond groups: ozone, nitric oxide, nitrogen dioxide, carbon monoxide, sulfur dioxide, nonmethane organic carbon, methane, ethane, propane, paraffins, ethene, olefins, formaldehyde, higher aldehydes, acetone, other ketones, toluene, xylene, benzaldehyde, and isoprene. Near-surface predictions of temperatures, RHs, and ozone were the most accurate parameters compared.

REFERENCES

Anthes, R.A., E.Y. Hsie, and Y.H. Kuo, Description of the Penn State/NCAR Mesoscale Model Version 4 (MM4). *NCAR Tech. Note, NCAR/TN-282+STR*, Natl. Cent. for Atmos. Res., Boulder, Colo., 1987.

Arakawa, A., Boundary conditions in limited-area models. Course notes, Dept. of Atmospheric Sciences, University of California, Los Angeles, 1982.

Arakawa, A., *Introduction to the UCLA General Circulation Model*. Course Notes, Atmospheric Sciences 281, Appendix to Chapter 5, Univ. of California, Los Angeles, 1995.

Arakawa, A. and W.H. Schubert, Interaction of a cumulus cloud ensemble with large scale environment, Part I, *J. Atmos. Sci.*, **31**, 674–701, 1974.

Arakawa, A. and V.R. Lamb, Computational design of the basic dynamical processes of the UCLA general circulation model, *Methods Comput. Phys.*, **17**, 174–265, 1977.

Arakawa, A. and V.R. Lamb, A potential enstrophy and energy conserving scheme for the shallow water equations, *Mon. Wea. Rev.*, **109**, 18–36, 1981.

Arakawa, A., and M.J. Suarez, Vertical differencing of the primitive equations in sigma coordinates, *Mon. Wea. Rev.*, **111**, 34–45, 1983.

Blackadar, Modeling the nocturnal boundary layer, *Preprints, Third Symp. on Atmospheric Turbulence, Diffusion, and Air Quality*, Raleight, N.C., Amer. Met. Soc., 46–49, 1976.

Blumenthal, D.L., Field study plan for the San Joaquin Valley Air Quality Study (SJVAQS) and the Atmospheric Utility Signatures, Predictions, and Experiments (AUSPEX) Program. Final Report to the California Air Resources Board, STI-98020-1241-FR. Sacramento, CA, 1993.

Briggs, G.A., Plume rise predictions. Chapter 3 of Lectures on Air Pollution and Environmental Impact Analyses, Sponsored by the American Meteorological Society, Boston, Mass, Sept. 29–Oct. 3, 1975.

Byun, D.W. and J.K.S. Ching, eds., *Science algorithms of the EPA Models-3 Community Multiscale Air Quality* (CMAQ) Modeling system, U.S. Environmental Protection Agency, Office of Research and Development, Washington, D.C., EPA/600/R-99/030, 1999.

Copeland, J.H., R.A. Pielke, and T.G.F. Kittel, Potential climatic impacts of vegetation change: A regional modeling study, *J. Geophys. Res.*, **101**, 7409–7418, 1996.

Damassa, J., S. Tanrikulu, K. Magliano, A.J. Ranzieri, and E. Niccum, Performance evaluation of SAQM in Central California and attainment demonstration for the 3–6 August 1990 ozone episode. California Air Resources Board, Sacramento, CA, 1996.

Dickinson, R.E., R.M. Errico, F. Giorgi, and G.T. Bates, A regional climate model for the western United States, *Clim. Change*, **15**, 383–422, 1989.

Ding, P. and D.A. Randall, A cumulus parameterization with multiple cloud-base levels. *J. Geophys. Res.*, 103(11), 341, 353, 1998.

Dudek, M.-P., X.-Z. Liang, and W.-C. Wang, A regional climate model study of the scale-dependence cloud-radiation interaction, *J. Clim.*, **9**, 1221–1234, 1996.

Food and Agricultural Organization (FAO), Soil Map of the World, Land and Water Development Division, FAO, Rome, Italy, 1995.

Giorgi, F. and G.T. Bates, On the climatological skill of a regional model over complex terrain, *Mon. Weather Rev.*, **117**, 2325–2347, 1989.

Giorgi, F. and M. R. Marinucci, Improvements in the simulation of surface climatology over the European region with a nested modeling system, *Geophys. Res. Lett.*, **23**, 273–276, 1996.

Giorgi, F. and L. O. Mearns, Introduction to special section: Regional climate modeling revisited, *J. Geophys. Res.*, **104**, 6335–6352, 1999.

Grell, G.A., J. Dudhia, and D. R. Stauffer, Description of the fifth generation Penn State/NCAR Mesoscale Model (MM5). *NCAR Tech. Note, NCAR/TN-398+STR*, pp. 121, Natl. Cent. for Atmos. Res., Boulder, Colo, 1994.

Jacobson, M.Z., Computation of global photochemistry with SMVGEAR II, *Atmos. Environ.*, **29**A, 2541–2546, 1995.

Jacobson, M.Z., Development and application of a new air pollution modeling system, part II, Aerosol module structure and design, *Atmos. Environ.*, **31**, 131–144, 1997a.

Jacobson, M.Z., Development and application of a new air pollution modeling system, part III, Aerosol-phase simulations, *Atmos. Environ.*, **31**, 587–608, 1997b.

Jacobson, M.Z., Studying the effects of aerosols on vertical photolysis rate coefficient and temperature profiles over an urban airshed, *J. Geophys. Res.*, **103**(10), 593, 604, 1998a.

Jacobson, M.Z., Improvement of SMVGEAR II on vector and scalar machines through absolute error tolerance control, *Atmos. Environ.*, **32**, 791–796, 1998b.

Jacobson, M.Z., *Fundamentals of Atmospheric Modeling*, Cambridge Univ. Press, New York, 656 pp., 1999a.

Jacobson, M.Z., Isolating nitrated and aromatic aerosols and nitrated aromatic gases as sources of ultraviolet light absorption, *J. Geophys. Res.*, **104**, 3527–3542, 1999b.

Jacobson, M.Z., Effects of soil moisture on temperatures, winds, and pollutant concentrations in Los Angeles, *J. Appl. Meteorol.*, **38**, 607–616, 1999c.

Jacobson, M.Z., Global direct radiative forcing due to multicomponent anthropogenic and natural aerosols, *J. Geophys. Res.*, **106**, 5385–5402, 2001a.

Jacobson, M.Z., Strong radiative heating due to the mixing state of black carbon in atmospheric aerosols, *Nature*, **409**, 695–697, 2001b.

Jacobson, M.Z., GATOR-GCMM: A global-through urban-scale air pollution and weather forecast model. 1. Model design and treatment of subgrid soil, vegetation, roads, rooftops, water, sea ice, and snow, *J. Geophys. Res.*, **106**, 5385–5402, 2001c.

Jacobson, M.Z., GATOR-GCMM: 2. A study of day- and nighttime ozone layers aloft, ozone in national parks, and weather during the SARMAP field campaign, *J. Geophys. Res.*, **106**, 5403–5420, 2001d.

Jacobson, M.Z., R. Lu, R.P. Turco, and O.B. Toon, Development and application of a new air pollution modeling system – Part I. Gas-phase simulations, *Atmos. Environ.*, **30**B, 1939–1963, 1996.

Jakobs, H.J., H. Feldmann, H. Hass, and M. Memmesheimer, The use of nested models for air pollution studies: An application of the EURAD model to a SANA episode, *J. Appl. Meteorol.*, **34**, 1301–1319, 1995.

Jones, R.G., J.M. Murphy, M. Noguer, and M. Keen, Simulation of climate change over Europe using a nested regional climate model, II, Comparison of driving and regional model responses to a doubling of carbon dioxide, *Q. J. R. Meteorol. Soc.*, **123**, 265–292, 1997.

Juang, H.-M. H. and M. Kanamitsu, The NMC nested regional spectral model, *Mon. Weather Rev.*, **122**, 3–26, 1994.

Kumar, N., M.T. Odman, and A.G. Russell, Multiscale air quality modeling: Application to southern California, *J. Geophys. Res.*, **99**, 5385–5397, 1994.

Lagarias, J.S. and W.W. Sylte, Designing and managing the San Joaquin Valley air quality study. *Journal of Air and Waste Management Assn.*, **41**, 1176–1179, 1991.

Laprise, R., D. Caya, M. Giguere, G. Bergeron, H. Cote, J.P. Blanchet, G. J. Boer, and N.A. McFarlane, Climate and climate change in western Canada as simulated by the Canadian Regional Climate Model, *Atmos. Ocean*, **36**, 119–167, 1998.

Leung, L.R. and S.J. Ghan, A subgrid parameterization of orographic precipitation, *Theor. Appl. Climatol.*, **52**, 95–118, 1995.

Leung, L.R. and S.J. Ghan, Pacific Northwest climate sensitivity simulated by a regional climate model driven by a GCM. Part I: Control simulation, *J. Climate*, **12**, 2010–2030, 1999.

Liang, J. and M.Z. Jacobson, Comparison of a 4000-reaction chemical mechanism with the Carbon Bond IV and an adjusted Carbon Bond IV-EX mechanism using SMVGEAR II, *Atmos. Environ.*, in press, 2000.

Lu, R and R.P. Turco, Air pollutant transport in a coastal environment. Part I: Two-dimensional simulations of sea-breeze and mountain effects, *J. Atmos. Sci.*, **51**, 2285–2308, 1994.

Lu, R. and R.P. Turco, Air pollutant transport in a coastal environment, II, Three-dimensional simulations over Los Angeles basin, *Atmos. Environ.*, **29**, 1499–1518, 1995.

Marinucci, M.R., F. Giorgi, M. Beniston, M. Wild, P. Tschuck, A. Ohmura, and A. Bernasconi, 1995, High resolution simulations of January and July climate over the western alpine region with a nested regional modeling system, *Theor. Appl. Climatol.*, **51**, 119–138, 1995.

Mathur, R., L.K. Peters, and R.D. Saylor, Sub-grid representation of emission source clusters in regional air quality modeling, *Atmos. Environ.*, **26A**, 3219–3238, 1992.

McGregor, J.J., Regional climate modeling, *Meteorol. Atmos. Phys.*, **63**, 105–117, 1997.

McGregor, J.J. and K.J. Walsh, Nested simulations of perpetual January climate over the Australian region, *J. Geophys. Res.*, **98**(23), 283, 290, 1993.

McHenry, J., N.L. Seaman, C.J. Coats, A. Lario-Gibbs, J. Vukovich, N. Wheeler, and E. Hayes, Real-time nested mesoscale forecasts of lower tropospheric ozone using a highly optimized coupled numerical prediction system. *Preprints, AMS Symposium on Interdisciplinary Issues in Atmospheric Chemistry*, January 10–15, 1999.

Mlawer, E.J., S.J. Taubman, P.D. Brown, M.J. Iacono, and S.A. Clough, Radiative transfer for inhomogeneous atmospheres: RRTM, a validated correlated-k model for the longwave, *J. Geophys. Res.*, **102**(16), 663, 682, 1997.

Morris, R.E., M.A. Yocke, T.C. Myers, and V. Mirabella, Overview of the variable-grid urban airshed model (UAM-V), Presentation at the 85th Annual Air & Waste Management Association Meeting & Exhibition, Kansas City, Missouri, June 21–26, 1992.

Odman, M.T. and C.L. Ingram, Multiscale Air Quality Simulation Platform (MAQSIP): Source code documentation and validation, ENV-96TR002-v1.0. Available from MCNC, 3021 Cornwallis Rd., Research Triangle Park, NC 27709, 1996.

Pielke, R.A., W.R. Cotton, R.L. Walko, C.J. Tremback, W.A. Lyons, L.D. Grasso, M.E. Nicholls, M.D. Moran, D.A. Wesley, T.J. Lee, and J.H. Copeland, *Meteorol. Atmos. Phys.*, **49**, 69–91, 1992.

Platt, C.M.R., A parameterization of the visible extinction coefficient of ice clouds in terms of the ice/water content, *J. Atmos. Sci.*, **54**, 2083–2098, 1997.

Pleim, J.E., J.S. Chang, and K. Zhang, A nested grid mesoscale atmospheric chemistry model, *J. Geophys. Res.*, **96**, 3065–3084, 1991.

Podzun, R., A. Cress, D. Majewski, and V. Renner, Simulation of European climate with a limited area model. Part II. AGCM boundary conditions. *Contrib. Atmos. Phys.*, **68**, 205–225, 1995.

Qian, Y. and F. Giorgi, Interactive coupling of regional climate and sulfate aerosol models over eastern Asia, *J. Geophys. Res.*, **104**, 6477–6799, 1999.

Qian, J.-H., F. Giorgi, and M.S. Fox-Rabinovitz, Regional stretched grid generation and its application to the NCAR RegCM, *J. Geophys. Res.*, **104**, 6501–6513, 1999.

Ranzieri, A.J. and R.H. Thuillier, SJVAQS and AUSPEX: a collaborative air quality field measurement and modeling program. Solomon, P.A. (ed.), Planning and Managing Regional Air Quality, Lewis Publishers, Florida, in conjunction with the Pacific Gas and Electric Company, 1994.

Roberts, P.T., T.B. Smith, C.G. Lindsey, and W.R. Knuth, Analysis of San Joaquin Valley Air Quality and Meteorology, Sonoma Technology Report STI-98101-1006-FR, Santa Rosa, California, 1990.

Smith, T.B., D.E. Lehrman, D.D. Reible, and F.H. Shar, The origin and fate of airborne pollutants within the San Joaquin Valley: Extended summary and special analysis topics. Report prepared for the California Air Resources Board, Sacramento, CA by Meteorology Research, Inc., Altadena, CA and the California Institute of Technology, Pasadena, CA, MRI FR-1838.

Solomon, P.A. and R.H. Thuillier, SJVAQS/SUSPEX/SARMAP 1990 Air Quality Field Measurement Project Volume II: Field measurement characterization. PG&E Cost reduction projects report 009.2-94.1, 1995.

Toon, O.B., C.P. McKay, T.P. Ackerman, and K. Santhanam, Rapid calculation of radiative heating rates and photodissociation rates in inhomogeneous multiple scattering atmospheres, *J. Geophys. Res.*, **94**(16), 287, 301, 1989.

Umeda, T. and P.T. Martien, Evaluation of a Data Assimilation Technique for a Mesoscale Meteorological Model used for Air Quality Modeling, *J. Appl. Meteorol.*, in review, 2000.

United States Geological Survey (USGS) / U. Nebraska, Lincoln / European Commission's Joint Research Center 1-km resolution global landcover characteristics data base, derived from Advanced Very High Resolution Radiometer (AVHRR) data from the period April 1992 to March 1993, 1999.

Walcek, C.J. and N.M. Aleksic, A simple but accurate mass conservative, peak-preserving, mixing ratio bounded advection algorithm with fortran code, *Atmos. Environ.*, **32**, 3863–3880, 1998.

Walmsley, J.L. and M.L. Wesely, Modification of coded parameterizations of surface resistances to gaseous dry deposition, *Atmos. Environ.*, **30**A, 1181–1188, 1996.

Welch, R.M., S.K. Cox, and J.M. Davis, *Solar Radiation and Clouds*, Meteorological Monograph 17, Am. Met. Soc., 1980.

Wesely, M.L., Parameterization of surface resistances to gaseous dry deposition in regional-scale numerical models, *Atmos. Environ.*, **23**, 1293–1304, 1989.

MODELING ATMOSPHERIC PARTICULATE MATTER IN AN AIR QUALITY MODELING SYSTEM USING A MODAL METHOD

FRANCIS S. BINKOWSKI*, SHAWN J. ROSELLE*, MICHELLE R. MEBUST[†],

AND BRIAN K. EDER*

1. Introduction. The Community Multiscale Air Quality (CMAQ) modeling system [*Byun and Ching,* 1999] was developed as a tool to deal with multiple air pollutants such as ozone, acidic deposition and aerosol particles or particulate matter (PM). We will give a short description here of the mathematical algorithms used within CMAQ to treat PM along with a preliminary comparison with selected observed values of atmospheric visual range.

The inclusion of aerosol particles in air quality models [*Seinfeld and Pandis,* 1998] may be done in several different ways. Dividing the size range of the particle size distribution into a set of size sections has been very popular [*Gelbard et al.,* 1980, *Wexler et al.,* 1994; *Kleeman et al.,* 1997; *Meng at al.,* 1998; *Jacobson,* 1997, 1999]. Another method [*Wright et al.,* 2000] predicts the first six moments of the size distribution, but not the size distribution itself. The method chosen for the aerosol component of the CMAQ model [*Byun and Ching,* 1999] is derived from that introduced in the Regional Particulate Model (RPM) [*Binkowski and Shankar,* 1995], an extension of the Regional Acid Deposition Model (RADM) [*Chang et al.,* 1987, 1990]. As in RPM, a modal representation is assumed here. Fine particles with diameters less than 2.5 μm (PM2.5) are represented by two subdistributions, or modes [*Whitby,* 1978], called the Aitken and accumulation modes. (Note: Aitken mode is used within CMAQ for the size range identified by Whitby as the "nuclei mode".) The Aitken mode includes particles with diameters up to approximately 0.1 μm for the mass distribution, and the accumulation mode covers the mass distribution in the range from 0.1 to 2.5 μm. *Whitby* [1978] also included a coarse mode covering the size range from 2.5 μm to 10μm. PM10 (particles with diameters less than 10 μm) mass is then represented by the sum of the masses in the Aitken, accumulation, and coarse modes.

The modal representation has also been used by *Ackermann et al.* [1998] who applied a model similar to RPM to Europe. *Pirjola et al.* [1998]

*Atmospheric Sciences Modeling Division, Air Resources Laboratory, National Oceanic and Atmospheric Administration, Research Triangle Park, NC 27711.

[†]Atmospheric Modeling Division, National Exposure Research Laboratory, U.S. Environmental Protection Agency, Research Triangle Park, NC 27711.

applied a monodisperse modal model to examine sulfate particle formation in the Arctic boundary layer.

2. Model description. In the CMAQ modeling approach, one calculates only three integral properties of the size distribution for each mode: the total particle number concentration, the total surface area concentration, and the total mass concentrations of the individual chemical components. The current approach differs from that taken by *Binkowski and Shankar* [1995] where the sixth moment was chosen as a third integral property, in place of the second moment. The sixth moment was chosen because of a mathematical simplification [see *Whitby and McMurry*, 1997] that allows analytical expressions to be used for the coagulation terms. The current approach uses numerical quadratures to calculate all of the coagulation terms. The numerical quadratures were compared with the analytical expressions exhibited in *Whitby et al.* [1991] and are accurate to at least six decimal places. The choice of using numerical quadratures was made because the coagulation integral for second moment, unlike the sixth moment, does not have an analytical form.

Conceptually within the fine particle group, the smaller Aitken (i) mode represents fresh particles either from nucleation or from direct emission, while the larger accumulation (j) mode represents aged particles. Primary emissions may also be distributed between these two modes. The two modes interact with each other through coagulation. Each mode may grow through condensation of gaseous precursors; each mode is subject to wet and dry deposition. Finally, the smaller mode may grow into the larger mode and partially merge with it. The chemical species treated in the aerosol component are fine species sulfate, nitrate, ammonium, water, anthropogenic and biogenic organic carbon, elemental carbon, and other unspecified material of anthropogenic origin. The coarse-mode species include sea salt, wind-blown dust, and other unspecified material of anthropogenic origin. Because atmospheric transparency, or visual range, is an important air quality related value, the aerosol component also calculates estimates of aerosol extinction coefficient and visual range.

The particle dynamics are described fully in *Whitby et al.* [1991] and *Whitby and McMurry* [1997]; therefore, only a brief summary of the method is given here. Discussion of chemical species and treatment of cloud processes are contained in *Binkowski* [1999] and *Binkowski and Roselle* [2001]. Note that in the following equations repeated subscripts are not summed.

Assuming a lognormal distribution for number concentration (N) for each mode defined as:

$$n\left(\ln D\right) = \frac{N}{\sqrt{2\pi \ln \sigma_g}} \exp\left[-0.5\left(\frac{\ln\left(D/D_g\right)}{\ln \sigma_g}\right)^2\right].$$

Particle diameter is denoted by D, the geometric mean diameter for the mode is denoted by D_g, and the geometric standard deviation for the mode is denoted by σ_g. The kth moment is obtained by integration as

$$M_k = \int_{-\infty}^{\infty} D^k n(\ln D)\, d(\ln D), \ \text{ with the result } \ M_k = M_0 D_g^k \exp\left(\frac{k^2}{2}\ln^2 \sigma_g\right).$$

M_0 is the total number N of aerosol particles within the mode, suspended in a unit volume of air. For $k = 2$, the moment is proportional to the total particulate surface area within the mode, per unit volume of air. For $k = 3$, the moment is proportional to the total particulate volume within the mode, per unit volume of air. The constant of proportionality between M_2 and surface area is π; the constant of proportionality between M_3 and volume is $\pi/6$. Note that the geometric standard deviation is the same no matter which moment is selected. M_3 is determined as follows from the total mass concentration within the mode:

$$M_3 = \sum_n \phi_n \left(\frac{\pi}{6}\rho_n\right)^{-1},$$

where the mass concentration for species n is ϕ_n with standard density ρ_n. Given values of number, second and third moment concentrations, the geometric standard deviation and the geometric mean diameter for each mode are diagnosed as

$$\ln^2(\sigma_g) = \frac{1}{3}\left[\ln M_0 + 2\ln M_3\right] - \ln M_2$$

and

$$D_g^3 = \frac{M_3}{N \exp\left[\frac{9}{2}\ln^2(\sigma_g)\right]}.$$

The history variables within the model are the modal number concentrations, the modal species mass concentrations, and for the Aitken and accumulation modes, the modal surface area concentrations. Analytic solutions to the governing differential equations are used, assuming the coefficients are constant over the internal time interval of calculation. This time interval is set by CMAQ and is called the *synchronization time step*.

The equation for predicting the number concentration in the Aitken mode is given by

$$\frac{\partial N_i}{\partial t} = c_i - a_i N_i^2 - b_i N_i N_j.$$

Here c_i is a source term representing an increase of number due to new particle formation and possible direct input of particles emitted from sources.

The intramodal coagulation coefficient is represented by a_i and intermodal coagulation coefficient is represented by b_i.

The corresponding prediction equation for the accumulation mode is

$$\frac{\partial N_j}{\partial t} = c_j - a_j N_j^2 \,,$$

where the source term c_j is now related only to possible direct input of emitted particles.

The solutions to these equations for the case with source terms are given by the following (subscripts eliminated for simplicity and $b = 0$ for the accumulation mode)

$$N(t) = \frac{r_1 + r_2 \gamma \exp(\delta t)}{a[1 + \gamma \exp(\delta t)]} \,, \quad \text{with} \quad \delta = \sqrt{b^2 + 4ac} \,,$$

and

$$r_1 = \frac{2ac}{b + \delta} \,, \qquad r_2 = -\frac{b + \delta}{2} \,.$$

The initial conditions enter through

$$\gamma = -\frac{r_1 - aN(t_0)}{r_2 - aN(t_0)} \,.$$

For the accumulation mode, if there is no source term ($c = 0$) the solution simplifies to

$$N(t) = \frac{N(t_0)}{1 + aN(t_0)t} \,,$$

which is known as the Smoluchowski solution.

Species mass in the Aitken mode is predicted by an equation

$$\frac{\partial \phi_i}{\partial t} = P_i - L_i \phi_i$$

with solution $\phi_i(t) = \dfrac{P_i}{L_i} + \left[\phi_i(t_0) - \dfrac{P_i}{L_i}\right] \exp(-L_i t)$.

Here the production coefficient P_i represents the mass accompanying new particle production, condensational growth, and any mass coming from source emissions. The loss coefficient L_i represents mass transfer to the accumulation mode by intermodal coagulation. For the accumulation mode the governing equation is

$$\frac{\partial \phi_j}{\partial t} = P_j$$

with solution $\phi_j(t) = \phi_j(t_0) + P_j t$.

The production term P_j represents mass transferred from the Aitken mode, the contribution from condensational growth, as well as any possible mass emitted from sources.

Turning to the prediction of surface area in the Aitken and accumulation modes, we first note that surface area is π times second moment M_2. Thus, the prediction equation for the second moment in each mode is of the form

$$\frac{\partial M_2}{\partial t} = P_2 - L_2 M_2$$

with solution $M_2\left(t\right) = \dfrac{P_2}{L_2} + \left[M_2\left(t_0\right) - \dfrac{P_2}{L_2}\right]\exp\left(-L_2 t\right)$.

For the Aitken mode, the production term P_2 represents the increase in second moment from new particle production, condensational growth, and the contribution from any source emissions. The loss term L_2 represents the loss of second moment by intramodal coagulation and by intermodal coagulation with the accumulation mode. For the accumulation mode, the production term represents the second moment transferred from the Aitken mode by intermodal coagulation plus the increase in second moment from condensational growth, and any second moment increase coming from source emissions. The loss term represents the loss of second moment by intramodal coagulation.

3. Representation of visual range. Visual range X_r is usually defined to mean the furthest distance one both can see and identify an object in the atmosphere. For a detailed presentation on the concepts of visibility, see *Malm* [1979]. In a perfectly clean atmosphere composed only of nonabsorbent gases, the only process restricting visibility during daylight is the scattering of solar radiation by the gas molecules. This is known as Rayleigh scattering, usually represented by a scattering coefficient. If in addition to scattering, absorption also occurs, an absorption coefficient may be defined. The sum of the scattering and absorption coefficients is called the extinction coefficient, β_{ext}. If absorption is not occurring, the extinction coefficient is equal to the scattering coefficient. The visibility in an atmosphere in which Rayleigh scattering is the only active optical process may be taken as a reference value. A useful index for quantifying the impairment of visibility by the presence of atmospheric aerosol particles is the deciview [*Pitchford and Malm*, 1994]. The deciview index *deciV* is given as

$$deciV = 10\ln\left(\frac{\beta_{ext}}{0.01}\right),$$

where 0.01 [km^{-1}] is the standard value for Rayleigh scattering. Thus, a *deciV* equal to zero means that there are no aerosol particles within the visual range.

The following table shows the relationships among $deciV$, X_r, and β_{ext}.

$deciV$	X_r [km]	β_{ext} [km^{-1}]
60	1.0	4.00
50	3.6	1.50
40	7.2	0.55
30	19.5	0.20
20	52.9	0.07
10	144.0	0.03
0	361.0	0.01

The aerosol extinction coefficient β_{ext} [km^{-1}] must be calculated from ambient aerosol characteristics as index of refraction, volume V concentration and size distribution. The extinction coefficient is usually obtained from a convolution of the size distribution with the Mie extinction efficiency [*Willeke and Brockmann*, 1977]. Here we calculate the extinction coefficient using a very efficient and reasonably accurate approximation to the Mie efficiency Q_{ext}, at a wavelength λ, by *Evans and Fournier* [1990]. The extinction coefficient is then given by

$$\beta_{ext} = \frac{3\pi}{2\lambda} \int_{-\infty}^{\infty} \frac{Q_{ext}}{\alpha} \frac{dV}{d\ln\alpha} d\ln\alpha \,,$$

where $\alpha = \dfrac{\pi D}{\lambda}$.

For a lognormal size distribution

$$\frac{dV}{d\ln\alpha} = V_T \left(\frac{A}{\pi}\right)^{1/2} \exp\left[-A\ln^2\left(\frac{\alpha}{\alpha_\nu}\right)\right], \qquad \alpha_\nu = \frac{\pi D_{g\nu}}{\lambda} \,,$$

and $A = \dfrac{1}{2\ln^2 \sigma_g}$.

$D_{g\nu}$ is the geometric mean diameter for a volume distribution with total volume V_T, and is related to the geometric mean diameter for the number distribution D_g

$$D_{g\nu} = D_g \exp\left(3\ln^2 \sigma_g\right).$$

Another useful visual range concept is the Koschmieder visual range equation which can be used to define an extinction coefficient

$$\beta_K = \frac{3.912}{X_r}$$

from a measured value of visual range, X_r, in km.

4. Preliminary results for a simulation of visual range. We turn now to results from a simulation of visual range from 00Z on July 11, 1995, to 00Z on July15, 1995 (00Z means 00 hours UTC). The model run was started with initial data for 00Z on July 6, 1995. The grid size is 36 km and the domain covers the eastern half of the US. The interval between the July 6 and July 10 is a "spin-up" period to allow the chemical species concentrations to develop properly. Visual range is observed routinely at major airports and military bases. Using the Koschmieder equation, the observed visual range was converted to an extinction coefficient and the deciview calculated. Only hourly observations collected under conditions with no precipitation and with relative humidities less than 90% were used. The data source for the extinction coefficients is the Center for Air Pollution Impact and Trend Analysis (CAPITA) at Washington University, Saint Louis, Missouri. The following table displays the observed and simulated mean, minimum and maximum *deciV*, along with the square of the correlation coefficient between simulated and observed values for each day at local noon. The total number of pairs of observed and simulated values is also indicated.

Day		Mean	Minimum	Maximum	r^2
July 11, 1995	Observed	25.4	11.0	31.8	0.17
($n = 155$)	Simulated	15.2	0.0	34.3	
July 12, 1995	Observed	26.5	13.8	34.0	0.30
($n = 157$)	Simulated	17.2	0.0	35.3	
July 13, 1995	Observed	27.4	11.0	38.4	0.50
($n = 162$)	Simulated	18.6	0.0	39.3	
July 14, 1995	Observed	27.4	11.0	36.6	0.29
($n = 158$)	Simulated	19.2	0.0	37.7	
July 15, 1995	Observed	26.8	11.0	38.5	0.29
($n = 159$)	Simulated	16.8	0.0	35.7	
5 Days	Observed	26.7	11.0	38.8	0.30
($n = 791$)	Simulated	17.4	0.0	34.7	

The model predicts the maxima reasonably well, but appears to underpredict the mean and the minima. There is a reason for underpredicting the minima. At any airport, there is a limit to how far an observer can see an object under ideal conditions. This distance is the maximum visual range or minimum possible deciew. So even if conditions were ideal, with molecular scattering as the dominant restriction to visual range, the deciview value would not be zero, but some larger value. For example, a value of 11 for observed deciview corresponds to an extinction coefficient of 0.03 and the (Koschmieder) visual range is 130 km. Very few eastern U.S. airports have objects that can be seen at that distance. For the five day episode the observed mean value is larger than the predicted mean because of the effect of many larger observed minima at any given airport. A more detailed comparison will be presented by *Mebust et al.* [2001].

REFERENCES

ACKERMANN, I.J., H. HASS, M. MEMMESHEIMER, A. EBEL, F.S. BINKOWSKI, AND U. SHANKER. Modal aerosol dynamics model for Europe: development and first applications. *Atmos. Environ.*, **32**, 2981–2999, 1998.

BINKOWSKI, F.S., Aerosols in Models-3 CMAQ, Chapter 10 of Byun and Ching, *Science Algorithms of the EPA Models-3 Community Multiscale Air Quality (CMAQ) Modeling System*, EPA/600/R-99/030, Office of Research and Development, United States Environmental Protection Agency, Washington, DC 1999.

BINKOWSKI, F.S. AND U. SHANKAR. The regional particulate model 1. Model description and preliminary results. *J. Geophys. Res.*, **100**, D12, 26191–26209, 1995.

BINKOWSKI, F.S. AND S.J. ROSELLE. Models-3 community multiscale air quality (CMAQ) model aerosol component. I: Description, *J. Geophys. Res.*, in review.

BYUN, D.W. AND J.K.S. CHING, *Science Algorithms of the EPA Models-3 Community Multiscale Air Quality (CMAQ) Modeling System*, EPA/600/R-99/030, Office of Research and Development, United States Environmental Protection Agency, Washington, DC, 1999.

CHANG, J.S., R.A. BROST, I.S.A. ISAKSEN, S. MADRONICH, W.R. STOCKWELL, AND C.J. WALCEK. A three-dimensional Eulerian acid deposition model: Physical concepts and formulation, *J. Geophys. Res.*, **92**, 14681–14700, 1987.

CHANG, J.S., F.S. BINKOWSKI, N.L. SEAMAN, D.W. BYUN, J.N. MCHENRY, P.J. SAMSON, W.R. STOCKWELL, C.J. WALCEK, S. MADRONICH, P.B. MIDDLETON, J.E. PLEIM, AND H.L. LANDSFORD. The regional acid deposition model and engineering model, NAPAP SOS/T Report 4, in National Acid Precipitation Assessment Program, Acidic Deposition: State of Science and Technology, Volume I, Washington, D.C., 1990.

EVANS, T.N. AND G.R. FOURNIER. Simple approximation to extinction efficiency valid over all size ranges, *Appl. Optics*, **29**, 4666–4670, 1990.

GELBARD, F., Y. TAMBOUR, AND J.H. SEINFELD. Sectional representations for simulating aerosol dynamics, *Jour.of Colloid and Interface. Sci.*, **76**, 541–556, 1980.

JACOBSON, M.Z. Development and application of a new air pollution modeling system-II. Aerosol module structure and design, *Atmos. Environ.*, **31**, 131–144, 1997.

JACOBSON, M.Z. *Fundamentals of Atmospheric Modeling*, Cambridge University Press, New York, p. 656, 1999.

KLEEMAN, M.J., G.R. CASS, AND A. ELDERING, Modeling the airborne particle complex as a source-oriented external mixture, *J. Geophys. Res.*, **102**, 21355–21372, 1997.

MALM, W.C., Considerations in the measurements of visibility, *J. Air Pollution Control Assoc.*, **29**, 1042–1052, 1979.

MENG, Z., D. DABDUB, AND J.H. SEINFELD, Size-resolved and chemically resolved model of atmospheric aerosol dynamics, *J. Geophys. Res.*, **103**, 3419–3435, 1998.

MEBUST, M.R., B.K. EDER, F.S. BINKOWSKI, AND S.J. ROSELLE. Models-3 Community multiscale air quality (CMAQ) model aerosol component. II: Model evaluation, *J. Geophys. Res.*, in review.

PIRJOLA, L., A. LAAKSONSEN, P. AALTO, AND M. KULMALA, Sulfate aerosol formation in the Arctic boundary layer, *J. Geophys. Res.*, **103**, 8309–8321, 1998.

PITCHFORD, M.L. AND W.C. MALM. Development and applications of a standard visual index, *Atmos. Environ.*, **28**, 1049–1054, 1994.

SEINFELD, J.H. AND S.N. PANDIS. *Atmospheric Chemistry and Physics from Air Pollution to Climatic Change*, Wiley, New York, 1998.

WEXLER, A.S., F.W. LURMANN, AND J.H. SEINFELD. Modeling urban and regional aerosols: I. Model development, *Atmos. Envion.*, **28**, 531–546, 1994.

WHITBY, K.T. The physical characteristics of sulfur aerosols, *Atmos. Environ.*, **12**, 135–159, 1978.

WHITBY, E.R. AND P.H. MCMURRY. Modal aerosol dynamics modeling, *Aerosol Sci. and Technol.*, **27**, 673–688, 1997.

WHITBY, E.R., P.H. MCMURRY, U. SHANKAR, AND F.S. BINKOWSKI. *Modal Aerosol Dynamics Modeling*, Rep. 600/3-91/020, Atmospheric Research and Exposure Assessment Laboratory, U.S. Environmental Protection Agency, Research Triangle Park, N.C., (NTIS PB91-161729/AS), 1991.

WILLEKE, K. AND J.E. BROCKMANN. Extinction coefficients for multimodal atmospheric particle size distributions, *Atmos. Environ.*, **11**, 995–999, 1977.

WRIGHT, D.L., R. MCGRAW, C.M. BENKOVITZ, AND S.E. SCHWARTZ. Six-moment representation for multiple aerosol populations in a sub-hemispheric chemical transformation model, *Geophys. Res. Lett.*, **27**, 967–970, 2000.

MODELING EMISSIONS AND CHEMISTRY OF MONOTERPENES FOR REGIONAL MODELS

R.J. BARTHELMIE[*] AND S.C. PRYOR[†]

Abstract. Organic compounds comprise a significant fraction of the total atmospheric fine particle loading and hence are associated with radiative forcing and health impacts. While a plethora of organic compounds have been found in the particle phase in most non-urban environments, a significant fraction are biogenically derived and part of this fraction are formed from condensation of the oxidation products of monoterpenes. Hence there is a need for numerical schemes to describe the emissions of monoterpenes and the oxidation and gas-particle partitioning of their products. Here we present semi-explicit chemical schemes developed for five monoterpene groups - α-pinene, β-pinene, Δ-3-carene representing the most important bicyclic monoterpenes, d-limonene representing unicyclic monoterpenes and finally ocimene representing acyclic monoterpene compounds. The chemistry of these compounds is based on experimental results (mainly from smog chambers) and known alkene chemistry. The semi-explicit schemes are evaluated in a box model configuration to simulate smog chamber studies of monoterpene oxidation. The chemical mechanisms have also been used in conjunction with speciated monoterpene emissions for different land use types of the United States to show regional variations in the resulting aerosol burdens. Combination of the emissions and chemistry models presented here offers the opportunity to represent these compounds (and their contribution to regional/global aerosol burdens) within atmospheric chemistry-transport models. Since the models provide specific product information, the results can be compared with field studies and used to identify candidate tracer compounds of monoterpene oxidation to assist the design of experimental studies.

1. Introduction. Global biogenic volatile organic compound (VOC) emissions are estimated as 1150 Tg C/yr of which 11% are monoterpenes (Guenther *et al.*, 1995). Monoterpene compounds and their derivatives are known to be present in the atmosphere although their emissions and chemistry are not fully understood. The main sources of monoterpenes are emissions from plants, in particular tree species (Lamb *et al.*, 1987). Monoterpenes are reactive in the atmosphere with a lifetime of approximately 0.1 day. Due to their structure and molecular weights, monoterpene degradation products have a relatively high aerosol yield (Grosjean *et al.*, 1993), (Pandis *et al.*, 1992b). Most secondary organic aerosols are mainly found in the fine size fraction which can be transported over relatively long distances and additionally is most active in producing health and visibility impacts. Biogenic compounds make a relatively large contribution to regional and global aerosol burdens (Griffin *et al.*, 1999a). Hence there is an ongoing need for more experimental and field studies and for improved modeling schemes for use in regional and global transport-chemistry models.

---

[*]Department of Wind Energy and Atmospheric Physics, Risø National Laboratory, 4000 Roskilde, Denmark; Email: R.Barthelmie@risoe.dk.

[†]Atmospheric Science Program, Department of Geography, Indiana University, Bloomington, IN 47405, USA; Email:spryor@indiana.edu.

Most regional or global chemistry models estimate monoterpenes emissions using PCBEIS2 (Birth and Geron, 1995) or similar emission schemes to produce bulk monoterpene emissions based on vegetation type and temperature. Recent experimental data have been combined with vegetation databases to allow estimates of speciated monoterpene emissions based on temperature and vegetation (Geron *et al.*, 2000). Thus it is now possible to estimate emission of individual monoterpene compounds (which are grouped by structure in this paper).

To estimate aerosol yields from monoterpene and other volatile organic compound (VOC) most current regional and global atmospheric chemistry-transport models use implicit schemes. These are implemented based either on single step aerosol yield or fractional aerosol coefficients. Gross estimates of aerosol yield of 762 μg m^{-3} per ppm for the monoterpenes are given by Pandis *et al.* (1992b) which is comparable to a fractional aerosol coefficient (aerosol yield in μg m^{-3}/initial VOC in μg m^{-3}) of 30% (Grosjean, 1992), (Grosjean and Seinfeld, 1989). A more advanced method is the absorption approach developed by Odum *et al.* (1996) and Hoffmann *et al.* (1997) which uses a two product model to estimate aerosol yields based on an empirical fit obtained through smog chamber experiments for individual compounds. Griffin *et al.* (1999a) used this approach to estimate secondary organic aerosol (SOA) yields from ozone (O$_3$) and hydroxyl radical (OH) oxidation. These implicit schemes are computationally efficient, however, these approaches are limited in their ability to correctly represent the time evolution of oxidation products (both gases and aerosols) and do not provide information regarding products which are likely to be found in the aerosol phase and are necessary for assessing likely water uptake and epidemiological studies.

In this paper, semi-explicit chemical schemes are developed for a number of monoterpene groups describing their oxidation paths which lead to condensable products. Mechanistic reaction sequences are expanded based on laboratory studies and alkene chemistry which include production of both condensable and gas products. The advantages of expanding these chemical schemes include:

- Improved oxidation chemistry. Some steps in the degradation of monoterpenes are oxidant consumers and some are oxidant producers. Detailed representation of oxidation chemistry is critical to correct treatment within numerical models.
- Carbon balance is retained. At each time step oxidation or degradation of monoterpene products in carbon mass which is partitioned into the aerosol phase.
- Concentrations of individual products which partition between the gas and aerosol phases are modeled rather than bulk aerosol yields.
- Time evolution of changes in aerosol concentration are predicted.
- Speciated aerosol concentrations can be produced allowing tracking of the source monoterpene for comparison with field experiments.

The emission speciation and chemical schemes are evaluated here using smog chamber experiments and applied in a box model to estimate the biogenic secondary organic aerosol yield for a number of different land use types.

2. Monoterpene emission fluxes. Many studies of monoterpene emission have been conducted (e.g. (Geron *et al.*, 1994), (Lamb *et al.*, 1987), (Lamb *et al.*, 1993), (Fuentes *et al.*, 1996)) which suggest that monoterpene emission fluxes are dependent on temperature. It has also been speculated that light plays a role for some monoterpene compounds and that humidity may also influence emission rates (Schade *et al.*, 1999). For regional and global scale modeling, these latter two factors are typically ignored due to large uncertainty in emission rates and dependency on these factors in the vegetation databases used to generate emission fluxes.

There are many different monoterpene compounds found in the atmosphere (Figure 1). Here we lump monoterpenes into five groups based on their emissions, structure and reactivity (Table 1). In non-urban areas, α-pinene, β-pinene and d-limonene are typically the most abundant monoterpenes (Isidorov *et al.*, 1985) with the importance of other monoterpenes such as camphene, myrcene, ocimene, Δ-3-carene and β-phellandrene varying (e.g. (Hagerman *et al.*, 1997), (Versino, 1997)). Experimental data from forests in North America suggest α-pinene, β-pinene, Δ-3-carene, d-limonene, camphene and α-terpinene dominate monoterpene concentrations (Geron *et al.*, 2000). However, because monoterpene reactivity with different oxidants varies over six orders of magnitude (Atkinson, 1997) for numerical modeling purposes emission rates rather than ambient concentrations are required. The five monoterpene groups treated herein are represented by α-pinene, β-pinene, Δ-3-carene, d-limonene and ocimene. These compounds have high emission fluxes compared to other monoterpenes and, as described below, represent structurally distinct classes of monoterpenes (with the exception of Δ-3-carene). For coniferous forest, this speciation grouping gives a relatively even distribution of the total monoterpene emission between α-pinene, β-pinene and d-limonene groups (Guenther *et al.* (1996) suggest these as the major components of monoterpene emissions). α-pinene has been indicated as the main component of coniferous forest emissions with significant contributions from d-limonene and β-pinene (Isidorov *et al.*, 1985). For deciduous forest and agricultural areas d-limonene and β-pinene appear to dominate mass emissions but the uncertainties are probably greater because of their greater diversity of vegetation and more limited observational data.

Monoterpenes can be divided into acyclic, monocyclic or bicyclic classes. In the current research bicyclic classes are represented by α-pinene, β-pinene and Δ-3-carene while d-limonene represents monocyclic compounds and ocimene is used to represent acyclic compounds. Clearly the number of rings and the position of the double bonds within cylic com-

TABLE 1
Monoterpene groups and their reaction rates.

Monoterpene	Oxidant	Reaction rate (cm^3 molecule^{-1} s^{-1})
α-pinene	OH	53.7×10^{-12}
	NO$_3$	6.16×10^{-12}
	O$_3$	86.6×10^{-18}
β-pinene	OH	78.9×10^{-12}
	NO$_3$	2.51×10^{-12}
	O$_3$	15×10^{-18}
Δ-3-carene	OH	88×10^{-12}
	NO$_3$	9.1×10^{-12}
	O$_3$	37×10^{-18}
d-limonene	OH	171×10^{-12}
	NO$_3$	12.2×10^{-12}
	O$_3$	200×10^{-18}
ocimene	OH	2.52×10^{-12}
	NO$_3$	22×10^{-12}
	O$_3$	540×10^{-18}

pounds are important both in terms of the reactivity of the compounds and their subsequent chemistry. Cleavage of a double bond may mean the loss of one or more carbon atoms depending on the position of the bond and the presence of one or more rings. This affects the aerosol formation potential of the oxidized compound since partitioning into the aerosol phase is strongly dependent on molecular weight (and functional groups). Other cyclic monoterpenes with relatively high emissions (see Table 2) were included by grouping them with α-pinene, β-pinene, Δ-3-carene or d-limonene based on structure, reaction rates and the number of carbon atoms in the major reaction product. For example, sabinene is a bicyclic compound with an external bond attached to a ring with a nine carbon reaction product which is represented most closely by β-pinene. All acyclic compounds are represented by ocimene and have been added to the mechanism although their oxidation products are expected to contribute less

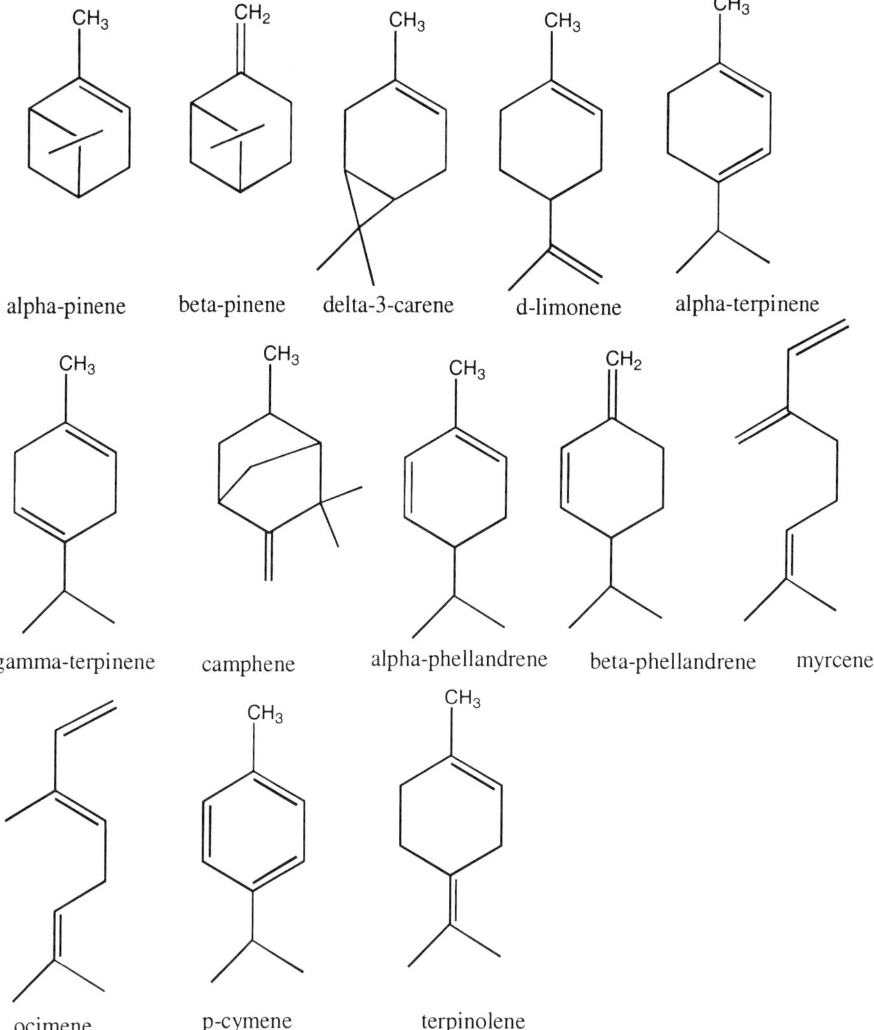

FIG. 1. *Structure of monoterpene compounds.*

to SOA concentrations than cyclic groups because cleavage of the double bonds produces an array of compounds most of which have less than seven carbon atoms. However, these compounds may contribute to the reactivity of the atmosphere since each possesses three double bonds.

3. Monoterpene mechanism.

3.1. Monoterpene degradation chemistry. The chemical schemes developed here describe the oxidation of monoterpenes by the hydroxyl radical (OH), nitrate radical (NO_3) and ozone (O_3). The mechanisms are

TABLE 2

α-pinene oxidation products and their estimated partitioning coefficients.

Description	Formula	Partitioning coefficient		
		(Barthelmie and Pryor, 1999)	(Yu *et al.*, 1999)	(Kamens *et al.*, 1999) (at 295K)
alpha-pinene	$C_{10}H_{16}$			
hydroxy alkyl radical	$C_{10}(OH)H_{16}$			
hydroxy peroxy radical	$C_{10}(OH)H_{16}OO$			
hydroxy alkoxy radical	$C_{10}(OH)H_{16}O$			
pinonaldehyde	$C_{10}(O)H_{16}O$	0.0010	0.0012	
Criegee intermediate	$C_{10}H_{16}O_3$			
nitrooxy alkyl radical	$C_{10}(ONO_2)H_{16}$			
nitrooxy peroxy radical	$C_{10}(ONO_2)H_{16}OO$			
nitrooxy alkoxy radical	$C_{10}(ONO_2)H_{16}O$			
nitrooxy hydroperoxide	$C_{10}(ONO_2)H_{16}OOH$			
nitrooxy peroxy nitrate	$C_{10}(ONO_2)H_{16}OONO_2$			
carbonyl nitrate	$C_{10}(ONO_2)H_{16}O$	0.0024		
hydroxy nitrate	$C_{10}(ONO_2)H_{16}OH$	0.0024		
hydroxy alkyl peroxynitrate	$C_{10}(OONO_2)H_{16}OH$			
pinonic acid	$C_{10}(O)H_{15}OOH$	0.0036	0.030	
pinic acid	$C_9(OOH)H_{12}OOH$	0.1093	0.028	0.051
dinitrate	$C_{10}(ONO_2)H_{16}ONO_2$	0.0038		
oxirane	$C_{10}(O)H_{16}$			

based on a combination of experimental results (mainly laboratory work e.g. (Kamens *et al.*, 1999), (Palen *et al.*, 1992), (Yokuchi and Ambe, 1985), (Pandis *et al.*, 1992a), (Yu *et al.*, 1999), (Hallquist *et al.*, 1999)) and alkene chemistry (Atkinson, 1997), (Stockwell *et al.*, 1997). The main objective is to provide a mechanism for describing the development of secondary organic aerosol (SOA) concentrations suitable for incorporation into global or regional scale numerical models. Hence, in order to ensure computational feasibility, the oxidation scheme for each monoterpene is limited to approximately 35–50 reactions (with additional reactions for d-limonene) and approximately 20 new products including radicals. The mechanism is designed to be integrated with either the Carbon Bond Mechanism (Gery *et al.*, 1989) or with the Regional Atmospheric Chemistry Mechanism (RACM) (Stockwell *et al.*, 1997). Given that many pathways and products of monoterpene oxidation remain unknown, it is clearly not possible to model all potential pathways and products of oxidation reactions. In the following sections, we provide an overview of the three major pathways of monoterpene oxidation in the troposphere with a focus on cyclic compounds. These are described in detail for α-pinene and β-pinene in Barthelmie and Pryor (1999) and for d-limonene and Δ-3-carene in Barthelmie and Pryor (2001).

Oxidation of monoterpenes by OH proceeds via production of a hydroxy alkyl radical which rapidly reacts with oxygen (O_2) to give a hydroxy peroxy radical. The initial reactions are assumed to be dominated by addition of OH on the terminal carbon. In the atmosphere, peroxy radicals are expected to react with nitric oxide (NO), nitrogen dioxide (NO_2) and NO_3, hydroperoxyl radical (HO_2), methyl peroxy radical (MO_2), or higher peroxy radicals (ACO_3) (Atkinson, 1997), (Stockwell *et al.*, 1997). Reaction with NO gives both hydroxy alkoxy radicals and NO_2 or hydroxy nitrates (Carter and Atkinson, 1989) while reaction with NO_2 is expected to produce hydroxy peroxy nitrates which rapidly decompose back to the reactants (Atkinson, 1997). Stockwell *et al.* (1997) suggest that reaction with NO_3 produces two alkene adducts, one of which decomposes and the other reacts to form carbon nitrates and HO_2 (incorporated into one step in the mechanism presented here). While reaction with HO_2 could be expected to proceed solely to form a hydroperoxide which decomposes, reactions with methyl or higher alkyl peroxides are anticipated to form mainly alkoxy radicals or alcohol plus ester combinations (Atkinson, 1997). To avoid addition of numerous large compounds, these reactions are assumed to produce the associated alkoxy radical and aldehyde/formaldehyde combinations including small amounts of condensable product. Reactions of the alkoxy radical are lumped into reaction with O_2, decomposition/isomerization, reaction with NO, and reaction with NO_2. These pathways are expected to give relatively similar condensable products except for possible nitrate formation if NO/NO_2 are present. Although alkoxy radicals produced from monoterpene oxidation also undergo isomerization (Atkinson and Carter, 1991), the final products of isomerization/decomposition are expected to be similar to those contained within the mechanism.

Monoterpene oxidation by O_3 proceeds via formation of an ozonide which gives two Criegee intermediates (carbonyl oxides) and two carbonyl compounds (Grosjean *et al.*, 1993). The Criegee intermediates are either stabilized by collision or decompose (Atkinson, 1997). Winterhalter *et al.* (2000) describe the pathways for Criegee intermediates as an ester route, a hydroperoxide route and a stabilzation route. Monoterpenes with double bonds internal to the primary ring (α-pinene, Δ-3-carene and d-limonene) react mainly (80–100%) via the hydroperoxide route (Winterhalter *et al.*, 1999) which is known to be a source of OH radicals (Atkinson, 1997). Formation of hydroperoxides in the mechanism is followed by their decomposition to carbonyl alkoxy radicals which are assumed to form mainly their dicarbonyl equivalents. This has been incorporated into one step. Stabilized carbonyl oxides can react with H_2O, formaldehyde (HCHO), or NO (Becker *et al.*, 1993). Grosjean and Grosjean (1997) suggest further reactions of carbonyl oxides which have been incorporated into one degradation step in these mechanisms.

The first step of oxidation by the NO_3 radical is production of an excited nitrooxy alkyl radical. This is stabilized and then reacts with oxygen to form a nitrooxy peroxy radical (Atkinson, 1997). An alternative minor route includes production of an oxirane and NO_2. Pathways for degradation of the nitrooxy peroxy radical are reaction with NO to give dinitrates and/or nitrooxy alkoxy radicals and NO_2, HO_2 to form hydroperoxide nitrates, RO_2 (or self-reactions) to form nitrooxy alkoxy radicals or oxygenated nitrates, and NO_2 to give unstable nitrooxy peroxynitrates (Barnes *et al.*, 1990). Main reactions of nitrooxy alkoxy radical are assumed to be reaction with O_2 or decomposition/ isomerization (incorporated into one step) or reaction with NO.

Reactions of acyclic compounds also follow these general pathways. The main differences between these reactions and those of cyclic or bicyclic compounds is that the compound fractures subsequent to attack on the double bond leading to products which are less likely to isomerize and have low molecular weights ($< 7C$) meaning that they are unlikely to be condensable.

3.2. Partitioning between gas and aerosol phase products. Although there is some experimental evidence that oxidation products of monoterpenes may nucleate in the remote troposphere (Kavouras *et al.*, 1998), (Kumala *et al.*, 1999), transfer of monoterpene oxidation products which have intermediate vapor pressures to the aerosol phase is most likely dominated by absorption into existing aerosols.

Here, the absorption model of Odum *et al.* (1996) and Hoffmann *et al.* (1997) (Equation (1)) is used to describe partitioning of condensable products between gas and aerosol phases. The organic aerosol concentration

M_0 at time t is predicted based on the change of reactive organic gas concentration ΔROG, the individual formation yield α_i, and the partitioning coefficient (absorption equilibrium constant) K_{om}.

$$(1) \qquad M_0(t) = M_{0,i} + \Delta ROG(t) M_{0,n-1} \sum_i \left(\frac{\alpha_i K_{om,i}}{1 + K_{om,i} M_{0,n-1}} \right).$$

Partitioning coefficients and yields for the two product approach were derived by Griffin *et al.* (1999b) for monoterpenes by fitting Equation (1) to aerosol mass and yields from smog chamber data. Typically, in this approach, one product has a relatively large yield with a low partitioning coefficient, while the second product has lower yield and higher partitioning coefficient. However, it should be noted that there are no unique solutions for Equation (1) and two products with identical yield and partitioning coefficients and many even number (two, four or higher) product models (with a large number of solutions for the yield and the partitioning coefficients) give good fits to the experimental results described by Odum *et al.* (1996) and Hoffmann *et al.* (1997).

Recent experimental work has begun to provide partitioning coefficients for individual monoterpene oxidation products (e.g. (Yu *et al.*, 1999), (Kamens *et al.*, 1999)). In the research presented here if partitioning coefficients for individual products were unavailable, they were estimated using the following procedure. The initial value of K_{om} was ascribed to each product assuming that the partitioning coefficient should lie between the two values given for partitioning coefficients for the two product model given in Hoffmann *et al.* (1997). Higher molecular weights products were assumed to have higher partitioning coefficients than lower molecular weight products and the functional groups of each product were also taken into account e.g., a dicarboxylic acid has a higher partitioning coefficient than a ketone. A best fit algorithm was then employed to reproduce the data of Odum *et al.* (1996) and Hoffmann *et al.* (1997) given the above constraints for each condensable product. As shown in Table 2, there is good agreement between measured partitioning coefficients and those estimated here with the exception of the value for pinic acid which is over-estimated by this approach.

Considerable uncertainty remains in the determination of partitioning coefficients (which can only be resolved through experimental research) and these values are critical to determining aerosol mass. A sensitivity study conducted to quantify how this uncertainty is manifest in aerosol yields presented in Barthelmie and Pryor (1999) indicates that aerosol yields are more sensitive to under-estimates than to over-estimates of partitioning coefficients. A further point which is not addressed herein is the temperature dependence of partitioning coefficients. Kamens *et al.* (1999) suggest that K_p varies by two orders of magnitude over a range of 35 °C. The smog chamber experiments from which the partitioning coefficients presented in

Odum *et al.* (1996) and Hoffmann *et al.* (1997) were derived were conducted at approximately 40 °C which is not representative of tropospheric conditions.

4. Mechanism evaluation and testing.

4.1. Use of smog chambers for evaluation of chemical mechanisms. While smog chamber experiments represent an efficient tool for providing kinetic and mechanistic information regarding relationships between precursor concentrations and reaction products in a systematic manner in a controlled environment, it is important to note that smog chamber studies are typically constrained to using concentrations much higher than found in either rural or urban atmospheres. This appears to impact the nitrate radical pathway of monoterpene oxidation most strongly. Smog chamber studies are also usually conducted at relatively high temperatures which, as discussed above, affects partitioning coefficients, and also the reaction pathways. Some reactions are temperature dependent, with OH and NO_3 radical rate constants for reaction with α-pinene decreasing with increasing temperature while that of O_3 increases (Atkinson, 1997). It should also be acknowledged that smog chambers are also subject to artifacts such as adsorption-desorption processes at the walls, and heterogeneous reactions on the walls which can result in net product of radicals (Sakamaki and Akimoto, 1988).

4.2. Simulation conditions. In order to evaluate the monoterpene mechanisms outlined above they are incorporated within the RACM mechanism which is operated in a box model configuration and applied to simulate smog chamber experiments. The chemical solver is the Euler backward iterative method described by Hertel *et al.* (1993) where the precision for all non-monoterpene concentrations is 0.005 and for all monoterpene compounds and their products is 0.5. Each simulation is initialized with the initial conditions specified for the smog chamber experiments as described in Odum *et al.* (1996), Hoffmann *et al.* (1997) and Griffin *et al.* (1999b). These smog chamber experiments were performed in the same chamber and are of similar design with temperatures of between 310-320 K, approximately 5% relative humidity and initial concentrations of dry ammonium sulfate seed particles of approximately 1×10^{-4} μg m^{-3}. While NO_x concentrations are given for each run separately the propene concentration is often not specified explicitly. The ratio of NO_x to propene is given in Hoffmann *et al.* (1997) and Odum *et al.* (1996) while Griffin *et al.* (1999b) indicate that the average ratio of hydrocarbon to NO_x was 5:1 with an initial concentration of propene up to 300 ppb. Propene is represented in the RACM mechanism by a terminal alkene with a carbon number of 3.8 (Stockwell *et al.*, 1997). The time step used in the model is 1 s.

4.3. Results for α-pinene. For α-pinene, six simulations were conducted with initial monoterpene concentrations of between 19.5 and 94.5

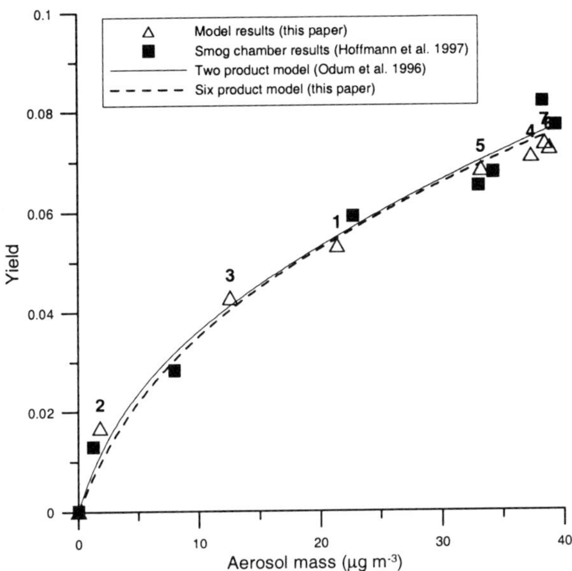

FIG. 2. *Comparison of aerosol mass and yield from the model described in this paper and with smog chamber results from Odum et al. (1996) and Hoffmann et al. (1997).*

ppb. These model runs simulate experiments described in Hoffmann *et al.* (1997). The length of the α-pinene smog chamber experiments are not explicitly stated by Hoffmann *et al.* (1997) and so the simulations are conducted until the change in aerosol yield is negligible (i.e. $dM_0/dt <$ 0.1 μg m^{-3} s^{-1}). Figure 2 shows aerosol mass and yield predicted by a six condensable product model conditioned as in Odum et al. (1996) used with the terminating yields obtained by the chemical mechanism. Also shown in this figure are the organic aerosol concentrations predicted by the two-product fit of Hoffmann *et al.* (1997) and Odum *et al.* (1996) based on the smog chamber data. As is evident, the organic aerosol concentrations given in Barthelmie and Pryor (1999) and partitioning coefficients given in Table 2 are similar to those of the two-product model. Aerosol yield in the mechanism is mainly dependent on the amount of monoterpene present. Within experimental uncertainty, the model results are in good agreement with the observations and are a good fit to the both the two-product model of Odum *et al.* (1996) and the six-product model (shown in Figure 2).

Table 2 also gives partitioning coefficients estimated by the method described above for individual oxidation products of α-pinene and those measured by Yu *et al.* (1999) and Kamens *et al.* (1999). There is good agreement for pinonaldehyde but the modeling technique gives higher values than measured for pinic acid.

In the simulations shown in Figure 2 pinonaldehyde is always the largest contributor to modeled aerosol yields despite its low partitioning coefficient and removal by secondary oxidation reactions. The next largest components of the aerosol mass in order of contribution; are hydroxy/carbonyl nitrate, dicarboxylic acid, carboxylic acid, nitrate, and dinitrate. These results are in good accord with a number of product studies conducted on α-pinene oxidation which have the presence of identified pinonaldehyde in yields of close to 30% (Arey *et al.*, 1990), (Hakola *et al.*, 1994) and multifunctional carbonyls and diacids (Glasius *et al.*, 1999). Noziere *et al.* (1999) identify pinonaldehyde as the major product of α-pinene oxidation in the presence of NO_x in the gas-phase with a molar yield of 87% and a yield of 18% for total gas-phase nitrates. For oxidation by ozone, Yu *et al.* (1999) also identified pinonaldehyde as the major product but found it present mainly in the gas-phase. In terms of molar yield, the major aerosol products determined by Yu *et al.* (1999) were pinic acid (1.8%), hydroxy pinonaldehydes (2.4%), pinonic acid (1.7%), norpinonic acid (2.1%) and hydroxy pinonic acid (2.1%). Since individual degradation pathways (NO_3, O_3 or OH) give different yields of various products, comparison between the model and the available laboratory measurements indicates a reasonable degree of agreement.

4.4. Results for β-pinene. Results of multiple β-pinene experiments are not reported by Hoffmann *et al.* (1997), but details of a single run are given including concentrations of some important gas-phase compounds. For comparison of the mechanism with this experiment, the model initialization is again carried out based on Hoffmann *et al.* (1997) where β-pinene is 95.0 ppb, NOxis 204 ppb, and the ratio of hydrocarbon to NOx is 9.1. There is reasonably good agreement between the measured and predicted aerosol concentrations both in terms of the onset of aerosol formation and the aerosol yield (Figure 3). While this does not provide a full evaluation of the model, it indicates that the mechanism can simulate the time evolution of the aerosol yield. The model simulations indicate that aerosol yields from β-pinene are higher than those from α-pinene. This is not indicated by some previous studies where aerosol yields from these two monoterpenes were approximately equal (e.g. (Zhang *et al.*, 1992) but appears to be a result of the importance of different oxidation pathways. Griffin *et al.* (1999b) indicate that with mixed oxidation pathways (either O_3, OH or NO_3) such as would occur in the smog chamber aerosol yields from oxidation of β-pinene are higher than those of α-pinene. However, when O_3 is the sole oxidant, aerosol yields from α-pinene are significantly higher. The simulations also suggest that aerosol yields from the nitrate pathway are higher for β-pinene.

4.5. Results for Δ-3-carene. Δ-3-carene contributes a significant fraction of total monoterpene fluxes over a range of vegetation types (Geron *et al.*, 2000). Although its reaction pathways are expected to be similar

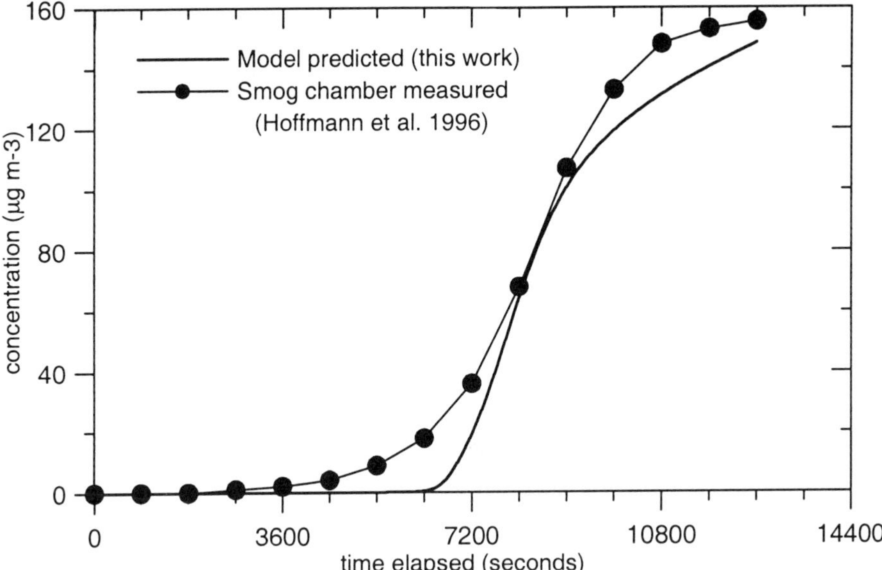

FIG. 3. *Model prediction of aerosol concentrations from oxidation of β-pinene in a smog chamber run in comparison with observed aerosol concentrations from Hoffmann et al. (1997).*

to that of α-pinene due to its internal double bond and bicyclic structure, Δ-3-carene is included as a second bicyclic class because its reaction rate with O_3 is only half that of α-pinene while reaction rates with the OH and NO_3 radicals are higher. Additionally aerosol yields from Δ-3-carene are higher than those from α-pinene but lower than those of β-pinene. In the modeling runs shown, measured partitioning coefficients from Yu *et al.* (1999) are employed if these are available for specific compounds. If partitioning coefficients are unknown (e.g. for products of nitrate oxidation) these have been estimated as described for α-pinene.

Figure 4 shows a number of aerosol mass and yields from smog chamber experiments for Δ-3-carene taken from (Griffin *et al.*, 1999b), (Hoffmann *et al.*, 1997), (Odum *et al.*, 1996), where the initial monterpene concentrations are 28 – 105 ppb. Also shown are the results of the numerical model simulations of these experiments, and the best fit yield and partitioning coefficients for the smog chamber experiments from the two-product model (given in Griffin *et al.* (1999b) as $\alpha_1 = 0.4454$, $K_1 = 0.0274$, $\alpha_2 = 0.0007$ and $K_2 = 0.0393$). As shown, at low aerosol mass/yield the model overpredicts both mass and yield. Since this is the region of most relevance to atmospheric conditions further investigation is required.

4.6. Results for d-limonene. Aerosol yields from d-limonene are high relative to other monoterpenes (Griffin *et al.*, 1999b). However, mod-

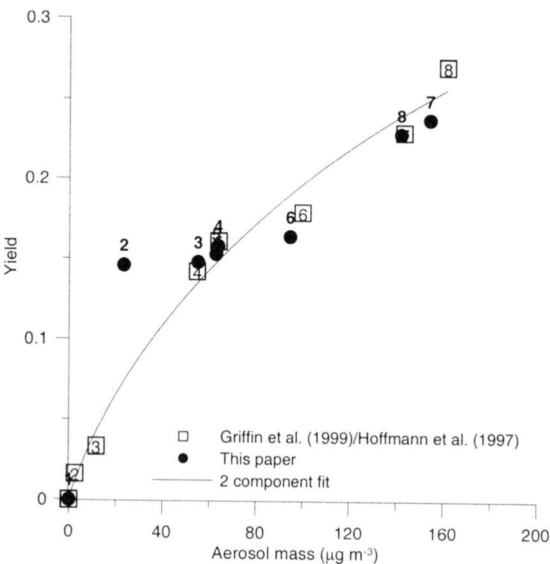

FIG. 4. *Modeled mass and aerosol yield for Δ-3-carene smog chamber experiments.*

eling of d-limonene oxidation is complicated by the possibility that both double bonds may be oxidized. Product analysis to date has indicated preferential oxidation of the internal double bond and that typically only one bond is oxidized giving gas phase products such as 4-acetyl-1-methyl-cyclohexene (AMCH) and endolim (Grosjean *et al.*, 1993), (Arey *et al.*, 1990), (Hakola *et al.*, 1994), (Calogirou *et al.*, 1997) and (Hallquist *et al.*, 1999). An outline of the oxidation pathways for d-limonene with ozone is given in Figure 5. As shown, here it is assumed that the oxidation proceeds mainly on one double bond.

Figure 6 is a mass/yield diagram for d-limonene oxidation based on smog chamber experiments for initial d-limonene concentrations of 20–65 ppb (Griffin *et al.* 1999b). Also shown are the model simulations for these experiments. There is relatively good agreement between the two product fit from Griffin *et al.* (1999b) and the six product model devised here and described in detail in Barthelmie and Pryor (2001). However, there is a large discrepancy in the mass/yield for experiment number 2 which has a relatively low initial d-limonene concentration of 20.5 ppb. As discussed above, the final mass/yield is highly sensitive to the availability of oxidation pathways so this discrepancy may be due in part to uncertainty with respect to the initialization of the smog chamber experiments.

4.7. Results for ocimene/myrcene. Relatively few studies have been conducted to examine ocimene or myrcene oxidation. As shown by

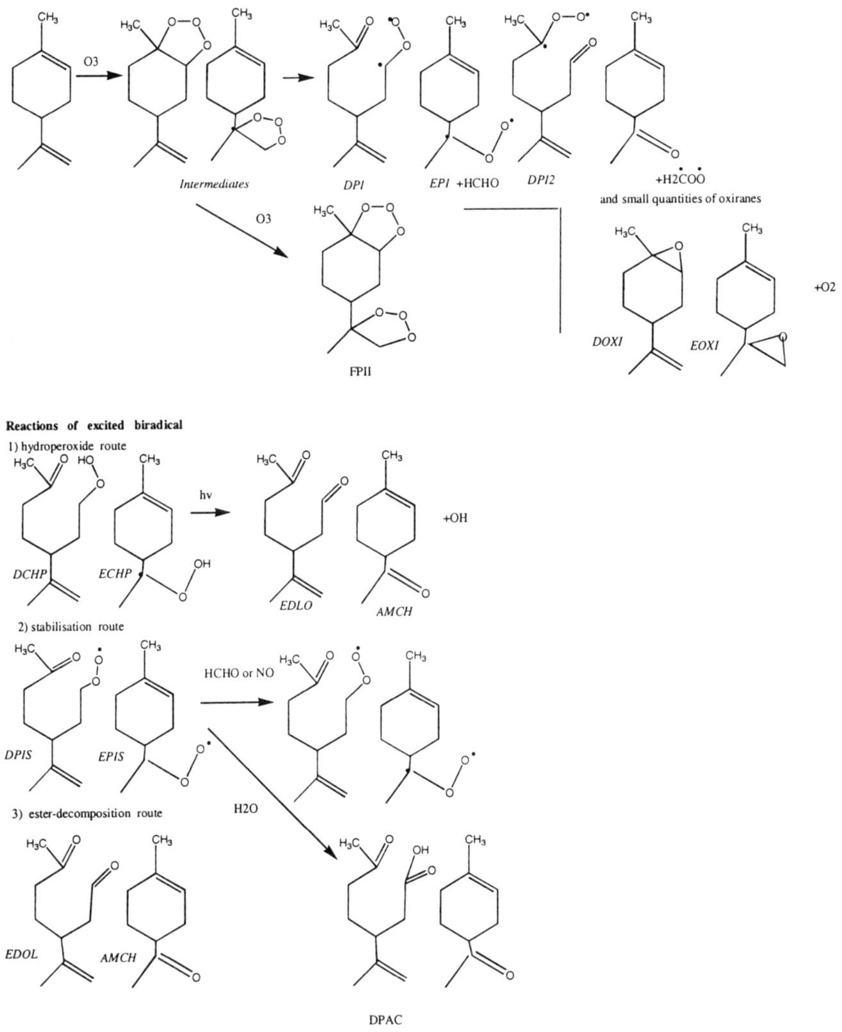

FIG. 5. *Outline of reaction pathways for d-limonene oxidation by ozone.*

Griffin *et al.* (1999b) acyclic monoterpenes have low aerosol yields compared with bicyclic and cyclic aerosol yields. However, these compounds are likely to play an important role in photochemical reactions in the same way that isoprene does. Table 3 shows details of aerosol concentrations as measured in ocimene and myrcene smog chamber experiments and modeled using the numerical mechanism. As with the other monoterpenes (with the exception of Δ-3-carene) there is reasonable agreement between predicted and modeled aerosol yields for low initial monoterpene concentrations but

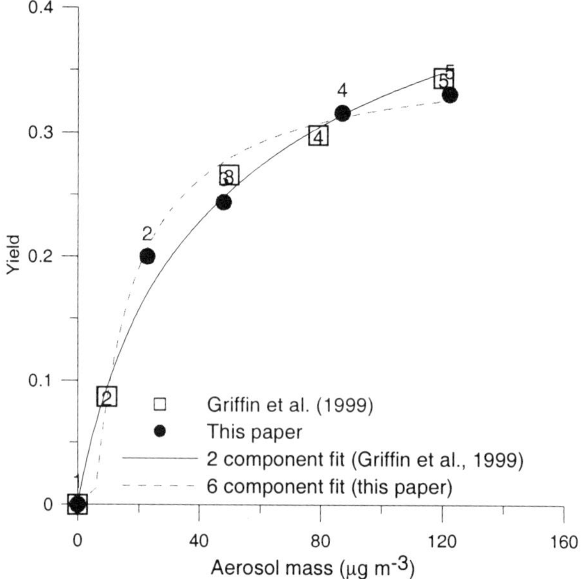

FIG. 6. *Modeled mass and aerosol yield for d-limonene smog chamber experiments.*

TABLE 3

Model and experimental results (Griffin et al., 1999b) for acyclic monoterpene oxidation.

Monoterpene	T (K)	Initial monoterpene concentration (ppb)	Measured aerosol concentration ($\mu g\ m^{-3}$)	Modeled aerosol concentration ($\mu g\ m^{-3}$)
Myrcene	311.1	9.8	3.5	2.9
Myrcene	311.9	77.5	57.5	9.8
Ocimene	313.2	45.6	7.3	8.2
Ocimene	315.7	76.5	17.6	15.6
Ocimene	315.7	142.0	50.3	17.3

the mechanism fails to capture high aerosol concentrations. One potential reason for this is that very high yields appear to be associated with nitrate radical oxidation. Nitrate concentrations can become high in smog chamber experiments. The model results are very highly sensitive to the concentrations of propene used to initialize the model (Barthelmie and Pryor, 2001) since this determines the oxidant spectrum. Nitrate concentrations in the box model may become as high as those in the smog chamber but this can occur after the monoterpene has been oxidized by the hydroxyl radical and ozone.

4.8. Discussion of model simulations for different monoterpenes. The β-pinene mechanism gives considerably higher aerosol yields than the α-pinene mechanism which agrees with Hoffmann *et al.* (1997) who reported aerosol yields of 6–7 % and 30 %, respectively for ~ 95 ppb of α-pinene or β-pinene. However, Hatakeyama *et al.* (1989) found higher aerosol carbon yields (18%) from α-pinene oxidation than from β-pinene (14%), and Zhang *et al.* (1992) suggest than aerosol yields for the two monoterpenes may be similar at completion. This apparent discrepancy could be a result of different experimental conditions relating to oxidant concentrations as described above. For all monoterpenes the aerosol yields generated by the chemical mechanisms described here are within the range reported previously in the literature. As in the smog chamber results presented in Griffin *et al.* (1999b) aerosol yields are highest for d-limonene, followed by β-pinene, Δ-3-carene and α-pinene. This hierarchy is also manifest in the chemical mechanisms presented here. However, considerable uncertainties remain with regard to the specific aerosol yields for atmospheric conditions.

Aerosol concentrations are dependent on the oxidants available, which is in accord with the different rates of oxidation and different yields of aerosol from different oxidation processes. Yields from nitrate oxidation in particular appear to be high and vary widely between different monoterpenes. Since smog chamber concentrations tend to be significantly higher than those in the atmosphere this adds to the uncertainty of applying these approaches for atmospheric conditions. Aerosol yields are also a reflection of the simulation conditions where, as in a smog chamber, the model is initialized with a set concentration of the monoterpene rather than using an emission flux as in the atmosphere.

5. Estimating atmospheric aerosol formation potential. Extending previous research which produced bulk monoterpene emissions, Geron *et al.* (2000) produced speciated monoterpene emissions from an extensive database of individual tree species emissions combined with a forestry database. Table 4 shows total monoterpene emission amounts for 13 areas defined by Geron *et al.* (2000).

At typical summer temperatures, highest monoterpene emissions are from coniferous forests while emissions from agricultural regions are much lower and mainly are associated with sparse forest cover. Emissions from crops are typically thought to be low although uncertainty remains particularly for the more reactive compounds like ocimene. Of the monoterpene compounds α-pinene dominates while β-pinene and d-limonene are important from almost all vegetation types. Figure 8 shows monoterpene emissions from Geron *et al.* (2000) lumped into the five monoterpene groups shown in Table 1 for one area with high emissions (West Coast) and one with low emissions (Agricultural Midwest). As shown, emissions are dominated by α-pinene for both land use types. β-pinene emissions are also significant but both d-limonene and Δ-3-carene make a larger contribution to agricultural monoterpene emissions than to the West Coast forest emissions.

To evaluate the aerosol formation potential of each emission area defined by Geron et al. (2000), a one dimensional model is employed for a 24 hour simulation. For each scenario, the initial monoterpene concentration (i.e. the concentration at midnight) is set as the concentration resulting from seven hours nighttime emissions with no loss. In each subsequent time step, monoterpene emissions into the cell are based on the fluxes given for each area in Table 4 (from Geron *et al.* (2000)) which have been temperature adjusted using the diurnal temperature profile shown in Figure 7. The bulk monoterpene emissions are then divided into the five classes in Table 1. The height of the cell (the mixed layer depth) varies with the diurnal profile of temperature (also shown in Figure 7). Application of a standard mixed layer depth and temperature profile clearly does not represent the average meteorological conditions for each area but using a uniform scenario (and neglecting deposition) presents an opportunity to compare the aerosol formation potential of each area. Oxidant chemistry (non-monoterpene VOC speciation and VOC and NO_x concentrations) are specified according to the 'Rural, Summer Surface' scenario number 1 taken from Stockwell *et al.* (1997). For each scenario (region) the model is run for a 24 hour period with a 1 second time step giving the terminating aerosol concentrations shown in Table 4. These concentrations are representative of those which would be found in a rural environment given the emission scenario specified if deposition of gases and aerosols was zero. Note also that since the partitioning between gas and aerosol phase depends on aerosol mass, the aerosol concentrations not accounting for deposition are higher than concentration totals calculated by simply subtracting deposition amounts.

The time series of monoterpene concentrations from the West coast scenario shown in Figure 7 illustrates the gradual accumulation of concentrations at night due to low mixed layer depth and lower oxidant concentrations. During the morning, the mixed layer depth grows and photochemical processes begin to generate oxidants decreasing monoterpene concentrations and increasing aerosol concentrations. During the afternoon and evening monoterpene concentrations begin to increase again and the depth of the mixed layer and concentrations of oxidants fall. Aerosol amounts continue to increase because the model does not account for deposition or advection.

As shown in Figure 8, ocimene (representing acyclic monoterpenes) is a small fraction of emissions from both agricultural and forest environments and makes a very small contribution to SOA concentrations. This is expected since oxidation of the double bonds leads to fractionation. While oxygenated products are more polar and therefore more likely to partition into the aerosol phase on average, loss of carbon to smaller molecules produces a lighter molecular weight product with a higher vapor

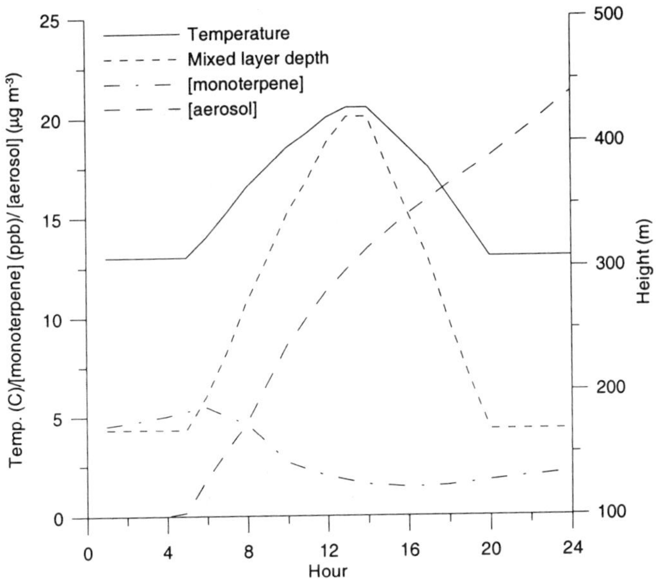

FIG. 7. *Time series of temperature, mixed layer depth, monoterpene and aerosol concentrations in the West Coast scenario.*

pressure. These products will probably remain in the gas-phase. While α-pinene, d-limonene and Δ-3-carene have reasonably high aerosol yields, the high aerosol yield from β-pinene reduces their percentage contribution to aerosol concentrations. The disproportionately large contribution made by β-pinene is a result of its high aerosol yield with nitrate.

In these simulations, the variability of reaction rates of individual monoterpenes with different oxidants is also important. small contribution from Δ-3-carene relative to d-limonene is thought to be mainly a result of its relatively slow reaction rate with O_3. In the first few hours of the simulations it is assumed to be night and the concentrations of both O_3 and NO_x are relatively low. As photolysis occurs, concentrations of the OH radical increase. Ocimene and then d-limonene have the fastest reaction rate with OH and will begin to be oxidized. The remaining monoterpenes may then have higher concentrations as concentrations of NO_3 increase. Hence the final aerosol yield is a combination of the reaction rate and the individual pathway aerosol yields.

Summary. A new chemical mechanism describing the chemistry of monoterpenes leading to secondary organic aerosol has been described. There are many advantages to using semi-explicit schemes such as those presented here; the possibility to speciate the resulting aerosol products, improved oxidant chemistry, balance of carbon, nitrogen and oxygen, mod-

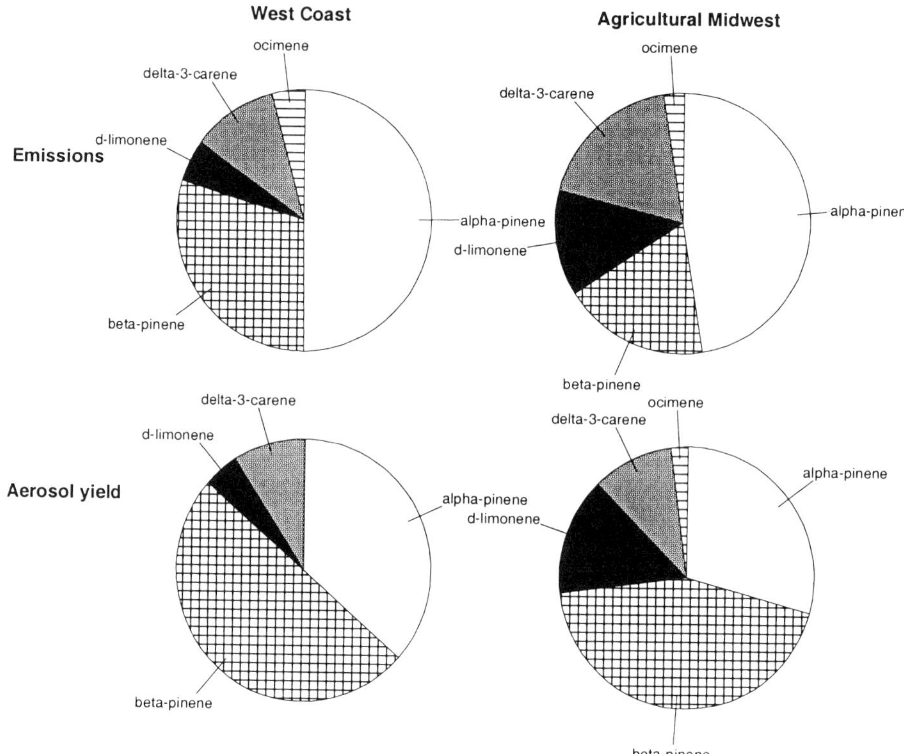

Fig. 8. *Comparison of speciated monoterpene emissions (in the five groups given in Table 1) and aerosol yield for two different land use types (West Coast and Agricultural Midwest).*

TABLE 4

Estimated monoterpene emissions from Geron et al. (2000) and estimated aerosol concentrations from the explicit mechanism.

	Total monoterpene emissions (μg C m^{-2} h^{-1}) at 30 °C	Aerosol concentrations predicted using the explicit mechanism (μg m^{-3})
1. West Coast	465	21
2. Great Basin	177	8
3. Northern Great Plains	21	1
4. Interior West	248	11
5. Northern Mixed Forest	159	7
6. Appalachia/Ozarks	178	7
7. Piedmont Southeast	410	18
8. Coastal Southeast	300	14
9. MidAtlantic/Midwest	185	7
10. Agricultural Midwest	39	1
11. Northern Coniferous Forest	825	42
12. Southern Great Plains	83	3
13. Alaska	381	18

eling the time evolution of aerosol concentrations, and fully coupled gas-aerosol chemistry. Within the mechanism presented here monoterpenes are represented within five groups reflecting their structure (number of rings and location of the double bond) and their importance in regional emissions. The final model has been integrated into a box model using the Regional Atmospheric Chemistry Mechanism giving nearly 500 reactions in total and about 250 products including radicals and condensables. Partitioning between the gas and particle phases uses the absorption approach. Comparison of the mechanisms are undertaken using smog chamber experiments. The box model is initialized as in the smog chamber with the same initial concentrations of propene, nitrogen oxides and monoterpene and run using the specified temperature and photolysis rates. In general the model gives good results in comparison with observed smog chamber results. The largest uncertainty exists in the nitrate oxidation pathway where there are fewest observations of products from laboratory studies and a lack of information about the range of partitioning coefficients to be expected for nitrated oxidation products from monoterpenes. This gives relatively high yields for some of the nitrate oxidation pathways, particularly for β-pinene.

The model is used to estimate aerosol formation potential from speciated monoterpene emissions from different land use types in the United States (Geron *et al.*, 2000). Highest monoterpene emissions and therefore highest aerosol potential is from coniferous forest types. α-pinene is always the largest emission and has a moderate aerosol yield giving it a relatively large contribution to SOA. β-pinene has high aerosol yields in the scheme adopted (from the nitrate pathway where the uncertainty is largest) and therefore contributes disproportionately to SOA. d-limonene also has relatively high aerosol yields, while Δ-3-carene has relatively small emissions and has the slowest reaction rate with ozone so its contribution to SOA is relatively small. Ocimene has both low emissions rates and relatively low aerosol yields so its contribution to SOA is small but it may have an important role in oxidant chemistry.

The development and testing of the monoterpene mechanism has illustrated that it is possible to describe the chemistry of biogenic volatile organic compounds which leads to products which partition between the gas- and aerosol- phases. Further evaluation and mechanism reduction is being undertaken for adoption of the mechanisms into regional air quality models.

Acknowledgments. We acknowledge support for this research from the National Science Foundation Atmospheric Chemistry Program grant ATM 9711755, from the Ford Research Centre at Aachen and from Environment Canada. We are grateful to Chris Geron of EPA for useful discussion and comments.

REFERENCES

Arey, J., Atkinson, R.A. and Aschmann, S.M., 1990: Product study of the gas-phase reactions of monoterpenes with the OH radical in the presence of NOx. *Journal of Geophysical Research*, **95(D11)**, 18539–18546.

Atkinson, R., 1997: Gas-phase tropospheric chemistry of volatile organic compounds: 1. Alkanes and Alkenes. *Journal of Physical and Chemical Reference Data*, **26**(2), 215–290.

Atkinson, R. and Carter, W.P.L., 1991: Reactions of alkoxy radicals under atmospheric conditions: the relative importance of decomposition versus reaction with O2. *Journal of Atmospheric Chemistry*, **13**, 195–201.

Barnes, I., Bastian, V., Becker, K.H. and Tong, Z., 1990: Kinetics and products of the reactions of NO3 with monoalkenes, dialkenes and monoterpenes. *Journal of Physical Chemistry*, **94**, 2413–2419.

Barthelmie, R.J. and Pryor, S.C., 1999: A model mechanism to describe oxidation of monoterpenes leading to secondary organic aerosol: 1 a-pinene and b-pinene. *Journal of Geophysical Research*, **104**, 23567–23571.

Barthelmie, R.J. and Pryor, S.C., 2001: A model mechanism to describe oxidation of monoterpenes leading to secondary organic aerosol: 2 d-limonene and delta-3-carene. *Journal of Geophysical Research*, **submitted**.

Becker, K.H., Bechara, J. and Brockmann, K.J., 1993: Studies on the formation of H2O2 in the ozonolysis of alkenes. *Atmospheric Environment*, **27A**(1), 57–61.

Birth, T.L. and Geron, C.D., 1995: *User's guide to the personal computer version of the biogenic emissions inventory system (PC-BEIS), version 2.0*, EPA, Research Triangle Park.

Calogirou, A., Duane, M., Larsen, B., Kettrup, A. and Kotzias, D., 1997: *Atmospheric oxidation of limonene: comparison between laboratory and field studies*. In: P.M. Borrell, P. Borrell, T. Cvitas and W. Seiler (Editors), Transport and transformation of pollutants in the troposphere: Proceedings of EUROTRAC Symposium 1996. Computational Mechanics Publications, Southampton, pp. 185–188.

Carter, W.P.L. and Atkinson, R.A., 1989: Alkyl nitrate formation from the atmospheric photooxidation of alkanes; a revised estimation method. *Journal of Atmospheric Chemistry*, **8**, 165–173.

Fuentes, J.D., Wang, D., Neumann, H.H., Gillespie, T.J., Den Hartog, G. and Dann, T.F., 1996: Ambient biogenic hydrocarbons and isoprene emissions from a mixed deciduous forest. *Journal of Atmospheric Chemistry*, **25**, 67–95.

Geron, C., Rasmussen, R., Arnts, R.R. and Guenther, A., 2000: A review and synthesis of monoterpene speciation from forests in the United States. *Atmospheric Environment*, **34(11)**, 1761–1781.

Geron, C.D., Guenther, A.B. and Pierce, T.E., 1994: An improved model for estimating emissions of volatile organic compounds from forests in the eastern United States,. *Journal of Geophysical Research*, **99(D6)**, 12773–12791.

Gery, M.W., Whitten, G.Z., Killus, J.P. and Dodge, M.C., 1989: A photochemical kinetics mechanism for urban and regional scale computer modeling. *Journal of Geophysical Research*, **94**(D10), 12925–12956.

Glasius, M., Duane, M. and Larsen, B., 1999: Determination of polar terpene oxidation products in aerosols by liquid chromatography - ion trap mass spectrometry. *Journal of Chromatography A*, **833**, 121–135.

Griffin, R.J., Cocker, D.R., Flagan, R.C. and Seinfeld, J., 1999a: Estimate of global atmospheric organic aerosol from oxidation of biogenic hydrocarbons. *Geophysical Research Letters*, **26**(17), 2721–2724.

Griffin, R.J., Cocker, D.R., Flagan, R.C. and Seinfeld, J., 1999b: Organic aerosol formation from the oxidation of biogenic hydrocarbons. *Journal of Geophysical Research*, **104**, 3555–3567.

Grosjean, D., 1992: In situ organic aerosol formation during a smog episode: estimated production and chemical functionality. *Atmospheric Environment*, **26A**(6), 953–963.

Grosjean, D. and Seinfeld, J.H., 1989: Parameterization of the formation potential of secondary organic aerosols. *Atmospheric Environment*, **23**(8), 1733–1747.

Grosjean, D., Williams, E.L., Grosjean, E., Andino, J.M. and Seinfeld, J., 1993: Atmospheric oxidation of biogenic hydrocarbons: reaction of ozone with b-pinene, d-limonene and trans-caryophyllene. *Environmental Science and Technology*, **27**, 2754–2758.

Grosjean, E. and Grosjean, G., 1997: The gas phase reactions of unsaturated oxygenates with ozone: carbonyl products and comparison with the alkene-ozone reaction. *Journal of Atmospheric Chemistry*, **27**, 271–289.

Guenther, A., Hewitt, C.N., Erickson, D. *et al.*, 1995: A global model of natural volatile compound emissions. *Journal of Geophysical Research*, **100**(D5), 8873–8892.

Guenther, A., Zimmerman, P., Klinger, L., Greenberg, J., Ennis, C., Davis, K., Pollock, W., Westberg, H., Allwine, G. and Geron, C., 1996: Estimates of regional natural volatile organic compound fluxes from enclosure and ambient measurements. *Journal of Geophysical Research (Atmospheres)*, **101**(D1), 1345–1359.

Hagerman, L.M., Aneja, V.P. and Lonneman, W.A., 1997: Characterization of nonmethane hydrocarbons in the rural southeast United States. *Atmospheric Environment*, **31**(23), 4017–4038.

Hakola, H., Arey, J., Aschmann, S.A. and Atkinson, R.A., 1994: Product formation from the gas-phase reactions of OH-radicals and O3 with a series of monoterpenes. *Journal of Atmospheric Chemistry*, **18**, 75–102.

Hallquist, M., Wangberg, I., Ljungstrom, E., Barnes, I. and Becker, K.-H., 1999: Aerosol and product yields from NO3 radical-initiated oxidation of selected monoterpenes. *Environmental Science and Technology*, **33**, 553–559.

Hatakeyama, S., Izumi, K., Fukuyama, T. and Akimoto, H., 1989: Reactions of ozone with a-pinene and b-pinene in air: yields of various gases and particulate products. *Journal of Geophysical Research*, **94**, 13013–13024.

Hertel, O., Berkowicz, R., Christensen, J. and Hov, O., 1993: Test of two numerical schemes for use in atmospheric transport-chemistry models. *Atmospheric Environment*, **27**, 2591–2611.

Hoffmann, T., Odum, J.R., Bowman, F., Collins, D., Klockow, D., Flagan, R.C. and Seinfeld, J.H., 1997: Formation of organic aerosols from the oxidation of biogenic hydrocarbons. *Journal of Atmospheric Chemistry*, **26**, 189–222.

Isidorov, V.A., Zenekevich, I.G. and Ioffe, B.V., 1985: Volatile organic compounds in the atmosphere of forests. *Atmospheric Environment*, **19**(1), 1–8.

Kamens, R., Jang, M., Chien, C.J. and Leach, K., 1999: Aerosol formation from the reaction of alpha-pinene and ozone using a gas-phase kinetics-aerosol partitioning model. *Environmental Science and Technology*, **33**(9), 1430–1438.

Kavouras, I.G., Mihalopoulos, N. and Stephanou, E., 1998: Formation of atmospheric particles from organic acids produced by forests. *Nature*, **395**, 683–686.

Kumala, M., Toionen, A., Makela, J. and Laaksonen, A., 1999: Analysis of the growth of nucleation mode particles observed in Boreal forest. *Tellus*, **50**, 449–462.

Lamb, B., Gay, D., Westberg, H. and Pierce, T., 1993: A biogenic hydrocarbon emission inventory for the USA using a simple forest canopy model. *Atmospheric Environment*, **27A**(11), 1673–1690.

Lamb, R., Guenther, A., Gay, D. and Westberg, H., 1987: A national inventory of biogenic hydrocarbon emissions. *Atmospheric Environment*, **21**, 1695–1705.

Noziere, B., Barnes, I. and Becker, K., 1999: Product study and mechanisms of the reactions of alpha-pinene and of pinonaldehyde with OH radicals. *Journal of Geophysical Research*, **104**, 23645–23658.

Odum, J.R., Hoffman, T., Bowman, F., Collins, D., Flagan, R.C. and Seinfeld, J.H., 1996: Gas/particle partitioning and secondary organic aerosol yields. *Environmental Science and Technology*, **30**, 2580–2585.

Palen, E.J., Allen, D.T., Pandis, S., Paulson, S., Seinfeld, J. and Flagan, R., 1992: Fourier transform infrared analysis of aerosol formed in the photo-oxidation of isoprene and b-pinene. *Atmospheric Environment*, **26**, 1239–1251.

Pandis, S., Paulson, S.E., Baltensperger, U., Seinfeld, J.H., Flagan, R.C., Palen, E. and Allen, D.T., 1992a: *Aerosol production during the photooxidation of natural hydrocarbons*. In: S.E. Schwarz and W.G.N. Slinn (Editors), Precipitation Scavenging and Atmosphere Surface Exchange., pp. 1641–1651.

Pandis, S.N., Harley, R.A., Cass, G.R. and Seinfeld, J.H., 1992b: Secondary organic aerosol formation and transport. *Atmospheric Environment*, **26A**(13), 2269–2282.

Sakamaki, F. and Akimoto, H., 1988: HONO formation as unknown radical source in photochemical smog chamber. *International Journal of Chemical Kinetics*, **20**, 111–116.

Schade, G., Goldstein, A. and Lamanna, M., 1999: Are monoterpene emissions influenced by humidity? *Geophysical Research Letters*, **26**(14), 2187–2190.

Stockwell, W.R., Kirchner, F. and Kuhn, M., 1997: A new mechanism for regional atmospheric chemistry modeling. *Journal of Geophysical Research*, **102(D22)**, 25847–25879.

Versino, B., 1997: *The BEMA project: biogenic emissions in the Mediterranean area*. In: P.M. Borrell, P. Borrell, T. Cvitas and W. Seiler (Editors), Transport and transformation of pollutants in the troposphere: Proceedings of EUROTRAC Symposium 1996. Computational Mechanics Publications, Southampton, pp. 213–217.

Winterhalter, R., Neeb, P., Grossmann, D., Kolloff, A., Horie, O. and Moortgat, G.K., 2000: Products and mechanism of the gas phase reaction of ozone and b-pinene. *Journal of Atmospheric Chemistry*, **35**(2), 165–197.

Winterhalter, R., Neeb, P. and Moorgat, G.K., 1999: General mechanism of gas phase monoterpene ozonolysis for application in tropospheric chemistry models, European Geophysical Society, The Hague.

Yokuchi, Y. and Ambe, Y., 1985: Aerosols formed from the chemical reaction of monoterpenes and ozone. *Atmospheric Environment*, **19**(8), 1271–1276.

Yu, J., Cocker, D., Griffin, R., Flagan, R. and Seinfeld, J., 1999: Gas-phase ozone oxidation of monoterpenes: gaseous and particulate products. *Journal of Atmospheric Chemistry*, **34**, 207–258.

Zhang, S.H., Shaw, M., Seinfeld, J.H. and Flagan, R.C., 1992: Photochemical aerosol formation from alpha- and beta-pinene. *Journal of Geophysical Research*, **97**(D18), 20717–20729.

LIST OF WORKSHOP PARTICIPANTS

- Ingmar Ackermann, Ford Forschungszentrum Aachen
- Javier Armendariz, Institute for Mathematics and its Applications
- Bruce Ayati, Institute for Mathematics and its Applications
- Rebecca Barthelmie, Department of Geography, Indiana University
- Francis S. Binkowski, Air Resources Laboratory Environmental Protection Agency
- Daewon Byun, Atmospheric Science Modeling Division, EPA
- Gregory Carmichael, Department of Chemical and Biochemical Engineering, Center for Global and Regional Environmental Research, University of Iowa
- John Chadam, Department of Mathematics, University of Pittsburgh
- Jason Ching, Atmospheric Modeling Division, US Environmental Protection Agency
- David Chock, Research Laboratory, Ford Motor Company
- Donald Dabdub, Mechanical and Aerospace Engineering, University of California - Irvine
- Dacian Daescu, Center for Global and Regional Environmental Research, University of Iowa
- Frank Dentener, Institute for Marine and Atmospheric Research IMAU, Utrecht University
- Fred Dulles, Institute for Mathematics and its Applications
- Alan Dunker, Chemical and Environmental Sciences Laboratory, General Motors Research and Development Center
- Yalchin Efendiev, Institute for Mathematics and its Applications
- Hendrik Elbern, Institute for Geophysics and Meteorology (EU-RAD), Universität Zu Köln
- Rodney Fox, Department of Chemical Engineering, Iowa State University
- Panos Georgopoulos, DECM, Environmental and Occupational Health Sciences Institute
- Giulio Giunta, Istituto di Matematica, Istituto Universitario Navale - Napoli
- Sun-ling Gong, ARQM, Atmospheric Environment Service
- Katharine Gurski, USRA/Center of Excellence in Space Data and Information, NASA - Goddard Space Flight Center
- Alan Hansen, Environment Department, EPRI

- Jiwen He, Department of Mathematics, University of Houston
- Mark Jacobson, Department of Civil and Environmental Engineering, Stanford University
- Seongjai Kim, Department of Mathematics, University
- Jana Milford, Department of Engineering-Mechanical, University of Colorado
- Willard Miller, Institute for Mathematics and its Applications
- Mike Moran, ARQI, Meteorolocical Service of Canada
- Ralph E. Morris, ENVIRON International Corporation
- Ed Nam, Ford Research Laboratory
- Talat Odman, Department of Civil Engineering, Georgia Tech
- Eduardo P. Olaguer, Health and Environmental Sciences, Dow Chemical Company
- Humberto Ortiz, Mechanical Engineering, University of Minnesota
- Spyros N. Pandis, Department of Chemical Engineering, Carnegie Mellon University
- John Pleim, Atmospheric Science Modeling Division, USEPA
- Sara Pryor, Amospheric Science Program, Department of Geography, Indiana University
- Betty K. Pun, Atmospheric and Environmental Research, Inc.
- S.T. Rao, Earth and Atmospheric Sciences, SUNY - Albany
- Angelo Riccio, Department of Chemistry, University of Naples - Federico II
- Roberto San Jose, Technical University of Madrid
- Adrian Sandhu, Computer Science Department, Michigan Technological University
- Fadil Santosa, MCIM, IMA and Minnesota Center for Industrial Mathematics
- Ananthakrishna Sarma, Science Applications International Corporation (SAIC)
- Uma Shankar, NCSC
- Michael Smiley, Department of Mathematics, Iowa State University
- Qingyuan Song, ARQM, Atmospheric Environmental Service
- Bruno Sportisse, Centre d'Enseignement et de Recherche sur l'Eau, la ville et l'environnement, École Nationale des Ponts et Chaussées (ENPC-CEREVE)
- William R. Stockwell, Division of Atmospheric Sciences, Desert Research Institute
- Henning Struchtrup, Department of Mechanical Engineering, University of Victoria
- Alison Tomlin, Department of Fuel and Energy, University of Leeds

- Gail Tonnesen, Center for Environmental Research and Technology, University of California at Riverside
- Itshushi Uno, Kyushu University
- Jan Verwer, CWI
- Sheng-Wei Wang, Atmospheric and Oceanic Princeton University
- Robert Yamartino, Sigma Research Division of Earth Tech
- Jeffrey Young, Atmospheric Modeling Division, U.S. Environmental Protection Agency
- Jesus Zepeda, Civil Engineering, University of Minnesota
- Yang Zhang, Atmospheric and Evironmental Research, Inc.
- Zahari Zlatev, Department for Atmospheric Environment, National Environmental Research Institute

IMA SUMMER PROGRAMS

1987 Robotics
1988 Signal Processing
1989 Robust Statistics and Diagnostics
1990 Radar and Sonar (June 18–29)
      New Directions in Time Series Analysis (July 2–27)
1991 Semiconductors
1992 Environmental Studies: Mathematical, Computational, and
      Statistical Analysis
1993 Modeling, Mesh Generation, and Adaptive Numerical Methods
      for Partial Differential Equations
1994 Molecular Biology
1995 Large Scale Optimizations with Applications to Inverse Problems,
      Optimal Control and Design, and Molecular and Structural
      Optimization
1996 Emerging Applications of Number Theory (July 15–26)
      Theory of Random Sets (August 22–24)
1997 Statistics in the Health Sciences
1998 Coding and Cryptography (July 6–18)
      Mathematical Modeling in Industry (July 22–31)
1999 Codes, Systems, and Graphical Models (August 2–13, 1999)
2000 Mathematical Modeling in Industry: A Workshop for Graduate
      Students (July 19–28)
2001 Geometric Methods in Inverse Problems and PDE Control
      (July 16–27)
2002 Special Functions in the Digital Age (July 22–August 2)

IMA "HOT TOPICS" WORKSHOPS

- Challenges and Opportunities in Genomics: Production, Storage, Mining and Use, April 24–27, 1999
- Decision Making Under Uncertainty: Energy and Environmental Models, July 20–24, 1999
- Analysis and Modeling of Optical Devices, September 9–10, 1999
- Decision Making under Uncertainty: Assessment of the Reliability of Mathematical Models, September 16–17, 1999
- Scaling Phenomena in Communication Networks, October 22–24, 1999
- Text Mining, April 17–18, 2000
- Mathematical Challenges in Global Positioning Systems (GPS), August 16–18, 2000
- Modeling and Analysis of Noise in Integrated Circuits and Systems, August 29–30, 2000
- Mathematics of the Internet: E-Auction and Markets, December 3–5, 2000
- Analysis and Modeling of Industrial Jetting Processes, January 10–13, 2001

- Special Workshop: Mathematical Opportunities in Large-Scale Network Dynamics, August 6–7, 2001
- Wireless Networks, August 8–10 2001
- Numerical Relativity, June 24–29, 2002

SPRINGER LECTURE NOTES FROM THE IMA:

The Mathematics and Physics of Disordered Media
   Editors: Barry Hughes and Barry Ninham
   (Lecture Notes in Math., Volume 1035, 1983)

Orienting Polymers
   Editor: J.L. Ericksen
   (Lecture Notes in Math., Volume 1063, 1984)

New Perspectives in Thermodynamics
   Editor: James Serrin
   (Springer-Verlag, 1986)

Models of Economic Dynamics
   Editor: Hugo Sonnenschein
   (Lecture Notes in Econ., Volume 264, 1986)

Forthcoming Volumes:

1999–2000: Reactive Flow and Transport Phenomena

> Confinement and Remediation of Environmental Hazards
> and Resource Recovery

> Dispersive Corrections to Transport Equations, Simulation
> of Transport in Transition Regimes/
> Multiscale Models for Surface Evolution and Reacting Flows

2000-2001: Mathematics in Multimedia

> Mathematical Foundations of Speech Processing and Recognition/
> Mathematical Foundations of Natural Language Modeling

> Mathematical Methods in Speech and Image Analysis/
> Image Processing and Low Level Vision/
> Image Analysis and High Level Vision

> Fractals in Multimedia

2001-2002: Mathematics in the Geosciences

> Time Series Analysis and Applications to Geophysical Systems

MTNS 2002 Conference

> Mathematical Systems Theory in Biology, Communication,
> Computation, and Finance